The Treatment of Liquid Aluminum-Silicon Alloys

John E. Gruzleski
Bernard M. Closset

The American Foundrymen's Society, Inc.

ISBN 0-87433-121-8

COPYRIGHT © 1990
American Foundrymen's Society, Inc.
Des Plaines, Illinois 60016-8399
U.S.A.

Printed in U.S.A.

Contents

Foreward, ix

Preface, xi

Acknowledgements, xiii

Chapter 1: Introduction

    1.1 Commercial Importance of Aluminum Foundry Alloys *1*

    1.2 Current Market for Aluminum Castings *5*

    1.3 The Future Potential for Aluminum Castings *9*

Chapter 2: Aluminum-Silicon Foundry Alloys

    2.1 Common Alloy Systems *13*

    2.2 Typical Applications *17*

    2.3 Cast Microstructure *18*

Chapter 3: Eutectic Modification

    3.0 Introduction *25*

    3.1 The Fundamentals of Modification *26*

    3.2 Chemical Modification by Sodium, Strontium or Antimony *31*

    3.3 Overmodification *44*

    3.4 Modifier Fading *46*

    3.5 The Effect of Phosphorus *50*

Chapter 4:   Modification and Porosity

4.0   Introduction *57*

4.1   Modification and Melt Hydrogen *58*

4.2   Modification and the Porosity Distribution *63*

4.3   Avoiding Porosity in Modified Castings *68*

Chapter 5:   The Properties of Modified Alloys

5.1   Tensile Properties *75*

5.2   Impact Properties and Fracture Toughness *85*

5.3   Fatigue Properties *89*

5.4   Thermal Shock Properties *90*

5.5   Machinability *90*

5.6   Foundry Properties *91*

Chapter 6:   Interactions of Modifiers

6.0   Introduction *95*

6.1   Strontium-Antimony Interactions *96*

6.2   Sodium-Antimony Interactions *99*

6.3   Sodium-Strontium Interactions *102*

6.4   Overcoming Negative Interactions *102*

Chapter 7:   Refinement and Modification in Hypereutectic Alloys

7.1   Structure of Hypereutectic Alloys *107*

7.2   Phosphorous Refinement *110*

7.3   The Effect of Refinement on Properties   *118*

7.4   Strontium Treatment   *120*

7.5   Sodium Treatment   *122*

7.6   Refinement by Other Elements   *125*

Chapter 8:   Grain Refinement

8.1   The Definition of a Grain   127

8.2   Principles of Grain Refinement   *128*

8.3   Grain Refinement by Chilling and the Effect on Dendrite Arm Spacing   *131*

8.4   Chemical Grain Refinement   *132*

8.5   The Effect of Grain Refinement on Properties   *137*

8.6   The Interrelationship between Chemical Grain Refinement, Cooling Rate and Modification   *141*

Chapter 9:   Gassing of Aluminum Foundry Melts

9.1   Hydrogen Solubility   *143*

9.2   Sources of Hydrogen   *146*

9.3   Gas and Shrinkage Porosity   *149*

9.4   Measuring the Hydrogen Concentration   *157*

Chapter 10:   Degassing Aluminum Foundry Melts

10.0   Introduction   *169*

10.1   Degassing Methods   *170*

10.2   Degassing Efficiency   *175*

10.3   Inert or Reactive Gas Treatment   *179*

10.4   Chemical Changes During Degassing   *181*

Chapter 11: Filtration

11.0 Introduction *185*

11.1 Inclusion Types and Origins *186*

11.2 Why Remove Inclusions? *186*

11.3 Filter Types and Placement *190*

11.4 How Filters Work *192*

11.5 Filter Efficiency *195*

11.6 Filter Sizing *196*

11.7 The Measurement of Melt Cleanliness *197*

Chapter 12: Treatment with Cover and Cleaning Fluxes

12.0 Introduction *203*

12.1 Cover Fluxes *204*

12.2 Cleaning Fluxes *205*

12.3 Flux Injection Process *209*

12.4 Flux Filtration *210*

12.5 Some Problems Related to Fluxes *210*

Chapter 13: Non-Destructive Microstructure Control

13.0 Introduction *213*

13.1 Principles of Thermal Analysis *214*

13.2 Thermal Analysis Control of Grain Size *219*

13.3 Thermal Analysis Control of Eutectic Modification *220*

13.4 Electrical Conductivity Control of Modification *225*

13.5 Electrical Conductivity Measuement of Dendrite Arm Spacing *229*

Contents

Chapter 14:   A Summary of Melt Treatment Processes, Techniques and Schedules

14.0   Introduction *233*

14.1   Melt Treatment Processes —A Summary *233*

14.2   Melt Addition Techniques *237*

14.3   Processing Schedules *244*

Index, *249*

Foreword

This is an excellent basic text which covers melt processing of aluminum silicon alloys. All principal processes—modification, degassing, fluxing, grain refining and filtration are covered in sufficient detail to enable a technically oriented foundryman to establish practices that will produce high quality melts.

While defect analysis *per se* is not a principal focus of the book, the knowledge conveyed on melt processes can assist in developing solutions for many metallurgically related defects, and the causes of less-than-optimum mechanical properties.

The reader will find this volume to be a very useful addition to their library on aluminum casting practices.

D.V.N./D.E.G.
Reviewers

Preface

This book is the result of ten years of collaboration between the authors on research related to the aluminum foundry industry. Our decision to write a book on liquid metal treatment arose from two factors. One of these was the spectacular growth in the aluminum casting industry since 1980, due to the general acceptance of cast aluminum alloys as high quality, light weight engineering materials. The other driving force behind our decision was the realization that while many facets of the aluminum casting business are quite advanced, most liquid metal treatments are less sophisticated. By and large, aluminum foundries know how to make beautiful and intricate molds. Significant time and money are expended on the molding process, while, until recently, only a limited effort went into controlling the quality of the liquid metal poured into those molds. We believe strongly that in order for the aluminum casting industry to prosper and to grow, much more attention must be paid to the quality of the liquid alloy.

Our purposes in writing this book were several. We wished to present the founder, in one volume, with a description of the various melt treatments that are available. We wanted to explain clearly why these treatments are performed, and their effects on the structure and properties of the final casting. It was also our desire to point out the strong and the weak points of melt treatments, along with some of the pitfalls, and to show that the same result can often be obtained by different routes.

The book is written primarily with the practicing foundry engineer in mind. While it will be useful to many students of metallurgy and foundry science, it is not intended as a detailed research textbook. Consequently, we have not attempted to present a comprehensive survey of the voluminous literature on aluminum foundry alloys. The state of the art in liquid metal processing is described, and the effects of liquid treatment on structure and properties are emphasized. Ultimately, it is casting quality which must be the prime concern of the foundry engineer. We sincerely hope that our efforts will help to achieve the desired quality level.

Acknowledgements

The authors are indebted to all of those societies and publishing houses that have kindly agreed to let us use material from their publications to illustrate the text. The source of all material used in the many figures is given in the references listed at the end of each chapter.

Much of the manuscript was written while one of the authors (JEG) was on sabbatical leave from McGill University. This leave was spent at the research center of La Société d'Aluminium Pechiney in Voreppe, France, and in the Department of Materials Science and Engineering, University of Arizona. The cooperation of each of these organizations in providing an atmosphere conducive to writing is gratefully acknowledged.

To Timminco Limited we owe a deep debt of gratitude. It was at Timminco that we first worked together. Their support of aluminum foundry research over the past decade has allowed many interesting and important questions to be raised and answered.

Finally, we would like to thank our two dedicated typists, Olga Gruzleski and Lucille LeBlanc who typed and re-typed the manuscript into its final form.

John Gruzleski
Montreal

Bernard Closset
Toronto

Chapter 1

Introduction

1.1   Commercial Importance of Aluminum Foundry Alloys

The casting of metal has evolved tremendously from its origin in prehistory. The earliest metal objects were wrought, casting having evolved as a fabrication process approximately 5000 years ago. Bronze, the first metal widely cast, was used to make bells, statues and guns. Like preceding epochs, the Industrial Revolution in Europe and North America is synonymous historically with the development of a casting process, in this case cast iron and steel. Tremendous expansion in the metal casting industry resulted from the need to produce new machinery of all types for the growing manufacturing and transportation industries.

Aluminum casting became affordable only after the invention of aluminum refining by the Hall-Heroult process. In the earlier part of the 20th Century, the application of aluminum castings was limited to decorative parts and cooking utensils. After World War II, a dramatic expansion of the aluminum casting industry occurred. New alloys were developed and casting processes were implemented to comply with engineering specifications, and to extend the range of commercial and technical applications. The recent "energy crisis" of the 1970s led to greater use of cast aluminum in many vehicles, because of its excellent strength-to-weight ratio.

A wide range of metals can be added to aluminum[1]. Among those regularly added and controlled as alloying elements are zinc, magnesium, copper, silicon, iron, lithium, manganese, nickel, silver, tin and titanium. The solid solubilities of these elements in aluminum vary considerably (Table 1.1). Some are used as solid solution strengtheners, while others are added because they form various desirable intermetallic compounds.

Aluminum alloys constitute a group of cast materials which, in tonnage terms, is second only to ferrous castings (Figures 1.1 and 1.2). In 1986, U.S. shipments of aluminum-based cast alloys were slightly above 1,000,000 tons, while shipments of copper- and zinc-based alloys were in the order of 250,000 tons each. Aluminum cast parts represented approximately 10% of the total tonnage of U.S. casting shipments in 1986. World-wide, approximately 20% of total aluminum production is, on average, converted into cast parts.

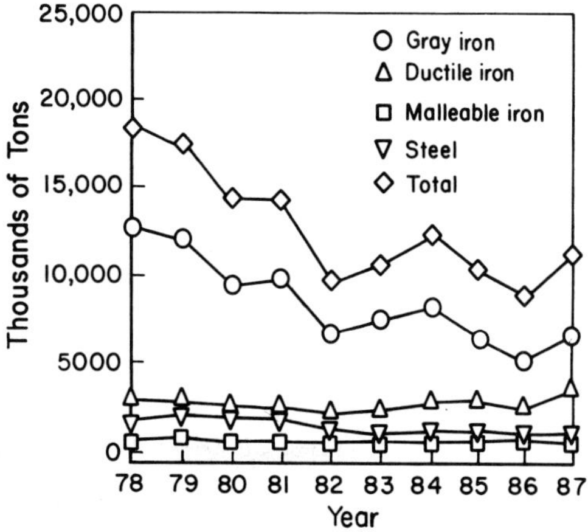

Figure 1.1. *U.S. ferrous casting shipments from 1978.*

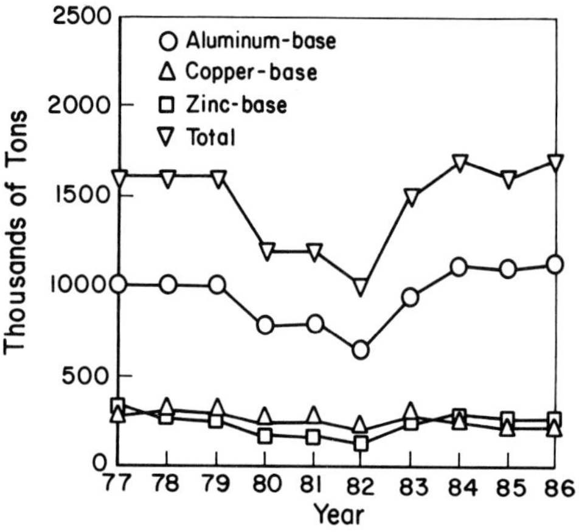

Figure 1.2. *U.S. nonferrous casting shipments from 1977 to 1986.*

Table 1.1. *Solid Solubility of Elements in Aluminum*

Element	Temperature (°C)	Maximum solid solubility (wt %)
Cadmium	649	0.4
Cobalt	657	<0.02
Copper	548	5.65
Chromium	661	0.77
Germanium	424	7.2
Iron	655	0.05
Lithium	600	4.2
Magnesium	450	17.4
Manganese	658	1.82
Nickel	640	0.04
Silicon	577	1.65
Silver	566	55.6
Tin	228	0.06
Titanium	665	1.3
Vanadium	661	0.4
Zinc	443	70.0
Zirconium	660.5	0.28

Table 1.2. Production of Cast Parts as a Percentage of Total Aluminum Products

Country	Percentage
United States	15 %
Britain	23 %
West Germany	23 %
Japan	27 %
France	29 %
Italy	37 %

Table 1.2 shows recent figures representing the percentage of aluminum production transformed into cast products. The large differences from country to country can mainly be attributed to variations in the amount of aluminum alloy used in vehicle applications. For example, in Europe and Japan, 60% and 75% respectively of all aluminum cast parts are used in the transport industry, compared to 44% in the U.S.

Casting Alloy Designation

Specifications for casting alloys are clearly distinguished from those of wrought alloys, and are defined by their chemical compositions. Foundry alloys are obtained either from electrolytic aluminum to which are added the constituent elements or from recycled aluminum metal. Presently it is estimated that more than 50% of aluminum cast parts are made from recycled metal.Each country has developed its own aluminum casting alloy nomenclature and designation, and so far no internationally accepted system has been adopted for identification. In the U.S. the Aluminum Association has adopted a four digit numerical system to identify aluminum casting alloys. The first digit indicates the major alloying element in the group, as follows:

1xx.x	Unalloyed composition; aluminum 99.0% or greater
2xx.x	Copper
3xx.x	Silicon with magnesium and/or copper
4xx.x	Silicon
5xx.x	Magnesium
6xx.x	Unused
7xx.x	Zinc
8xx.x	Tin
9xx.x	Unused

In the 1xx.x group, two digits to the left of the decimal denote the minimum of aluminum content. For example a 190.x designation corresponds to an aluminum of a 99.90% grade purity. The digit on the right of the decimal point indicates the product form, 0 and 1 respectively for castings and ingots.

In the alloy groups from 2xx.x to 9xx.x, the second two digits have no specific significance and serve only to identify the different alloys in the group. The last digit on the right of the decimal point identifies the product form.

Aluminum Cast Alloy Properties

Aluminum alloys are characterized by their low specific gravity which can vary slightly above and below the specific gravity of pure aluminum depending on the major alloying elements. In addition to their light weight, other advantages of aluminum casting alloys include relatively low melting temperatures, negligible gas solubility with the exception of hydrogen, excellent castability especially near the eutectic composition of 11.7%, good machinability and surface finishing, good corrosion resistance, and good electrical and thermal conductivity. A volumetric shrinkage of between 3.5% and 8.5% occurring during solidification constitutes the major drawback of aluminum castings. The shrinkage coefficient should be taken into account during mold design in order to obtain dimensional accuracy and to avoid hot tearing and shrinkage porosity. While the mechanical properties are usually inferior to those of wrought products, the heat treatment of some alloys considerably improves the mechanical properties, and will be discussed in Chapter 2.

The selection of an alloy composition for a particular application is based on three parameters:
- castability (a complex property depending on drossing tendency, mold geometry and alloy solidification characteristics);
- mechanical properties;
- usage properties.

The castability of aluminum alloys is determined by using specific sample molds which are able to evaluate the fluidity, hot tearing and shrinkage characteristics. Foundry properties depend mainly on the alloy composition and the solidification interval which can vary from 0C (eutectic alloys) to 140C (B390 alloy).

The best mechanical properties are generally obtained with heat treatable alloys which include eutectic alloys (A356.0, A357.0) and solid solution alloys (201.0). In the aeronautic and aerospace industry, the demand for greater casting quality and integrity has led to the concept of premium quality castings. Quality assurance for premium quality castings signifies that a specified minimum level of mechanical properties be obtained within various locations in a casting. Through a strict control of melting and pouring practices, impurity level control, grain size refinement and eutectic modification, mechanical properties far superior to those previously available can be obtained. Many commercial castings are approaching the premium quality level as the state of foundry technology improves. The nominal level of mechanical properties is generally well above that obtainable as recently as 15 years ago. The current goal is to control the quality and to maintain it at a high level 100% of the time.

Casting alloys for general use are selected according to such characteristics as machinability, corrosion resistance, hardness and mechanical properties. Alloys for special purpose applications are selected for their unique properties, such as high temperature resistance, low thermal expansion coefficient (390.0), or bearing properties (high Sn alloys).

Casting Processes

In general, aluminum castings can be produced by more than one process. Quality requirements, technical limitations and economic considerations dictate the choice of a casting process. The three main casting processes are as follows:

- sand casting: large castings (up to several tons), produced in quantities of from one to several thousand castings;
- permanent mold casting (gravity and low pressure) : medium size castings (up to 100 kg); in quantities of from 1000 to 100,000;
- high pressure die casting: small castings (up to 50 kg); in large quantities (10,000 to 100,000).

These castings and production sizes are typical, but of course exceptions are always possible. Other casting processes include: investment casting (lost wax), lost foam casting, plaster molding, ceramic molding, centrifugal casting, and new and emerging processes such as squeeze casting, and semi-solid casting.

1.2   Current Market for Aluminum Castings

The current market for aluminum castings is supplied by the most commonly used processes which are sand casting, permanent mold casting and high pressure die casting.   Table 1.3 shows the tonnage of aluminum castings shipped in the U.S. in recent years. In 1986, a total of 1,080,000 tons of aluminum castings were produced by the three major processes. High pressure die casting accounted for the largest share of shipments with 76.4%, while permanent mold casting and sand casting were used for 15.7% and

7.8% respectively. It is estimated that in the U.S., casting by the three major processes represents 90% of all aluminum cast parts production. Table 1.4 shows the casting production in Europe (France and West Germany) by the three main casting processes. In West Germany, castings poured by processes other than the three major ones represent less than 1% of the total tonnage. High pressure die casting in Europe accounts for 40% to 60% of aluminum parts production with significant variations from country to country.

Since 1945, the growth of aluminum casting in the U.S. has been mainly due to the expansion of the permanent mold and high pressure die casting industries (Table 1.3). Between 1955 and 1986 production of high pressure die castings increased from 178,000 tons to 825,000 tons which corresponds to an average annual growth rate of 11.7%. In other developed countries, growth in the high pressure die casting industry was also very substantial as shown in Table 1.5. The rapid expansion of high pressure die casting production can be directly related to the automotive industry especially in a country like Japan where the growth rate over the last 17 years was 13.8% on an average annual basis.

Aluminum Casting in the Automotive Industry

The sharp oil price increase in the early 1970s resulted in a trend to lighter, smaller and more economical automobiles. Aluminum is one of the materials being substituted in automobile production in order to reduce weight. Others are magnesium, plastics, and high strength, low alloy (HSLA) steels. Cast aluminum substitutions for cast iron have proved to be very cost effective, especially for cylinder heads.

Ford and Alusuisse have published estimates of fuel savings through the use of aluminum. Ford has shown that a fuel reduction of 1 liter every 120 km is obtained by 200 kg weight reduction. This amounts to 1000 liters over the life of the vehicle. A study by Alusuisse demonstrates that the replacement of steel components by 50 kg of aluminum results in an 850 liter fuel saving over a ten year period.

The data presented in Table 1.6 illustrates the steady rise in the aluminum content of American automobiles. Between 1975 and 1985, the proportion of aluminum contained in a complete automobile more than doubled from 2% to 5%. By 1982, castings accounted for the majority (60%) of this content.

In Europe and Japan, automobiles are generally smaller and hence lighter. The average weight of an automobile in Europe in 1982 was only 900 kg which is considerably less than the average weight of 1147 kg for a U.S. vehicle in 1985. The aluminum content in a European automobile is close to 4%. In Japan, the aluminum content of a typical vehicle has increased from 3% to 4-5% from the early- to mid-1980s.

Table 1.3. U.S. Aluminum Casting Shipments (1000 Tons Per Year)

Year	Sand Casting (Tons)	%	Permanent Mold Casting (Tons)	%	High Pressure Die Casting (Tons)	%	Total (Tons)
1945	98	52.5	54	29.0	34	18.5	186
1955	83	20.0	149	36.5	178	43.5	410
1963	72	15.0	150	32.0	254	53.0	476
1983	83	9.4	151	17.0	653	73.6	887
1986	85	7.8	170	15.7	825	76.4	1080

Table 1.4. Aluminum Casting Production in Europe

Country	Total (Tons)	Sand (Tons)	(%)	Permanent Mold (Tons)	(%)	High Pressure Die Casting (Tons)	(%)
West Germany(1)	448,600	59,400	13	148,100	33	241,100	54
France(2)	170,000	15,000	9	80,000	48	75,000	43

(1) in 1988
(2) in 1984

Table 1.5. Aluminum High Pressure Die Casting Production (1000 Tons Per Year)[2]

	1970	1980	1985	1986	1987	1988
U.S.A.	427.6	427.1	565.0	748.9	676.4	-
Japan	157.7	369.1	490.0	503.4	527.9	587.5
West Germany	97.1	153.2	205.5	215.8	231.8	241.1
Italy	102.0	168.0	177.0	186.0	203.0	213.0
France	48.4	85.3	75.7	82.4	91.0	100.2

Table 1.6. Average Aluminum Content of Automobiles in the U.S.A. (kg per vehicle)[2][5]

Year	Castings	Wrought Products	Total
1965	24	8	32
1970	26	9	35
1975	27	11	38
1977	29	15	44
1979	33	20	53
1982	36	24	60
1984	37*	25*	62
1987	40*	26*	66
1989	43*	28*	71

*Estimated

Automobile vehicle production is presently concentrated in three major economic blocks comprising North America, Western Europe, and Japan, producing approximately 30 million units annually. It is estimated that between 60% to 70% of all aluminum castings produced are destined for the transportation industry. In the U.S. and Canada, the automobile industry consumes well over half of all aluminum castings made annually. Aluminum is used in a variety of applications in the automobile including engine blocks, cylinder heads, intake manifolds, pistons, wheels, compressor parts, brake calipers, and transmission and steering systems.

Automotive parts are generally produced by permanent mold and sand casting mainly because of the large numbers involved. Safety parts such as wheels and brake calipers which require good mechanical properties are usually cast by the permanent mold process (gravity and low pressure).

Other Market Segments

Both military and civilian aircraft have traditionally been large consumers of aluminum castings. Aircraft companies are encouraging sand foundries, in particular, to implement quality assurance programs in order to supply premium quality structural castings, and considerable research is being carried out to determine the fracture properties of their castings. All predictions are for a strong world aerospace industry through the 1990s. Excellent opportunities should exist for premium quality sand and investment castings in this area. In addition to the automotive and aircraft industries, major users for aluminum castings are found in computers, and office and communications systems. These account for 10% of total U.S. aluminum casting production. Other important markets include: industrial machinery and equipment, power tools, household appliances, and motors and generators.

1.3   The Future Potential for Aluminum Castings

Predicting the future use of aluminum castings is complicated by shifts in the world economy. The U.S. production of aluminum castings has followed a cyclical pattern over the past 20 years, reflecting the influence of economic factors. (Fig. 1.3). During that period, a growth of 2-3% was maintained, due mainly to the expansion of the die casting industry, which grew at a much faster pace than the sand and permanent mold sector, which showed a

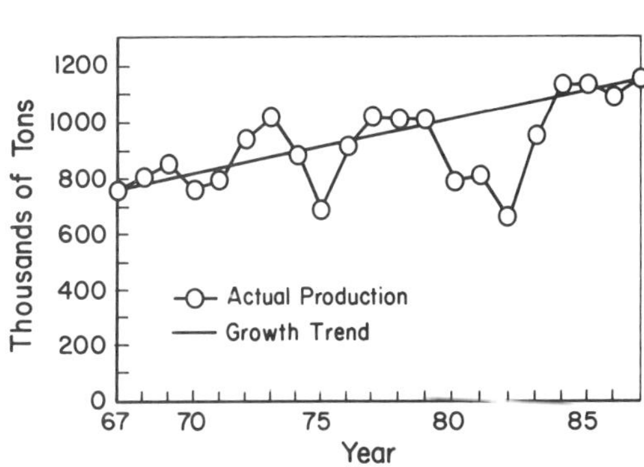

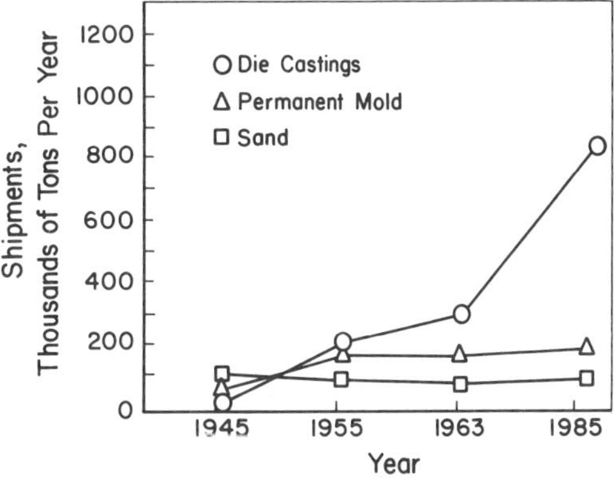

Figure 1.3. U.S. aluminum casting production [3].   *Figure 1.4. Aluminum casting shipments per year in the U.S.A.*

reduced rate of growth over the same period (Fig. 1.4). During the last decade, shipments of aluminum castings in the U.S. have risen dramatically to an annual average of more than 950,000 tons. By comparison, between 1945 and 1955, the annual average was only in the order of 270,000 tons.

Market opportunities in the U.S. for aluminum casting producers are expected to grow through the next decade. Demand for aluminum die castings is projected to grow to over 1,000,000 tons per year by the end of the 1990s. In the same period, aluminum sand and permanent mold castings may reach 700,000 tons per year. These opportunities will not be met by a business-as-usual attitude. Improvements in quality, advances in materials, and developments in processes will determine the success rate. A Japanese-style approach towards quality and productivity in the mass production of automotive components could also influence market growth for aluminum castings. The single most important market opportunity will be in the transportation sector, especially the automobile and commercial vehicle segments of this industry.

Future of Aluminum in the Automotive Industry

The energy crisis of the 1970s, and more recent environmental concerns, have virtually guaranteed that the trend toward lighter weight automobiles will continue. Palazzo, in a 1977 study[4], showed a direct correlation between vehicle weight and fuel consumption. Aluminum ranks high among the light weight substitutes for denser metals such as iron, copper and zinc, and this accounts for its already extensive use in the transportation industry (Table 1.7). At present, all indicators point to a continued increase in the use of aluminum castings in transportation as the quest for lighter, more efficient and less polluting automobiles continues.

Table 1.7. Aluminum Castings in Transportation

Country	Percentage of Production
United States	60
West Germany	67
Italy	67
France	75
Japan	75

Overall the automotive sector offers the best opportunity for growth over the next decade with possible new uses for aluminum in intake manifolds, cylinder heads, engine blocks, wheels, pistons, transmission casings and housings, brake drums, and power stirring components. A large number of these components are currently manufactured from ferrous metals, especially cast iron.

At present only one-third of automotive engine cylinder heads are produced in aluminum casting alloys. Gravity or low pressure permanent mold casting are the methods of choice to cast 319.0 type alloy. It is projected that by the mid-1990s, nearly 100% of all cylinder heads will be converted to aluminum, requiring an additional 100,000 tons of castings per year. Additional production capacity is being built in the U.S.A., Canada, Mexico, Europe, Korea, and Japan. In some of the new plants, the lost foam process is being applied to the casting of cylinder heads.

Another growth opportunity for aluminum is aluminum wheels. Initially utilized as design elements, aluminum wheels are today installed because of their superior heat dissipation during braking, improved road handling and decreased tire wear. In addition, a weight savings of about 25 kg can be made if aluminum is used in place of steel wheels. It is estimated that slightly more than 30% of original equipment wheels are aluminum cast.

In recent years, new production facilities have been built in North

America, and production capacity has been expanded in existing plants in Europe, Japan and North America. New foundries have also opened in developing countries such as Taiwan, Malaysia, Korea and South America, and additional production capacity is planned in aluminum producing countries such as Canada, Australia and Venezuela. Today, it is estimated that 30 million aluminum cast wheels are produced. A production level of 50 million wheels per year by the late 1990s is possible with a market penetration of 50%.

The largest single opportunity for increased use of aluminum castings is provided by engine blocks. In the U.S.A. approximately 5% of all engines are made of cast aluminum. If the current expansion trend continues, 20-25% of the engines could contain cast aluminum blocks by the late 1990s requiring an increased 100,000 tons of metal per year. At present no all-aluminum engines are made in the U.S.A. In Europe, aluminum engines are common in West Germany, France and England, especially in up-scale passenger cars. In one English automobile, a 30% engine weight saving was made by replacing a cast-iron block/aluminum cylinder head with an all-aluminum engine.

Aluminum engine blocks are cast in 380.0 and 390.0 alloy using the low pressure aluminum mold or the high pressure die casting processes. A new alloy known as 3HA was developed and patented in Australia and offers better properties than existing high-silicon aluminum alloys used for cylinder heads and blocks. The increased strength and hardness of the 3HA alloy results from the addition of strontium to an alloy containing 15% silicon and significant amounts of copper, magnesium, nickel and iron. In the U.S.A. alone this alloy has the potential of consuming more than 200,000 tons of aluminum per year.

Aluminum pistons have been made for many years in North America, Europe and Japan. Continued growth for this application is foreseen as new engines which have recently been developed make use of aluminum pistons. Another possible aluminum-based material for pistons is metal-matrix composites. In addition to engines and pistons substantial growth in the use of aluminum cast parts is forecast for such automotive items as brake calipers, suspension arms, engine driveshafts and transmission cases.

Possible competition to aluminum for the manufacturing of cylinder heads and blocks could come from magnesium or cast iron. A new casting process known as the FM process has been developed in France and allows the production of thin walled cast iron. Magnesium engine blocks have been cast in the U.S.A. and weigh only 19 kg compared to 49 kg for the cast iron block. In Sweden an engine with 50 kg of magnesium components made possible a 200 kg weight savings and dramatically increased the fuel efficiency to 35 km/liter (81 mpg).

References

1. Polmear, I.J. *Light Metals*, E. Arnold Publishers, London (1981).
2. *The Economics of Aluminum*, Third Ed., Roskill Information Services Ltd, London (1988 p. 353.
3. "Metal Casting Census Index,"*Foundry Management and Technology*, Penton Publishing (1987).
4. Palazzo, F. "The Future of Aluminum in the Automotive Industry," (Organization of European Aluminum Smelters General Meeting, Florence 1977 Association of Light Alloy Refiners Ltd., London, p.17.
5. *Aluminum Statistical Review*, The Aluminum Association (1988).

Chapter 2

Aluminum-Silicon Foundry Alloys

2.1   Common Alloy Systems

In the United Sates, the Aluminum Association has identified and classified into 7 series (see Chapter 1 and Table 2.1) close to 240 compositions of aluminum casting alloys. This large number is due to the fact that alloy systems have usually been developed to serve individual processes and needs. Some alloys differ only in impurities or small amounts of alloying elements and are identified by a different prefix letter, for example 356.0 and A356.0. Many important foundries in North America, Europe and Japan have developed in-house casting alloys having proprietary compositions.

Aluminum castings can be produced generally by more than one process. High pressure die casting, sand mold and permanent mold are the preferred processes for large production quantities. Smaller casting quantities are produced by using plaster or investment molding.

Aluminum alloy compositions have been designed to make the castings either heat treatable or non heat treatable. In the as-cast condition, the castings are identified by a suffix F following the alloy number. The heat treatment of castings results in improved mechanical properties. Temper designations such as 0, T4, T5, T6 and T7 identify the type of heat treatment cycle. High pressure die castings are not normally heat treated because of the occurrence of blistering at solution temperatures in the 500-550C range. Several excellent general references dealing with aluminum castings can be found. These are given at the end of this chapter [1-4].

Aluminum-Silicon Alloys

Aluminum-silicon castings constitute 85% to 90% of the total aluminum cast parts produced. The most common aluminum casting alloys are listed in Table 2.1 and their properties are summarized in Table 2.2. Aluminum alloys containing silicon as the major alloying element offer excellent castability, good corrosion resistance, and can be machined and welded.

Binary eutectic or hypoeutectic aluminum-silicon alloys are characterized by good castability and corrosion resistance. From the phase diagram in Figure 2.1 it is seen that the 413.0 alloy (approximately 12% Si) contains a predominant eutectic phase, and therefore must be modified with either strontium or sodium to ensure adequate tensile strength and ductility. For high pressure die casting, alloy 413.0 has better castability than alloy 443.0. All casting processes can be used to cast 443.0 parts when ductility, corrosion resistance and pressure tightness are more important than strength.

Strengthening of Al-Si alloys is achieved by adding small amounts of Cu, Mg or Ni. In this family of alloys of hypoeutectic composition, silicon provides good casting properties and copper improves tensile strength, machinability and thermal conductivity at the expense of a reduction in ductility and corrosion resistance. Alloy 319.0 is used extensively for sand and permanent mold casting, while the 380.0 has been used for many years as the principal high pressure die casting alloy. Generally 319.0 and 380.0 alloys are supplied in the as-cast temper, but strength and machinability of a 319.0 alloy can be improved by T6 or T5 heat treatments.

Worldwide, the high pressure die casting industry uses the aluminum-silicon-copper alloy known as 380.0 in the U.S.A. Table 2.3 shows the different international designations which differ not only by the alloying elements such as silicon and copper, but also by the levels of the main impurities such as iron, zinc and manganese. Higher iron contents are preferred to avoid the risk of welding on with dies, but can promote the formation of brittle plates of α-AlFeSi or other complex intermetallics in the

Table 2.1. Composition of Common Aluminum-Silicon Casting Alloys(a)

Alloy	Method(b)	Si	Cu	Mg	Fe	Zn	Others
319.0	S, P	6.0	3.5	<0.10	<1.0	<1.0	
332.0	P	9.5	3.0	1.0	1.2	1.0	
355.0	S, P	5.0	1.25	0.5	<0.06	<0.35	
A356.0	S, P	7.0	<0.20	0.35	<0.2	<0.1	
A357.0	S, P	7.0	<0.20	0.55	<0.2	<0.1	0.05 Be
380.0	D	8.5	3.5	<0.1	<1.3	<3.0	
383.0	D	10.0	2.5	0.10	1.3	3.0	0.15 Sn
384.0	D	11.0	2.0	<0.3	<1.3	<3.0	0.35 Sn
390.0	D	17.0	4.5	0.55	<1.3	<0.1	<0.1 Mg
413.0	D	12.0	<0.1	<0.10	<2.0	-	
443.0	S, P	5.25	<0.3	<0.05	<0.8	<0.5	

(a) Remainder: Aluminum and other impurities
(b) S, Sand Casting; P, Permanent Mold Casting;
    D, High Pressure Die Casting

Table 2.2. Characteristics of Aluminum-Silicon Casting Alloys(1)

Alloy	Casting Method	Resistance To Tearing	Pressure Tightness	Fluidity	Shrinkage Tendency	Corrosion Resistance	Machin- ability	Weldability
319.0	S,P	2	2	2	2	3	3	2
332.0	P	1	2	1	2	3	4	2
355.0	S,P	1	1	1	1	3	3	2
A356.0	S,P	1	1	1	1	2	3	2
A357.0	S,P	1	1	1	1	2	3	2
380.0	D	2	1	2	-	5	3	4
390.0	D	2	2	2	-	2	4	2
413.0	D	1	2	1	-	2	4	4
443.0	P	1	1	2	1	2	5	1

(1) Ratings: 1, best ; 5, worst

Table 2.3. Different International Designations for 380.0 Alloy Type

Country	Specification	Designation	Cu	Si	Fe (max)	Zn (max)	Mn (max)
International	ISO DIS 3522	AlSi8Cu3Fe	2.5-4.0	7.5-9.5	1.3	1.2	0.6
Belgium	NBN 436	DG AlSi8Cu3Fe	2.5-4.5	7.0-9.5	1.3	1.0	0.6
Canada	HA3	SC 84N	3.0-4.0	7.5-9.5	0.6	0.1	0.1
		SC 84R	3.0-4.0	7.5-9.5	1.2	1.2	0.5
Denmark	DS 3002	4254	2.0-4.0	7.5-10.0	1.1	3.0	0.5
Finland	SFS 568	G-AlSi9Cu3Fe	2.0-4.0	7.5-10.0	1.25	1.2	0.5
France	NF A 57-703	A-S9U3A-Y4	2.5-4.0	7.5-10.0	1.3	1.2	0.5
W. Germany	DIN 1725/2	G-AlSi8Cu3 (226)	2.0-3.5	7.5-9.5	0.8	1.2	0.5
		GD-AlSi8Cu3 (226D)	2.0-3.5	7.5-9.5	1.3	1.2	0.5
Italy	UNI 3601	G-AlSi8.5Cu	3.0-4.0	7.5-9.5	0.8	0.05	0.5
	UNI 5075	GD-AlSi8.5 Cu3.5Fe	3.0-4.0	8.0-9.5	1.1	1.0	0.5
Japan	JIS H 5302	ADC 10	2.0-4.0	7.5-9.5	1.3	1.0	0.5
		ADC 12	1.5-3.5	10.5-12.0	1.3	1.0	0.5
Netherlands	NEN 6022	AlSi8Cu3	2.5-4.5	7.0-9.5	0.7	1.0	0.6
Norway	NS 17530	AlSi9Cu3	2.0-4.0	7.5-10.0	1.0	1.3	0.5
Spain	UNE 38-203-76	L2630	2.5-4.0	7.5-10.0	1.0	3.0	0.5
Sweden	SIS 144251	4251	2.0-3.0	6.0-8.0	0.7	2.0	0.5
	SIS 144252	4252	2.0-4.0	7.5-10.0	1.1	1.2	0.5
Switzerland	VSM 10895	G-AlSi8Cu3	2.0-3.5	7.5-9.5	1.3	1.2	0.5
UK	BS1490	LM24	3.0-4.0	7.5-9.5	1.3	3.0	0.5
USA	ASTM B179-80	A380.0	3.0-4.0	7.5-9.5	2.0	3.0	0.5

presence of manganese. The magnesium level is usually specified as below 0.3% to avoid the formation of Mg_2Si, which results in deterioration of the tensile strength.

Age hardenable aluminum-silicon alloys containing magnesium constitute another important alloy group. After solution treatment (T4) and

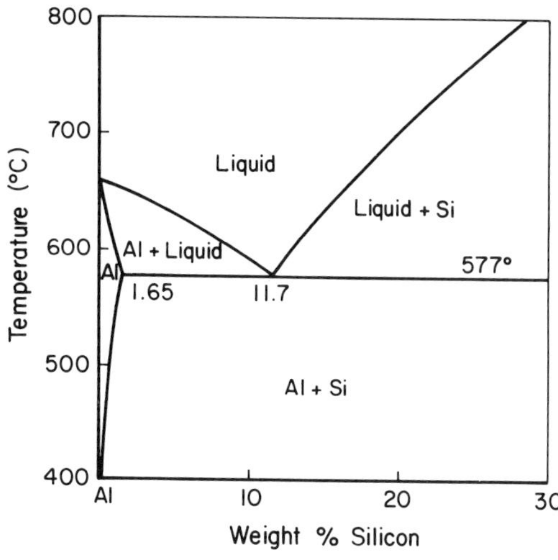

Figure 2.1. Equilibrium binary Al-Si phase diagram.

quenching, aging results in a uniform distribution of Mg_2Si precipitates throughout the aluminum dendrites. The aluminum-silicon-magnesium system is represented by a pseudo-binary Al-Mg_2Si diagram in Figure 2.2. Alloys of the aluminum-silicon-magnesium silicide group are cast in sand or permanent molds and have excellent castability, pressure tightness and corrosion resistance.

Structure control through eutectic modification and heat treatment provides a wide range of properties. Increased iron levels and slower solidification rates have a negative influence on the mechanical properties. Porosity is detrimental to tensile strength and elongation. Alloy 357.0 can be distinguished from a 356.0 alloy by its higher magnesium level. Consequently, heat treated 357.0 alloys have higher tensile strength than 356.0 alloys. A356.0 and A357.0 alloys are higher purity versions of 356.0 and 357.0 alloys. Lower iron levels and a small amount of beryllium in A357.0 result in improved ductility.

Alloy 355.0 of the aluminum-silicon group contains both magnesium and copper. The presence of these two elements results in higher tensile strength but to the detriment of ductility and corrosion resistance. The 332.0 alloy contains higher silicon and copper levels, and is used in internal combustion engines because of its thermal stability and lower coefficient of expansion.

Hypereutectic alloys such as 390.0 and 393.0 containing 15 to 25% silicon exhibit excellent wear resistance and low thermal expansion. Machinability improves as the silicon particles become finer and more evenly distributed by the phosphorous treatment. The 390.0 is low pressure cast in Europe; both high pressure die cast and permanent mold cast in the U.S.A.

Aluminum-Copper Alloys

The use of aluminum-copper alloys has declined in recent years due to their replacement by aluminum-silicon alloys which exhibit better casting properties. Generally aluminum-copper alloys contain less then 5.7% copper and are typical solid solution alloys. Figure 2.3 shows that after a solution treatment, $CuAl_2$ particles are precipitated from the quenched alloy. Generally, aluminum-copper alloys have a greater tendency to hot tearing and microshrinkage formation. Also, fluidity is reduced and good risering and gating is required to ensure casting soundness. A recently developed Al-4.5%Cu-0.25%Mg alloy with a small amount of silver (0.7%) is heat treatable and exhibits higher tensile strength than any aluminum-silicon-magnesium casting alloy.

Aluminum-Magnesium Alloys

Aluminum-magnesium alloys constitute the 5xx.x series and are characterized by high corrosion resistance, good machinability and excellent appearance when anodized. These alloys are very sensitive to Si contamination and are generally difficult to cast, since they freeze as solid solutions over a wide temperature range (Fig. 2.4). They require care in gating, good risering and greater chilling to ensure sound castings. Magnesium increases the tendency for oxidation of the molten alloy and small additions of Be are often made to control drossage.

2.2 Typical Applications

Aluminum-Silicon Based Alloys

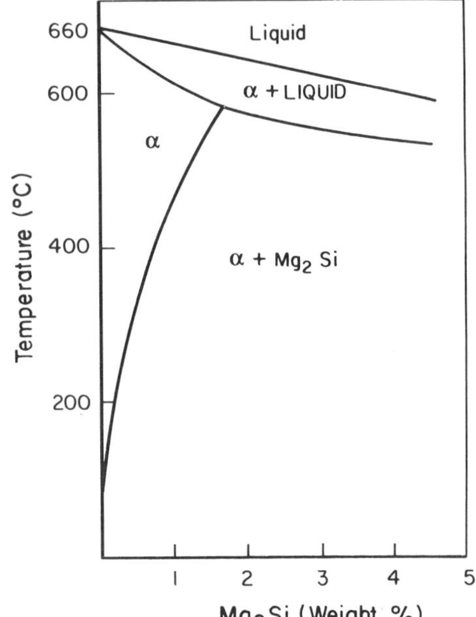

Figure 2.2. Pseudo Binary Al-Mg₂Si phase diagram [1].

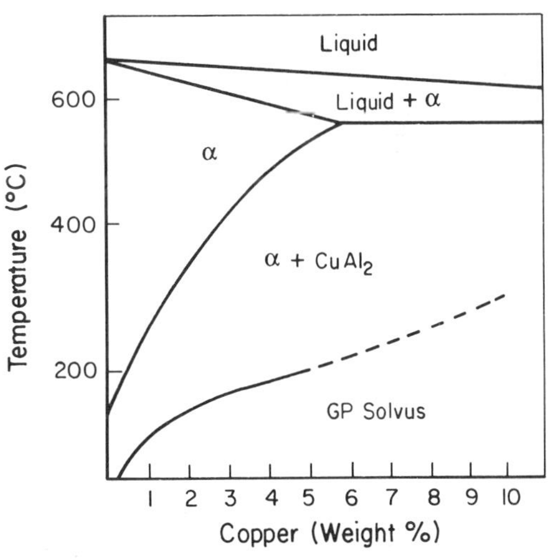

Figure 2.3. Al-Cu phase diagram.

Sand and permanent mold cast binary aluminum-silicon parts are the materials of choice for many automotive, domestic food and pump castings. Other applications include food handling components and castings exposed to marine atmospheres.

Alloy 319.0 is obtained from recycled metal. Parts made with this alloy include cylinder heads and intake manifolds. Properties of A356.0 and A357.0 alloys are very attractive for many automotive and aircraft part applications. In Australia and North America, A356.0 cast wheels are normally modified with strontium to obtain maximum tensile strength and

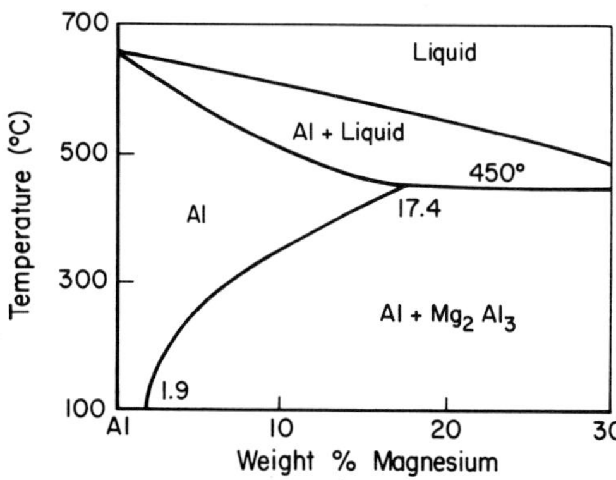

Figure 2.4. Equilibrium binary Al-Mg phase diagram.

ductility after heat treatment. Premium quality A356.0 and A357.0 alloys are usually specified for military and aircraft applications. Heat treated high purity C355.0 alloys are cast to produce tank engines, pump parts, high speed rotating parts and impellers.

Automobile gas and diesel pistons are made of permanent mold cast 332.0 alloys. The addition of nickel improves elevated temperature properties by forming stable intermetallics. In Europe and Japan, hypereutectic aluminum-silicon-copper-magnesium-nickel alloys are used for pistons and cylinders for diesel engines.

The hypereutectic 390.0 alloy was first high pressure die cast for engine blocks without iron liners. A 0.01% addition of phosphorous is recommended to obtain a refined primary silicon. European car manufacturers use low pressure die casting to produce 390.0 engine blocks. Today, 380.0 engine blocks with cast-in iron liners are high pressure cast in Europe and Japan. In the U.S.A. and Canada, engine casings, transmission parts and various other automotive parts are cast in 380.0 alloys.

Aluminum-Copper Alloys

Alloy 201.0 and A206.0 are used for premium-quality aerospace castings mainly because of their excellent tensile strength, and are currently finding increased applications in automotive castings.

2.3   Cast Microstructure

Much of the remaining part of this book is concerned with the use of melt treatments to control cast microstructure. Here, by way of introduction, we present a quick summary of the most common factors which determine the microstructure of cast aluminum alloys. For more detail, the reader is referred to the appropriate Chapters. A small collection of microstructures is also given here in order to illustrate the structure of the more common alloys. In view of the importance of eutectic modification to microstructure control, we have included typical modified microstructures. These specific examples are obtained by strontium modification. However, use of the proper level of sodium will result in similar microstructures.

The microstructures of aluminum-silicon alloys depend strongly both on the composition and the casting process. The rapid cooling in pressure die casting causes a fine eutectic structure, small dendritic cells and arm spacing, and reduced grain size. Slower cooling rates encountered in permanent mold and sand casting necessitate the use of eutectic modifiers such as strontium or sodium to obtain a finely dispersed eutectic silicon. In hypereutectic alloys, phosphorous is added to control the primary silicon. Grain refiners are added to aluminum-silicon, aluminum-copper and aluminum-magnesium alloys to produce a fine equiaxed grain structure.

Grain refinement improves resistance to hot tearing, decreases porosity and increases mass feeding. As a result, a grain refined cast part is more homogeneous with better casting soundness and increased mechanical properties. Aluminum-silicon alloys (A356.0, 413.0) exhibit a resistance to hot cracking index of 1 in Table 2.2, compared to aluminum-copper (201.0) and aluminum-magnesium (A535.0) alloys which have a hot cracking resistance index of 4. Therefore grain refining is of a paramount importance in obtaining sound aluminum-copper and aluminum-magnesium castings. In the case of aluminum-silicon casting alloys, the presence of a large volume fraction of liquid eutectic renders the effectiveness of grain refinement, especially at faster cooling rates (permanent mold) more difficult to substantiate.

Chemical modification dramatically alters the morphology of the eutectic silicon. Even at the rapid cooling rates encountered with high pressure die casting, the eutectic is changed from acicular or lamellar to fibrous with the addition of strontium or sodium.

Microstructures of Binary Aluminum-Silicon Eutectic Alloys

Binary aluminum-silicon alloys close to the eutectic composition (11.7% Si) exhibit, in the unmodified state, an acicular or a lamellar eutectic silicon well dispersed throughout the aluminum matrix (Fig. 2.5a). Because of non-equilibrium cooling and slight variations in composition, the presence of polyhedric primary silicon particles is also observed frequently. The addition of low levels of modifier, e.g., 0.005% to 0.01% Sr, results in a partially modified structure as shown in Figure 2.5b where both lamellar and fibrous eutectic is evident. Concentrations of 0.02% Sr are sufficient to bring about full modification with a completely fibrous structure (Fig. 2.5c). Modification causes the alloy to become slightly hypoeutectic with the result that aluminum dendrites appear more frequently in the microstructure.

Aluminum-Silicon-Magnesium Alloys

The unmodified A356.0 alloy shows a coarse eutectic phase located between the aluminum dendrites (Fig. 2.6a). The addition of 0.008% strontium changes the acicular eutectic to a finely dispersed fibrous eutectic (Fig. 2.6b).

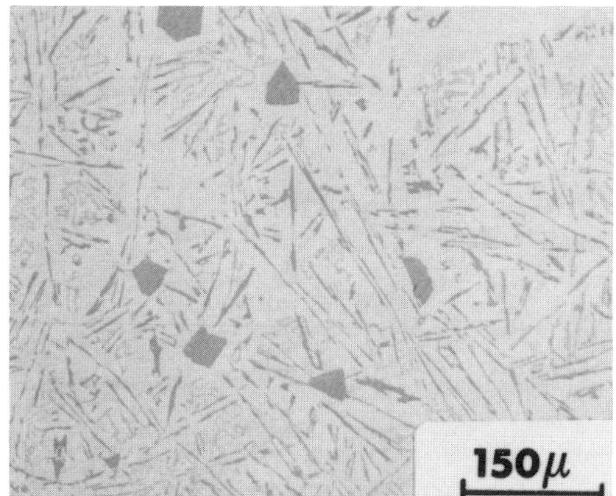

(a) unmodified

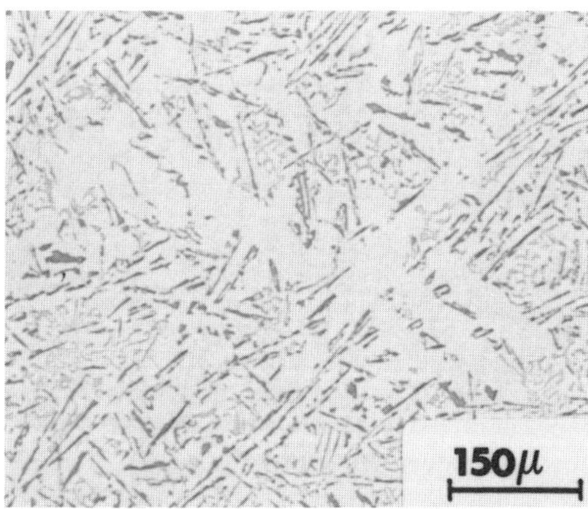

(b) partially modified with 0.01% Sr

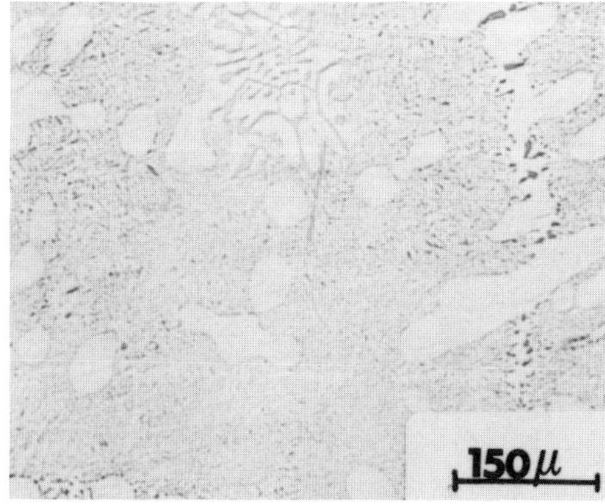

(c) fully modified with 0.02% Sr

Figure 2.5. Microstructures of 413 alloy (x100).

When the magnesium level is increased to produce an A357.0 alloy (0.55% Mg), the non-modified eutectic is lamellar and slightly finer than that found in non-modified A356.0 alloys (Fig. 2.7a). The addition of 0.013% strontium is sufficient to modify the eutectic as illustrated in Figure 2.7b.

When the magnesium content is raised to 0.65%, $Al_8FeMg_3Si_6$ intermetallics form which are difficult to dissolve during solution treatment at 540C (Fig. 2.8). A high iron level (0.4%) leads to the formation of Al_5FeSi and $Al_8FeMg_3Si_6$ intermetallics which cannot be changed by heat treatment (Fig. 2.9).

Aluminum-Silicon-Copper Alloys

Al-Si alloys containing various amounts of copper as an alloying element are readily modified with strontium. Figure 2.10 shows a non-modified and a modified structure of an 319.0 alloy. A 0.02% strontium level is usually adequate to obtain a fibrous eutectic silicon (Fig. 2.10b).

Addition of strontium also modifies the microstructure of 380.0 alloys. Figure 2.11 exhibits the non-modified structure cooled at a rate simulating sand casting. The addition of 0.018% strontium (2.11b) causes a fibrous eutectic silicon to form. At high pressure die cast cooling rates strontium (2.12a) is still an effective modifier, and changes the coarse microstructure to a fibrous one (Fig. 2.12b).

Hypereutectic Casting Alloys

The primary silicon in hypereutectic alloys is usually very coarse and imparts poor properties to these alloys. Phosphorus additions are normally made to refine the primary silicon resulting in the typical microstructure of a 390.0 alloy shown in Figure 2.13. The subject of refinement of primary silicon is covered in detail in Chapter 7.

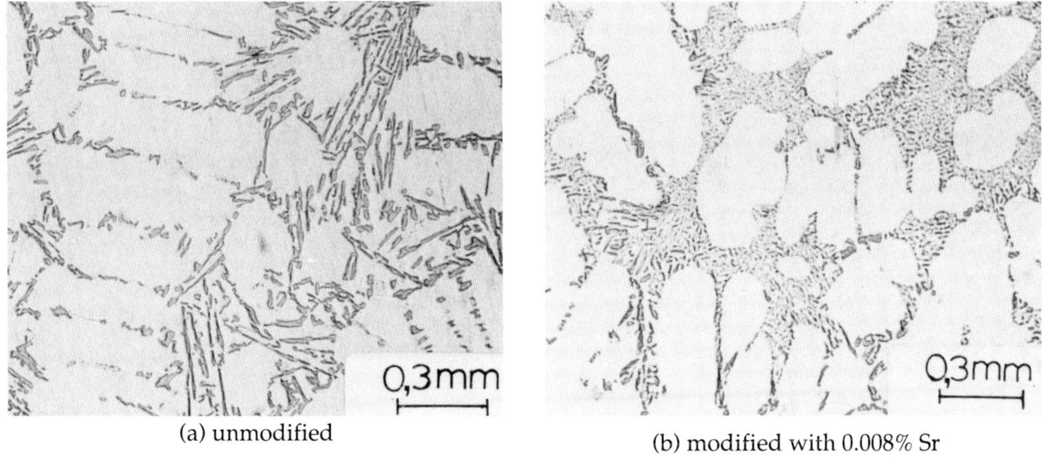

(a) unmodified　　　　　　　　(b) modified with 0.008% Sr

Figure 2.6. Microstructures of A356.0 alloy (x100).

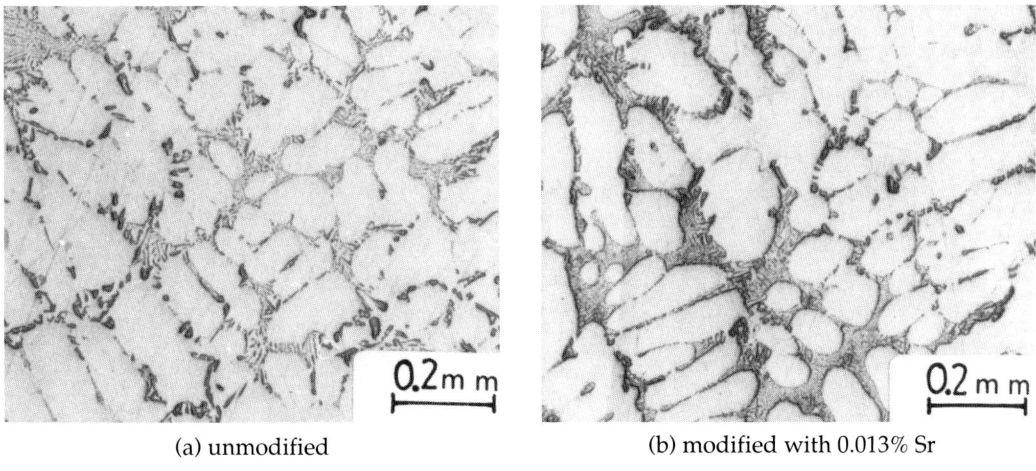

(a) unmodified　　　　　　　　(b) modified with 0.013% Sr

Figure 2.7. Microstructures of A357.0 alloy (x100).

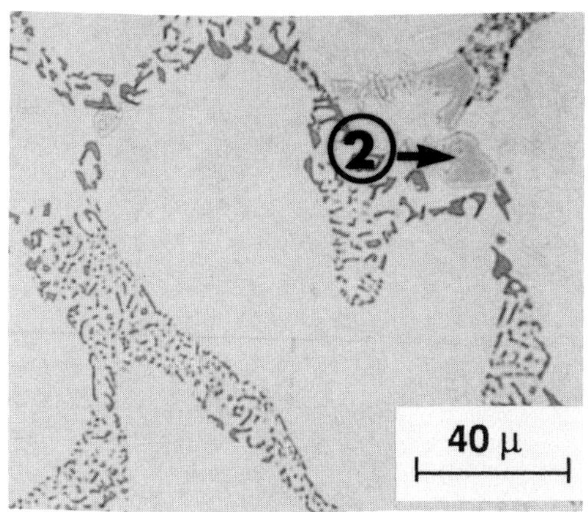

Figure 2.8. Microstructure of A357.0 alloy containing 0.65% magnesium (x100).
② $Al_8Fe\,Mg_3Si_6$

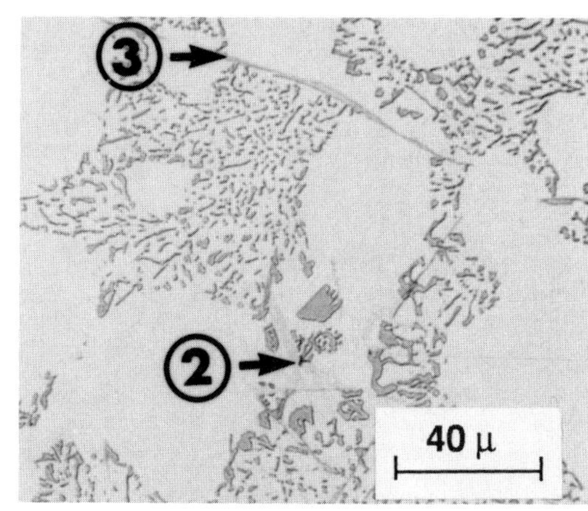

Figure 2.9. Microstructure of A357.0 alloy containing 0.35% iron and 0.60% magnesium (x100).   ② $Al_8Fe\,Mg_3Si_6$ ③ $Al_5Fe\,Si$

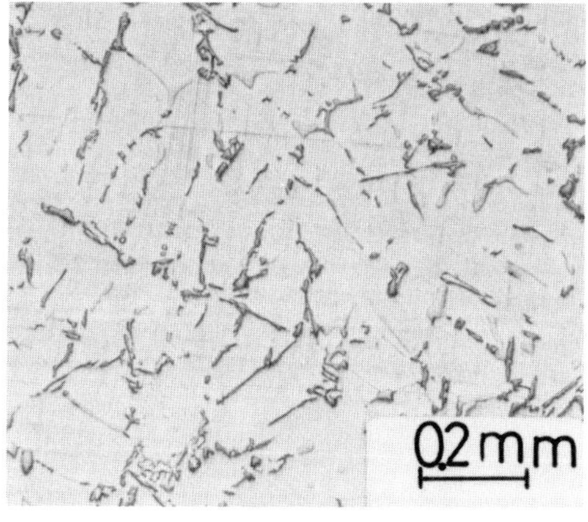

(a) unmodified

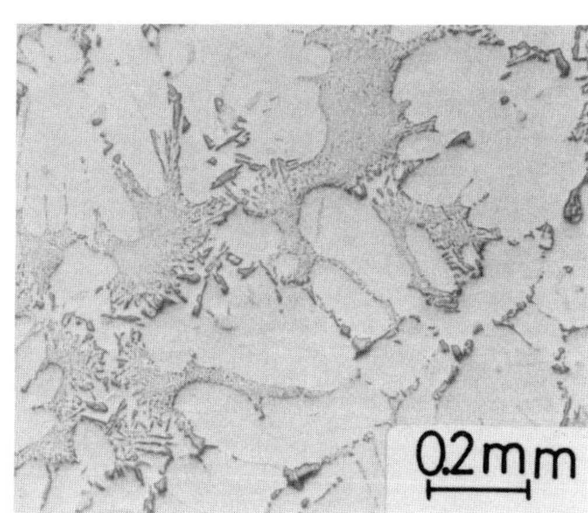

(b) modified with 0.019% Sr

Figure 2.10. Microstructures of 319.0 alloy (x100).

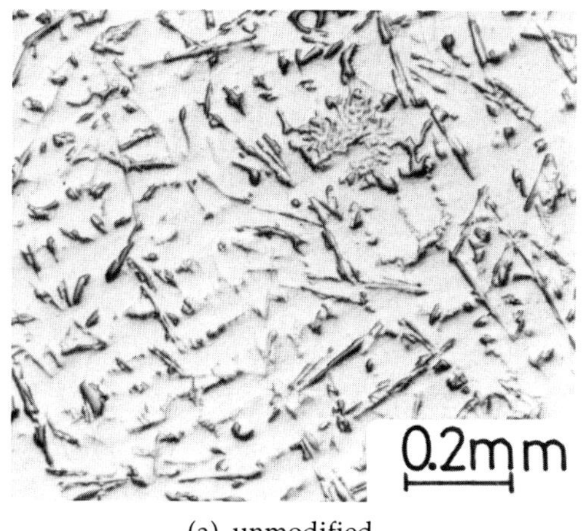

(a) unmodified

(b) modified with 0.018% Sr

Figure 2.11. Microstructures of 380.0 alloy cast at a low cooling rate ($1°C\ sec^{-1}$) (x100).

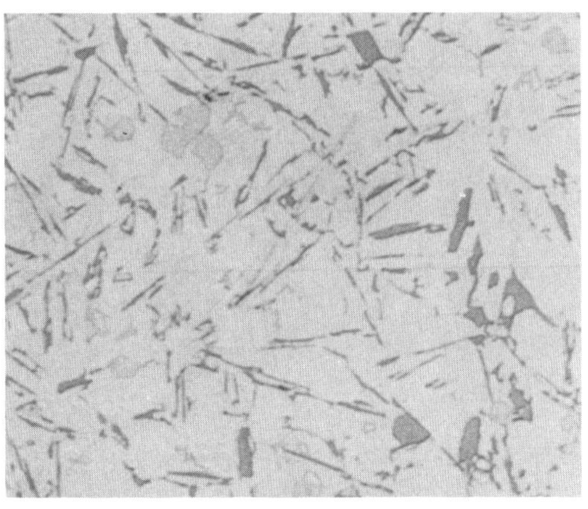

(a) unmodified

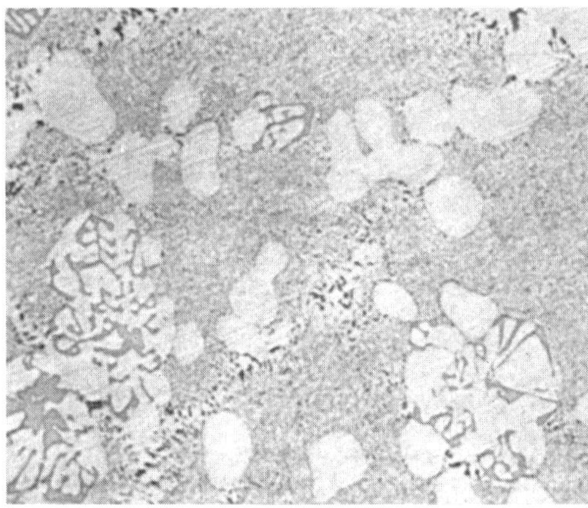

(b) modified with 0.025% Sr

Figure 2.12. Microstructures of 380.0 alloy cast by high pressure die casting (x100).

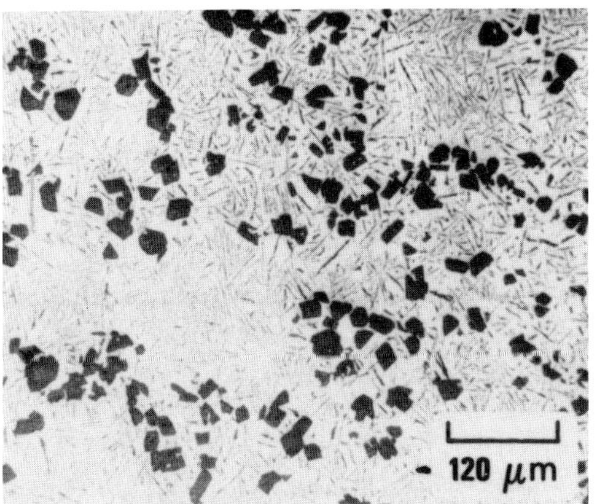

Figure 2.13. Microstructure of 390.0 alloy type, chill cast and treated with phosphorous.

References

1. *Aluminum Casting Technology*, American Foundrymen's Society, Inc., Des Plaines, Illinois (1986).
2. Polmear, I.J. *Light Alloys*, Edward Arnold Publishers, London (1981).
3. *Metals Handbook*, Ninth Edition, Vol 15, ASM International, Metals Park, Ohio (1988).
4. Hatch, J.E. *Aluminum-Properties and Physical Metallurgy*, ASM International, Metals Park, Ohio (1984).

Chapter 3

Eutectic Modification

3.0   Introduction

It was in 1921 that Aladar Pacz was granted a United States Patent on the discovery that Al-Si alloys containing between 5% and 15% silicon could be treated with alkali fluoride (preferably sodium fluoride) fluxes to yield alloys of improved ductility and machinability[1] . The discovery of Aladar Pacz is undoubtedly one of the most significant related to aluminum foundry alloys, since it placed these alloys in the family of modern engineering materials with the light weight, high strength and predictable fracture properties that are so important in today's transportation and defense industries.

Metallographic techniques were well known in the 1920s and it soon became evident that alloys treated with sodium salts possessed very different microstructures from untreated material. Figure 3.1 illustrates the important difference. Untreated alloy contains the silicon phase in the form of large plates with sharp sides and ends (acicular silicon). The addition of small amounts of sodium through Pacz's fluxes caused the eutectic silicon to solidify with a fine, apparently globular, morphology. Although the scanning electron microscope was unknown in Pacz's time, its development has led us to new insight into the nature of these structures. The scanning electron micrographs of Figure 3.1 show clearly the plate-like nature of the silicon in the alloy to which no sodium was added. The globules of the sodium treated structure are seen, in fact, to be the ends of silicon fibers which form an interconnected network. It is this transformation from acicular to fibrous silicon that is termed modification, and which is responsible for the enhanced properties associated with modified Al-Si castings.

Since the original discovery of modification, there have been many developments in its science and technology, although most of these have occurred only since about 1970. It is with these developments that this and the following three chapters are concerned. Although the central theme will be how modification relates to the casting process and to the properties of the cast alloy, the question is often asked by foundrymen, "Why does modification occur?" To begin to answer this, we must understand the basic science behind modification.

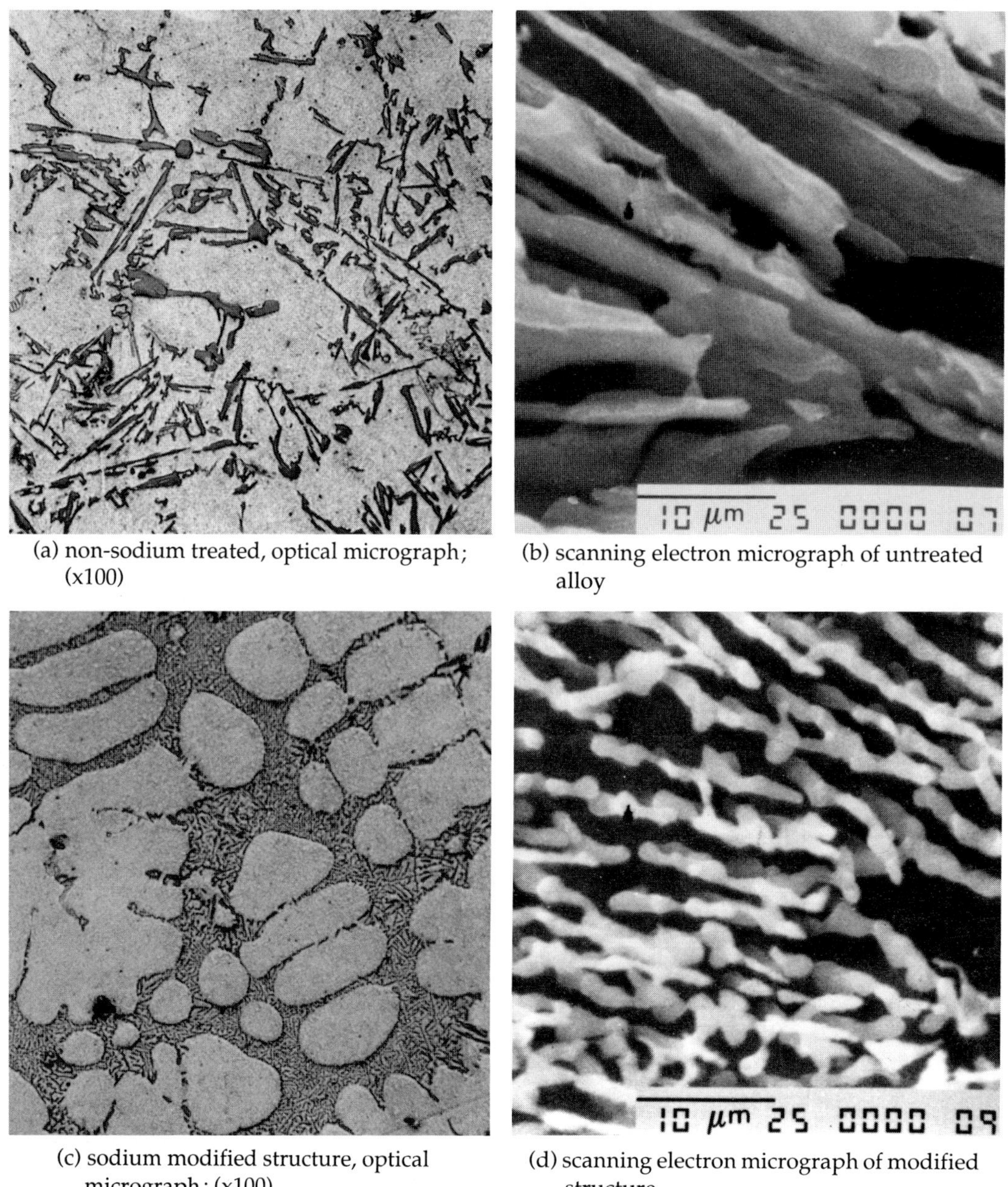

(a) non-sodium treated, optical micrograph; (x100)

(b) scanning electron micrograph of untreated alloy

(c) sodium modified structure, optical micrograph; (x100)

(d) scanning electron micrograph of modified structure

Figure 3.1. The as-cast microstructure of A356 alloy.

3.1 The Fundamentals of Modification

Of the several hundred scientific papers which have been written about modification since the 1920s, the vast majority have been concerned with explaining how it is possible to change eutectic silicon from coarse plates to fine fibers. Most of the explanations met with little success until very recently, when electron microscope techniques allowed the study of crystallographic defects in the silicon phase itself. Any workable explanation must deal with the following facts:

1) several elements are known to cause modification. These include some groups IA, IIA, and rare earth elements. Of all of these, sodium is the most effective in producing a fine, uniform, fibrous structure;
2) modifiers are effective at very low concentration levels, typically 0.01% to 0.02%;
3) a modified structure can be produced without any addition by very rapid solidification (quench modification).

The Solidification of Silicon

Of the two phases which take part in the freezing of an Al-Si eutectic, it is the silicon phase which plays the critical role in modification. In fact, our knowledge to date indicates that the aluminum solid solution exerts only a very minor influence on the process. Silicon is a nonmetal, and as such freezes in what is called a faceted manner, i.e., it forms crystals which are bounded by definite crystallographic planes, and it is usually able to grow in only certain very specific crystallographic directions.

The unmodified plate-like form of silicon exhibits the crystallography illustrated in Figure 3.2. This crystal can grow easily only in the <112> direction, and when the crystal structure of silicon is taken into account, this implies that the large flat faces of the crystal are <111> planes. A very important feature of silicon crystallization is that twins are easily formed. These crystallographic defects, illustrated schematically in Figure 3.3, occur when large groups of atoms uniformly shift position across a plane in the crystal structure, known as a twin plane. In silicon, twins form readily

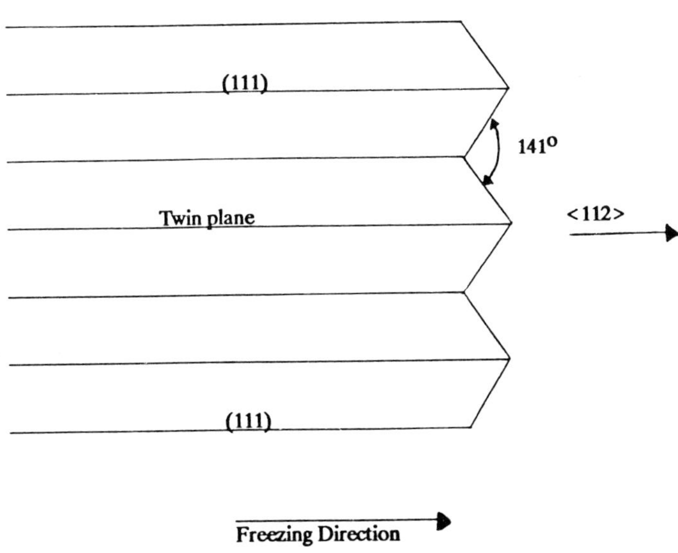

Figure 3.2. Schematic representation of the growth of an acicular silicon crystal from the melt.

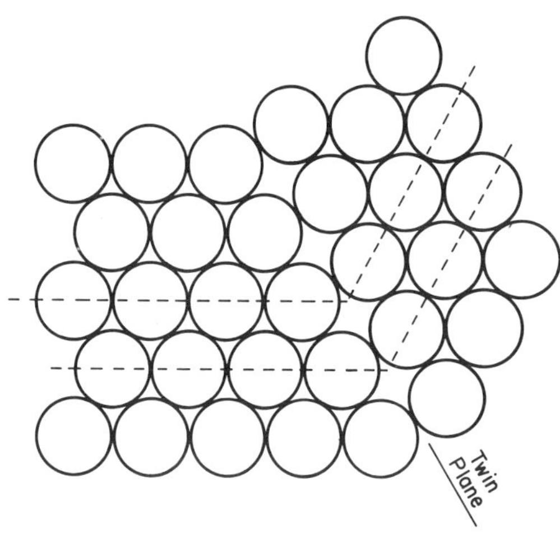

Figure 3.3. Twinning in a crystal. Note the conti-
nuity of the atom planes across the twin plane.

across the <111> planes, and this has the effect of producing a self-perpetuating groove of 141 degrees at the solid liquid interface (Fig. 3.2). Crystallization of the silicon takes place by the addition of atoms to form steps which move across the solid-liquid interface. These steps (Fig. 3.4) originate at the twins, and since the twin planes always intersect the solidifying interface, there is a constant supply of growth sites at which freezing of silicon can occur.

It is evident that the situation just described, and illustrated in Figure 3.2, is not one which favours branching of the silicon crystal, and hence unmodified silicon occurs in essentially an unbranched, flat-plate morphology. It is equally evident that the modified fibrous structure must arise from a very different type of growth, one which allows a free and easy branching to occur. This difference appears to lie in the number of twins found in unmodified and modified silicon. Careful studies by transmission electron microscopy have revealed that modified silicon fibers contain orders of magnitude more twins than do unmodified silicon plates (Fig. 3.5), and that the surface of the fibers is microfaceted and rough as a consequence of the intersection of myriads of twin planes with it. Silicon fibers are very crystallographically imperfect, and each surface imperfection is a potential site for branching to occur should it be required by the solidification conditions. As a result, fibers in the chemically modified eutectic are able to bend, curve and split to create a fine microstructure; the plates of the unmodified structure are inhibited by their relative crystallographic perfection and can do little but form in a coarse acicular fashion.

This remarkable difference in twin density is caused by the addition of only a fraction of one weight-percent of modifier. One possible explanation for the effect is that atoms of the modifier absorb onto the growth steps of the silicon solid-liquid interface. If the modifier atom radius has the correct size with respect to the atomic radius of silicon ($r_{modifier} : r_{silicon} = 1.646$) a growth twin will be caused at the interface[2] as illustrated in Figure 3.6. This phenomenon has been named, impurity induced twinning, and it is supported by the observation that modifiers become concentrated in the silicon, not the aluminum, phase.

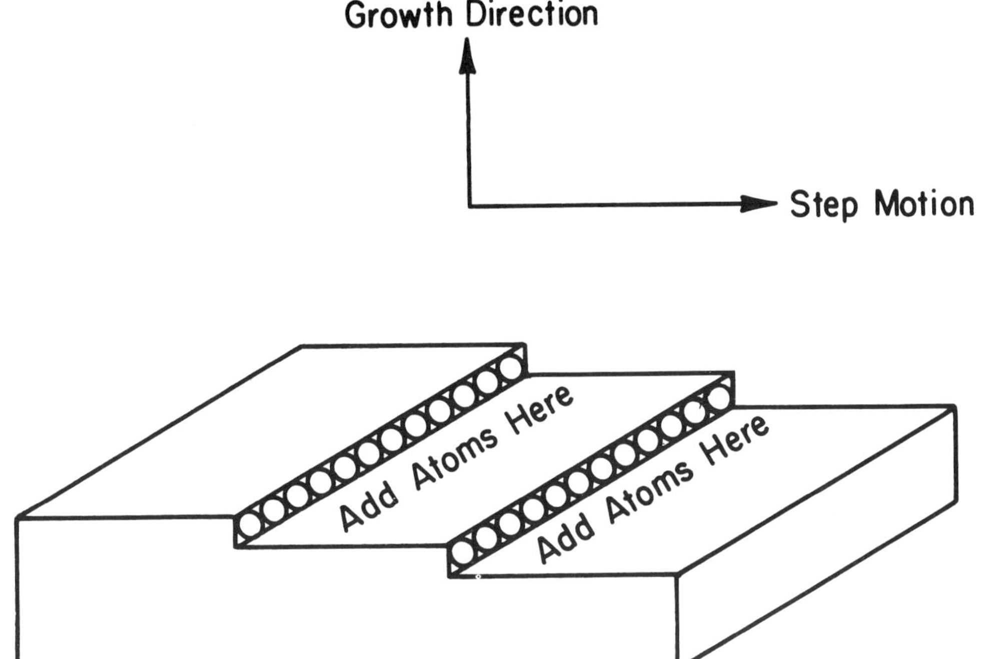

Figure 3.4. Schematic illustration of the solid-liquid interface of a solidifying silicon crystal. Growth of the crystal takes place by the movement of steps across the interface. This movement is caused by the addition of atoms at each step.

Modifying Efficiency

The effectiveness of various elements as modifiers is apparently related to the number of twins produced in modified silicon fibers. Measured values of twin spacings are presented in Table 3.1 for five elements which are known to modify. Clearly, sodium produces the greatest number of twins; it also produces the finest modified structures at the lowest concentrations. It might be expected therefore that sodium is the ideal modifier. Other factors do come into play, such as ease of dissolution, vapor pressure and stability in the melt. These, combined with the crystallographic factors discussed here, determine the choice of a modifier. Further discussion of the strong and weak points of the various modifiers will be found throughout this and subsequent chapters.

Quench Modification

A fibrous eutectic structure can be obtained, in the absence of chemical modifiers, by rapid solidification in the growth rate range 400 to 1000 µm per second. Quench modified structures appear optically identical to impurity modified material. However, electron microscopy has revealed that the silicon is similar to the unmodified form in that it contains no, or very low, levels of twinning. This structure is simply an exceedingly fine form of the unmodified eutectic occasioned by very rapid solidification. In practical

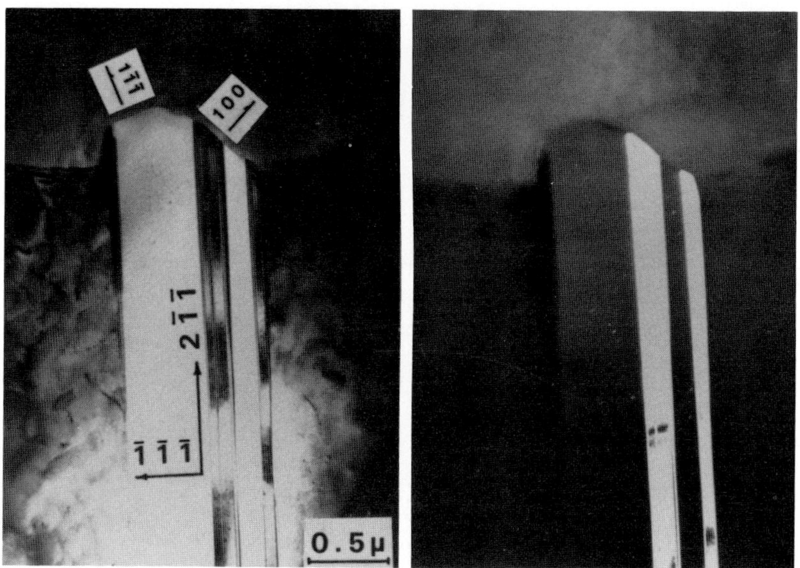

(a) acicular silicon showing a very low twin density

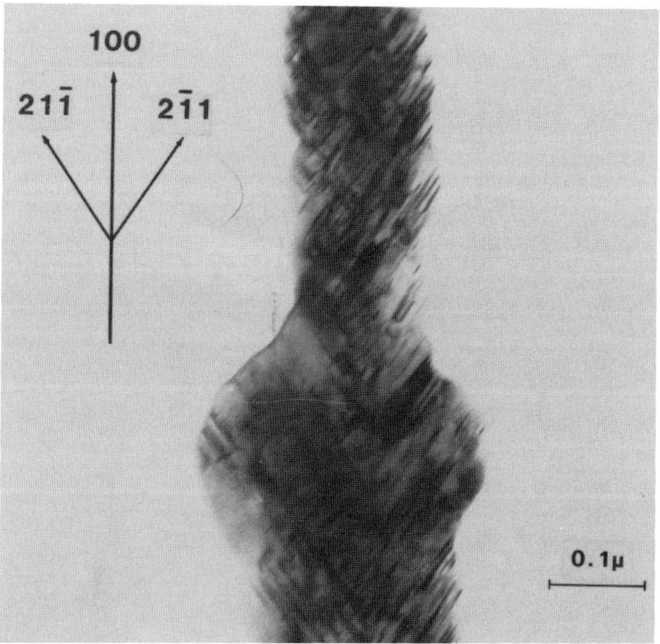

(b) a sodium-modified silicon fiber with very high twin density
and a very rough surface caused by multiple impurity induced
twinning[2]

Figure 3.5. Transmission electron micrographs of silicon.

terms it is of little consequence, as commercial casting processes, with the possible exception of die casting, do not operate at solidification rates sufficient to cause quench modification.

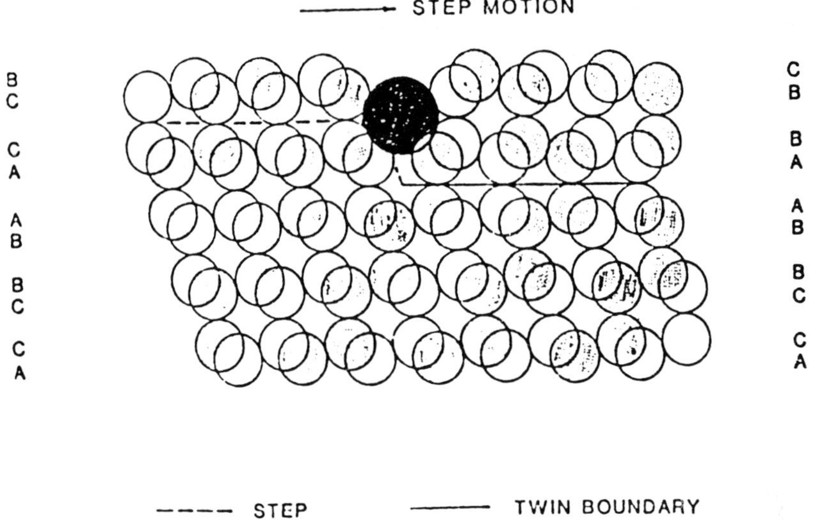

Figure 3.6. Impurity induced twinning. The modifier atom (black) absorbs onto the solid liquid interface of silicon at a growth step. As a result, the crystal structure of silicon is disturbed, and a twin results in the silicon which crystallizes after absorption of the foreign atom [2].

It is well known that chemical modifiers are more effective at higher freezing rates. For example, it is much easier to modify a chill casting than a heavy section sand casting. This is to be regarded as fundamentally different from quench modification. In the presence of a chemical modifier, both the twinning frequency and the angle of branching increase with freezing rate. Both of these promote modification and lead to finer structures.

3.2 Chemical Modification by Sodium, Strontium or Antimony

Several elements have been found to produce the fibrous modified structure. Among these are sodium, potassium, rubidium, cerium, calcium, strontium, barium, lanthanum and ytterbium. In addition, a few elements act to produce a finer version of the coarse acicular structure. This lamellar silicon can be produced by the addition of arsenic, antimony, selenium or cadmium to the melt. There are almost certainly other elements which are capable of changing the form of eutectic silicon in aluminum casting alloys, but for the moment, only sodium, strontium and antimony find any significant industrial use. The amount of each element required depends somewhat on the alloy composition—a higher silicon content requires more modifying agent. Typical retained sodium levels are in the range 0.005-0.01%; strontium amounts of 0.02% are sufficient to modify a 7% Si alloy such as 356, but up to 0.04% is needed for a eutectic alloy such as 413. Antimony which is used frequently in Europe and Japan is effective in amounts of 0.1% and greater.

Table 3.1. Measured Twin Spacings in Several Modified Systems[2]

Structure	Modifier	Twin Spacing at Constant Freezing Rate (nm)
acicular	None	400
fiber	Na	5
fiber	Sr	30
fiber	Ba	30
fiber	Ca	100
fiber	Yb	50

The chemical and physical properties of these three melt treatment agents are very different, and because of this, they are added to the melt in quite different forms. Sodium can be added as a flux in the same way as the original Pacz discovery, but many modern applications use elemental sodium, vacuum packed in small aluminum cans. The high reactivity of sodium requires a packaging technique which will minimize its oxidation and hydrogenation. Elemental sodium stored under kerosene has been used in the past, but this technique is messy and kerosene can add hydrogen to the melt. Because sodium has a very low solubility in aluminum (about 0.01%), the manufacture of aluminum base master alloys is not practical for the purposes of sodium modification.

Unlike the case with sodium, the manufacture of aluminum-strontium master alloys is entirely feasible, since the Al-Sr phase diagram contains several intermetallic compounds (Fig. 3.7). Some common master alloy compositions for strontium addition are:

Al-3.5%Sr
Al-10%Sr
Al-10%Sr-14%Si
90%Sr-10%Al

Pure strontium has sometimes been used as a modifier, but it is reactive with air and water vapor, and within a few minutes becomes covered with a mixture of SrO, SrO_2, $Sr(OH)_2$ and $(CaSr)NO_3$. This layer can completely prevent dissolution unless it is mechanically removed, and for this reason, pure metallic strontium is not the preferred way to add this element to Al-Si alloys. Master alloys which contain less than about 45% Sr are not reactive in air, and no special precautions are necessary for their packaging and storage. Alloys with 20% to 60% strontium have melting points in excess of 900C (1652F), and while they are stable in air, they are of no practical use. The 90%Sr-10%Al alloy does contain some elemental strontium in its microstructure, and hence it is somewhat reactive in air, although much less so than pure strontium. For this reason, it is, like sodium, packaged in sealed containers. These usually contain an inert gas rather than a vacuum, reflecting the lesser reactivity of the strontium alloys compared to elemental sodium. Some common sodium and strontium additives are shown in Figure 3.8.

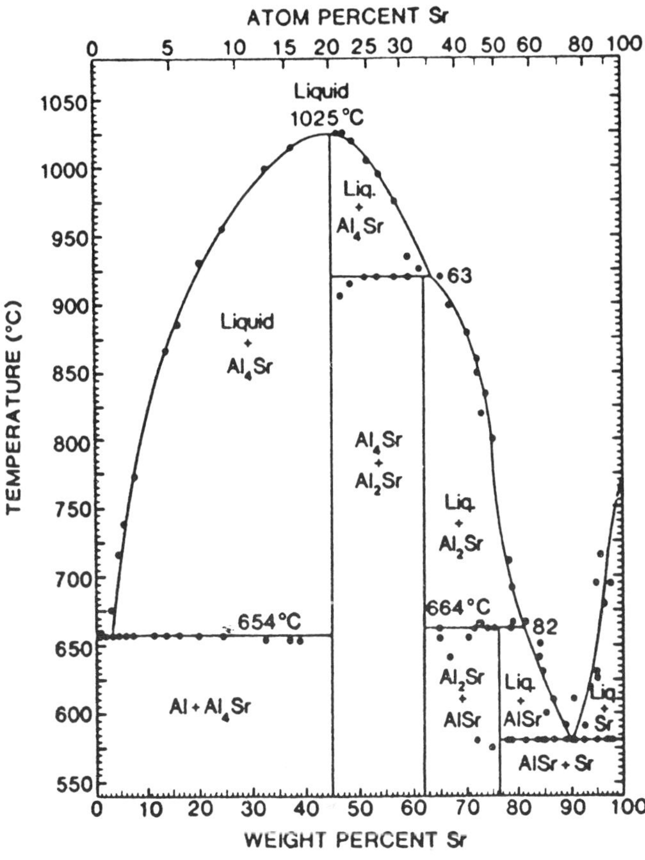

Figure 3.7. The Al-Sr phase diagram [3].

The melt treatment process itself is simple and can be done by plunging the additive held in a perforated cup or bell below the surface of the melt. A gentle stirring action improves the dissolution rate but this should not be so severe as to unduly agitate the bath surface, otherwise hydrogen gas pickup will be facilitated. The addition of sodium is accompanied by a violent reaction which in itself usually causes severe agitation and can result in increased hydrogen levels. Strontium treatment is quiescent and hydrogen pickup is not a problem. The question of hydrogen and modification will be treated more fully in Chapter 4. Dissolution of sodium is usually complete within five minutes. When strontium is the modifier it is important to match the addition temperature to the master alloy composition. Dissolution times of less then five minutes can be achieved if, for example, a 90%Sr-10%Al master alloy is used within the optimum temperature range, e.g., 680C-720C (1056F-1328F) for A356 alloys.

Both sodium and strontium premodified ingot is available from alloy

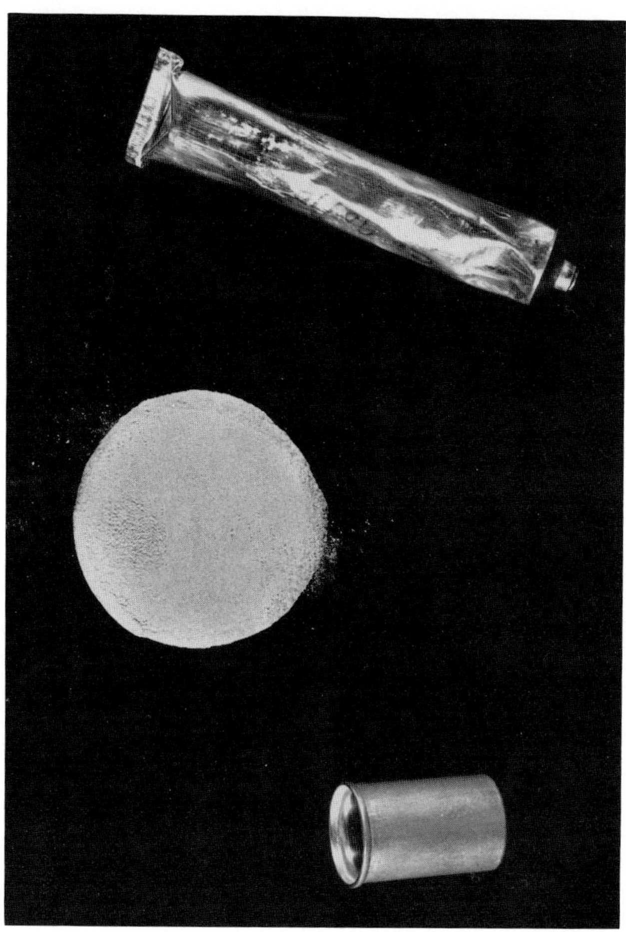

Figure 3.8. Common commercial modifying agents. From top to bottom: 90% Sr-10% Al alloy packed under inert gas; sodium flux tablet; vacuum-packed metallic sodium.

producers. In principle, such ingot needs to be only remelted and cast, but significant losses of modifier, particularly sodium, occur on remelting. Hence, precise control of modifier amounts is often difficult, and it is usually necessary to "top-up" the sodium or strontium concentration after remelting using one of the methods just described.

Antimony is a toxic material and in addition it can react with hydrogen dissolved in the liquid aluminum to form deadly stibine gas according to the reaction:

$$Sb + 3\underline{H} \longrightarrow SbH_3.$$

For these reasons, antimony is not added to melts in the foundry. Antimony treated alloy is purchased as premodified ingot from primary aluminum suppliers and is simply remelted and cast. As antimony is very stable in the melt, losses are virtually nil, and no extra additions are required.

Dissolution of Sodium and Strontium

The process of adding a modifier to a melt involves a dissolution of the modifying agent and its subsequent dispersal throughout the body of the melt. When sodium was the only modifier used, this was a very easy process. Sodium melts at 98C (207F) and thus will enter readily into melts which are normally treated in the range 775C to 800C (1425F to 1475F). Dissolution of sodium is practically instantaneous at these temperatures, but it has a very high vapor pressure (0.2 atm. at 730C, 1345F), and hence large amounts of the sodium which are added boil off almost immediately. Sodium recoveries are therefore poor (20% to 30% of the addition), despite the excellent dissolution characteristics of this element. A lowering of the melt temperature is of little use as the dissolution slows significantly below 700C (1290F). Thus sodium is characterized by an easy dissolution above 700C, but a poor and somewhat unpredictable recovery.

Strontium, on the other hand, can be added so as to give high (about 90%) and very reproducible recoveries, but its dissolution characteristics are more complex than those of sodium, and its successful application requires more understanding on the part of the user. Consideration of the Al-Sr phase diagram given in Figure 3.7 shows that high strontium containing master alloys, such as 90%Sr-10%Al, have microstructures consisting of elemental strontium and the compound AlSr. The low strontium, high aluminum master alloys, Al-10%Sr for example, consist of almost pure aluminum co-existing with the intermetallic compound Al_4Sr. These different structures of the master alloy impart very different dissolution characteristics.

High strontium containing master alloys dissolve by a process known as reactive dissolution. The elemental strontium (and perhaps the AlSr intermetallics as well) react with the liquid Al-Si alloy into which they are immersed, to produce new intermetallic compounds. Not all of these have been identified, but $SrAl_2Si_2$ is certainly one of them. Within a certain temperature range this reaction is highly exothermic, and the master alloy addition can be heated up to 100C above the bath temperature. Figure 3.9(a) shows the temperature distribution inside a piece of 90%Sr-10%Al master alloy immersed in a liquid A356 melt at 650C (1200F). It is seen that the master alloy quickly reaches the melt temperature and then surpasses it by almost 100C as the exothermic reaction proceeds. This reaction is complete after about 60 seconds and the thermocouple temperature diminishes to the bath temperature.

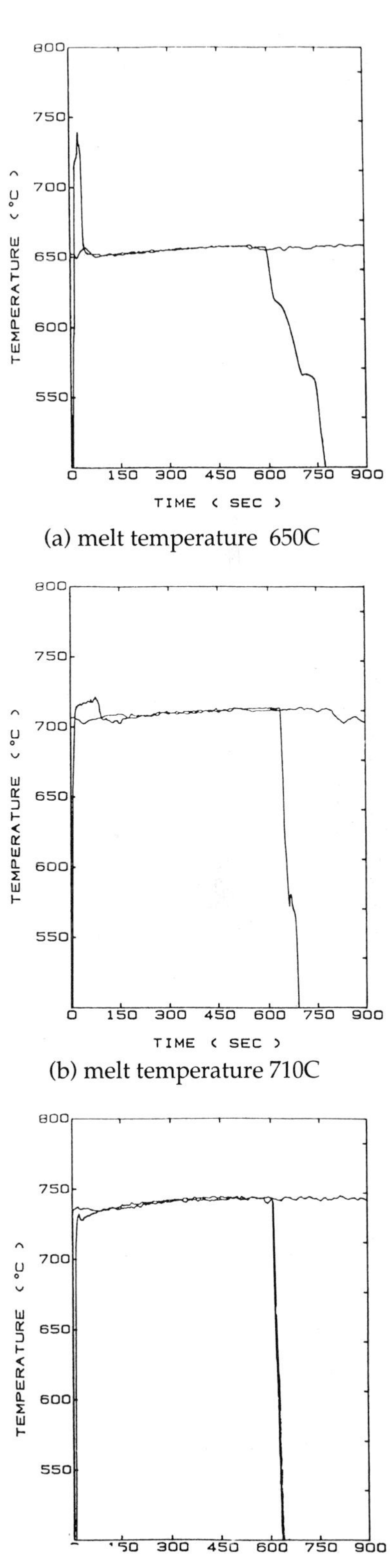

(a) melt temperature 650C

(b) melt temperature 710C

(c) melt temperature 740C

Figure 3.9. Temperature at the center of a 90%Sr-10%Al master alloy dissolving in an A356 bath **[4]**.

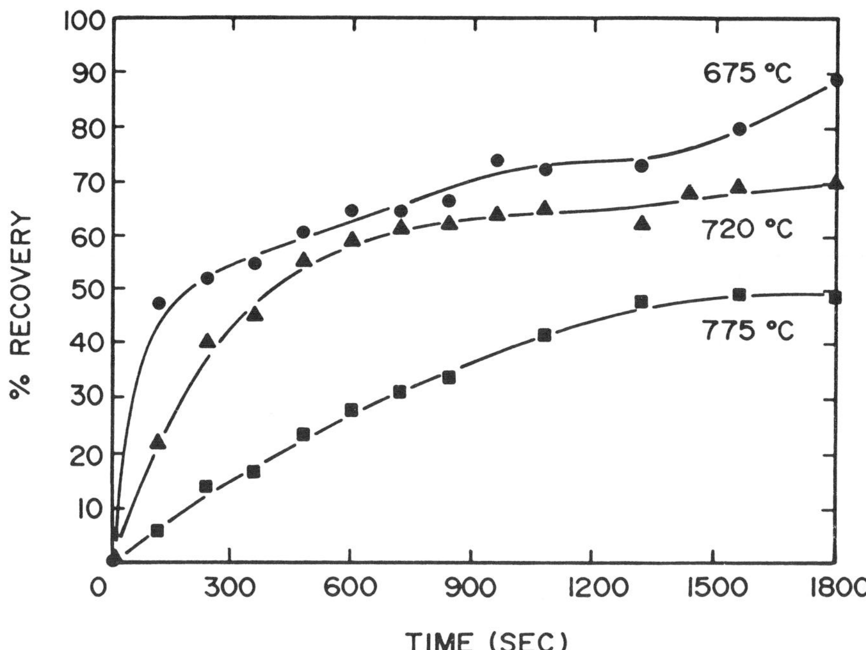

Figure 3.10. The recoveries of strontium added as a 90%Sr-10%Al master alloy to A356 melts held at various temperatures [4].

A characteristic of exothermic reactions is that their intensity diminishes as the temperature is increased. This is also the case with the dissolution of high strontium master alloys. Figures 3.9(b) and (c) illustrate the temperature behavior at melt temperatures of 710C (1310C) and 740C (1365F). The amount of exothermicity is much less at 710C than at 650C, and at 740C it has ceased altogether. The presence of an exothermic reaction seems to be necessary in order to achieve a high recovery of strontium when using these alloys. In Figure 3.10 are given measured recovery curves in A356 for a 90%Sr-10%Al alloy. Best recovery is achieved under conditions which promote the greatest exothermic reaction, i.e., at low temperatures. In the absence of an exothermic reaction, strontium does dissolve in the melt but at a much slower rate, see for example the behavior at 775C (1425F).

High strontium containing master alloys therefore dissolve best at low rather than high temperatures and should be added at the lowest practical temperature. Early users of strontium frequently reported difficulties when making additions to the melt, due, no doubt, to the fact that they attempted to perform the melt treatment at too high a temperature. In cases where lowering the melt temperature is not practical, a small ladle of melt can be removed and allowed to cool. The master alloy when added to this liquid will readily dissolve, and the contents of the ladle can be poured back.

Low strontium master alloys behave quite differently and exhibit classical dissolution behavior. Their dissolution improves as the temperature increases. Microstructures of two popular low strontium alloys are shown in

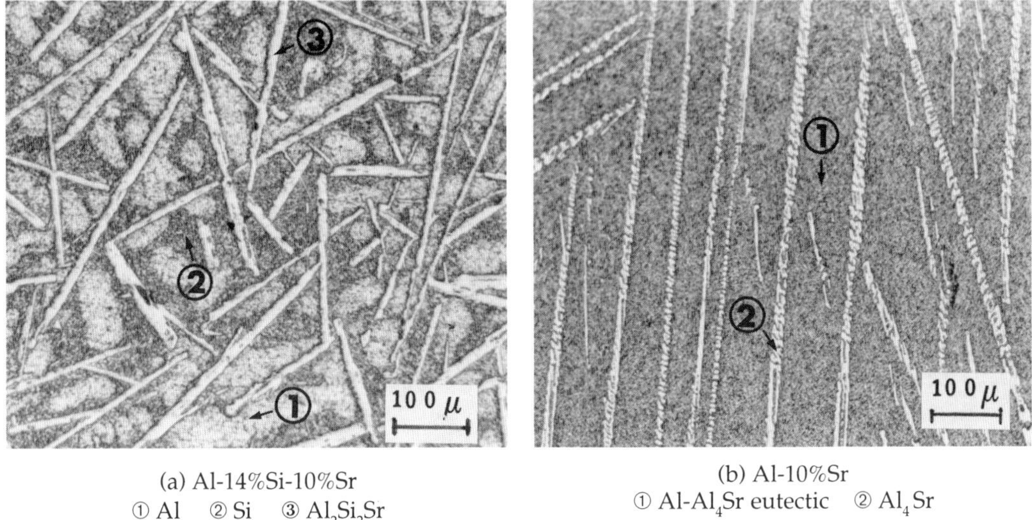

(a) Al-14%Si-10%Sr
① Al ② Si ③ Al$_2$Si$_2$Sr

(b) Al-10%Sr
① Al-Al$_4$Sr eutectic ② Al$_4$Sr

Figure 3.11. Microstructures of two common low-strontium, high-aluminum master alloys.

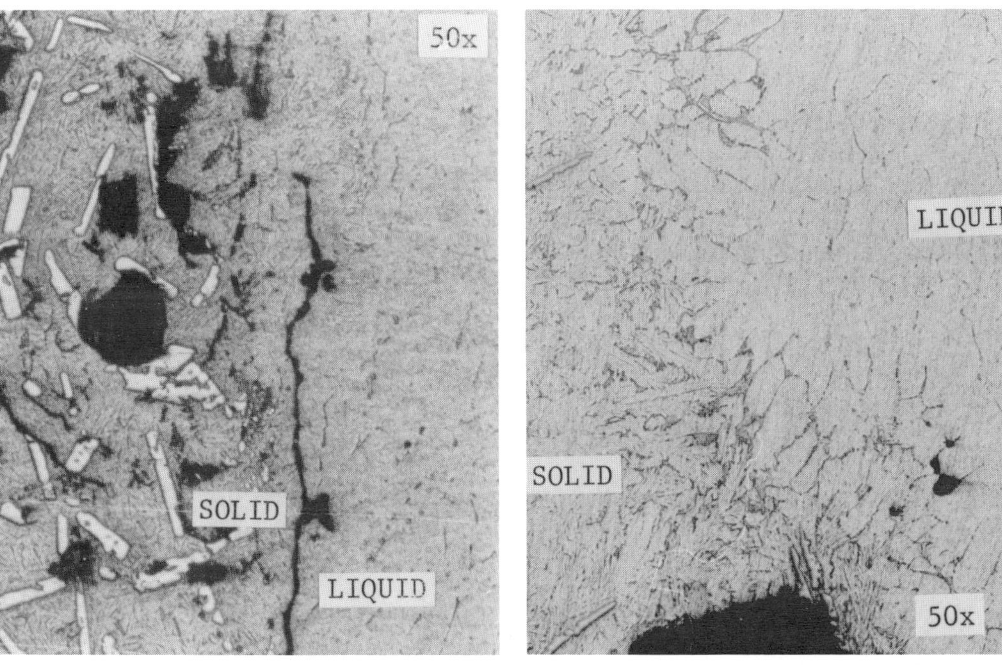

(a) dissolving interface 600 sec. after immersion showing break-up of oxide film on the master alloy

(b) after 775 sec. dissolution takes place from the interface and Al$_4$Sr phase gradually dissolves into the liquid[5]

Figure 3.12. Microstructures of dissolving 10%Sr-90%Al alloy. Samples were immersed in pure liquid aluminum and then quenched in liquid helium when partially dissolved.

(Previously published in the *Canadian Metallurgical Quarterly*, Vol. 28 [1989]. Reprinted with permission of the Canadian Institute of Mining and Metallurgy.)

Figure 3.11. The Al-10%Sr alloy contains the Al-Al$_4$Sr eutectic and large plates of Al$_4$Sr intermetallic. The Al-10%Sr-14%Si alloy is composed of Al-Si eutectic and large particles of Al$_2$Sr$_2$Si. In each case most of the strontium is locked-

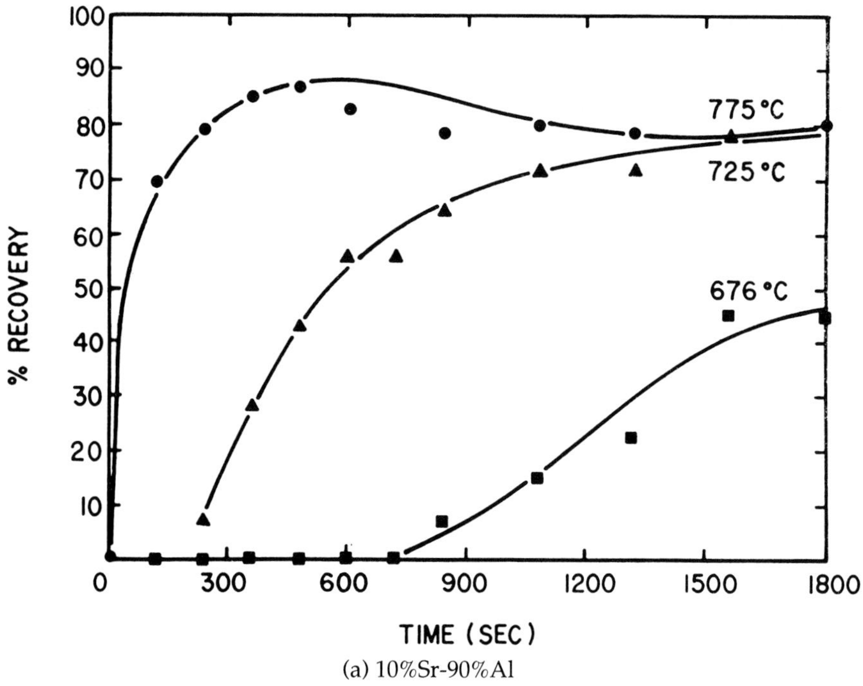

(a) 10%Sr-90%Al

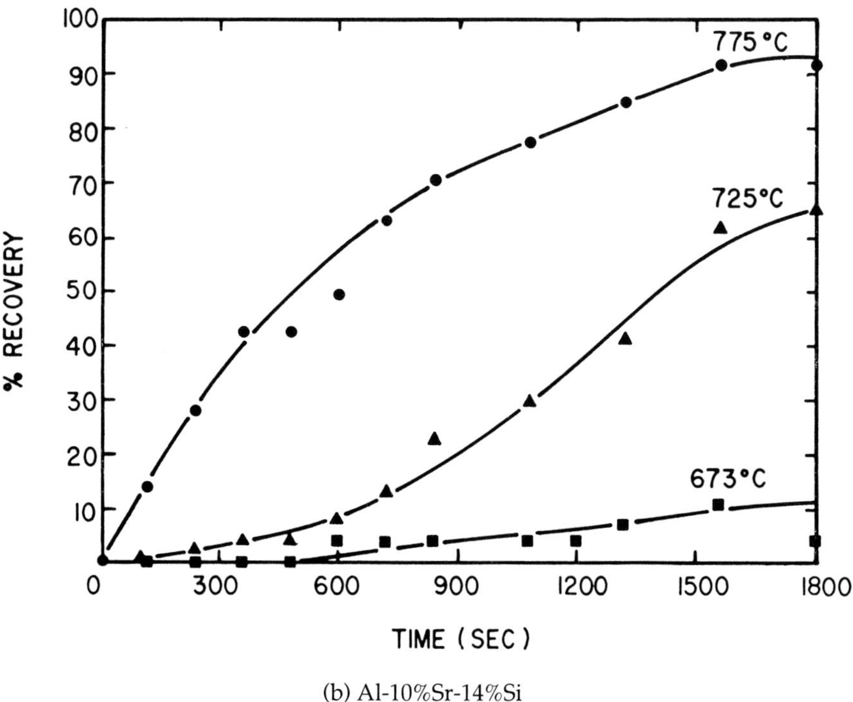

(b) Al-10%Sr-14%Si

Figure 3.13. Recovery of low-strontium, high-aluminum master alloy in A356 [5]. (Previously published in the *Canadian Metallurgical Quarterly*, Vol. 28 [1989]. Reprinted with permission of the Canadian Institute of Mining and Metallurgy.)

up in intermetallic compounds and the process of adding strontium to the melt is one of dissolving these compounds. This has been shown to proceed by simple dissolution [5] with the intermetallics gradually dissolving into the melt (Fig. 3.12). As a result, strontium recoveries are higher at higher melt temperatures, and are very poor if the temperature is too low.

Figure 3.13 presents some experimental data on the dissolution of Al-10%Sr and Al-10%Sr-14%Si alloys in A356. As is the case with high strontium alloys, recoveries of 90% can be achieved under optimum conditions. If the temperature is too low this can fall to as low as 10%, and be accompanied by a significant period during which no measurable strontium dissolution occurs. At all temperatures the Al-10%Sr alloy dissolves twice as fast as the Al-10%Sr-14% alloy[5]. This is presumably a reflection of the faster dissolution rate of the Al_4Sr phase compared to Al_2Si_2Sr.

Effects of Modification on the Microstructure

The microstructural change from acicular to fibrous silicon is not a sharp one and castings with an inadequate amount of either sodium or strontium will exhibit a mixed structure—one containing regions of fibrous silicon, lamellar silicon and acicular silicon. Modification with strontium is often less uniform than with sodium, and of course antimony will produce only a lamellar and never a fibrous structure.

The entire range of microstructures seen in the modification of a hypoeutectic alloy is shown in Figure 3.14 at both a relatively low and at a high magnification. These structures have been divided into six classes[6] with well modified structures falling into class 5, undermodified into classes 2-4, and lamellar into class 2. Formation of the very fine structure, sometimes called supermodified, of class 6, is not well understood at present. It can be produced in the laboratory by rapid quenching of partially solidified samples, and although it is sometimes observed in commercial castings, its production seems to be non-repeatable. Thus, the vast majority of modified castings will have structures of the 1 to 5 type.

In assessing the efficiency of modification, it is sometimes useful to quantify the microstructure. This can be readily accomplished by examining a polished section under a microscope and assigning to each class the proportion of the sample surface which has a particular type of modification. For example, suppose a given sample contains roughly 20% class 3, 50% class 4 and 30% class 5. Its modification rating (M.R.) would then be calculated as:

M.R. = (0.2x3)+(0.5x4)+(0.3x5) = 4.1

and the sample could be said to be reasonably well, but not perfectly, modified.

Five variables determine the exact microstructure which will form:

200x   800x

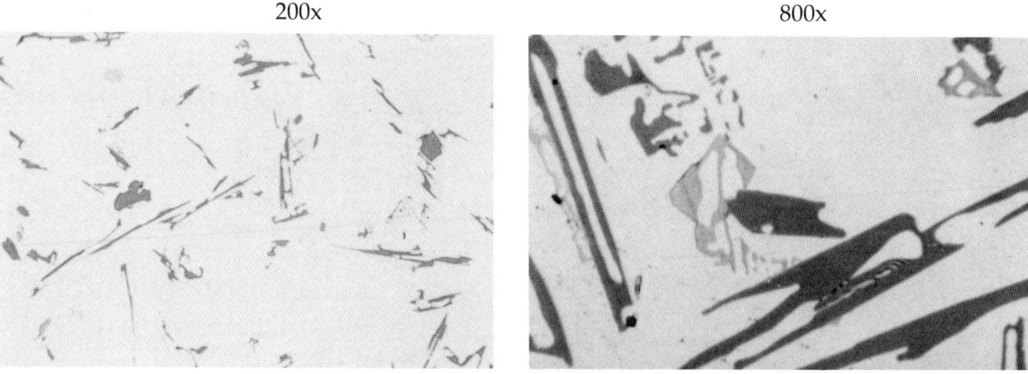

Class 1. Fully Unmodified Structures

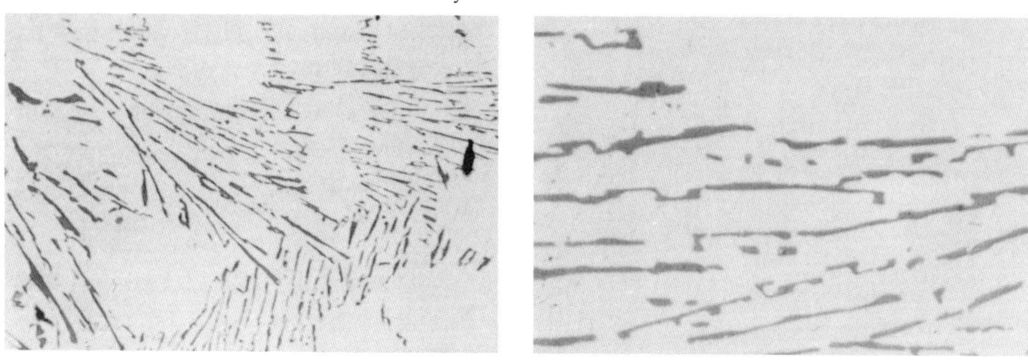

Class 2. Lamellar Structures

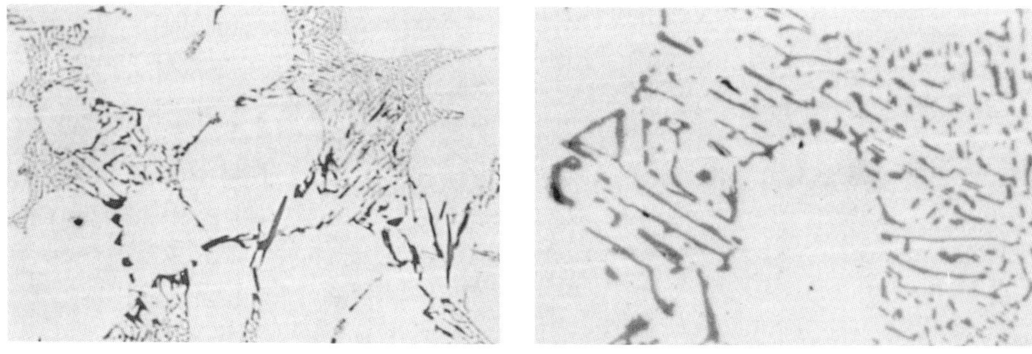

Class 3. Partial Modification

Figure 3.14. Rating system for modified microstructures[6].

- type of modifier used;
- impurities present in the melt;
- amount of modifier used;
- freezing rate;
- silicon content of the alloy.

These five variables interact in a highly complex manner, and our present state of knowledge is insufficient to allow an exact quantitative prediction of microstructure, say by modification rating, for any specific set of the five. We

200x 800x

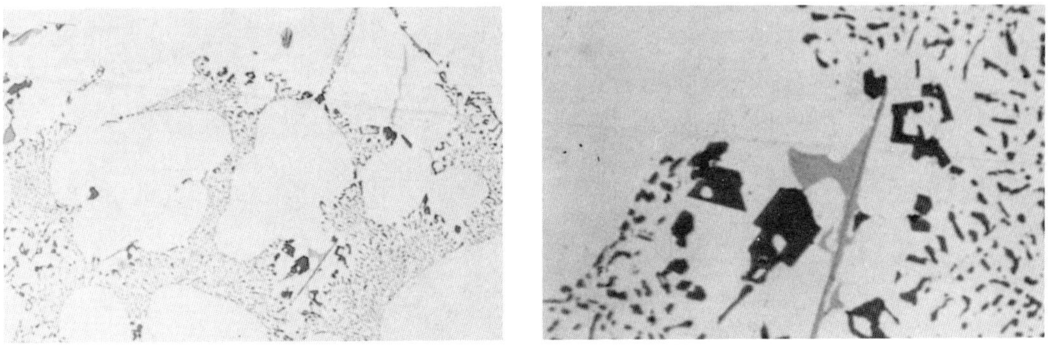

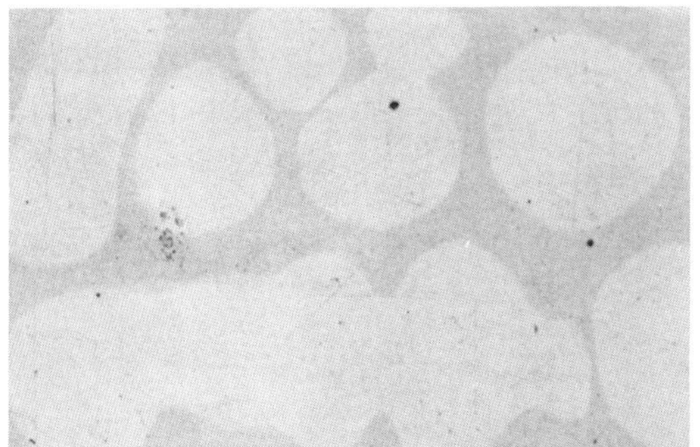

Class 4. Absence of Lamellar Structures

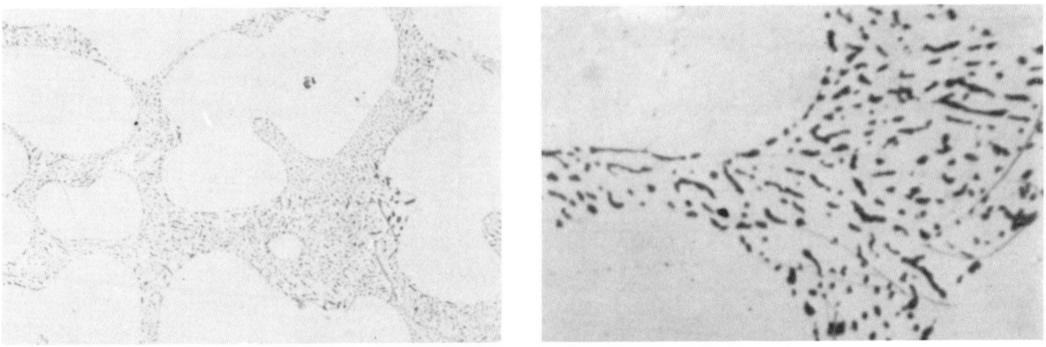

Class 5. Fibrous Silicon Eutectics

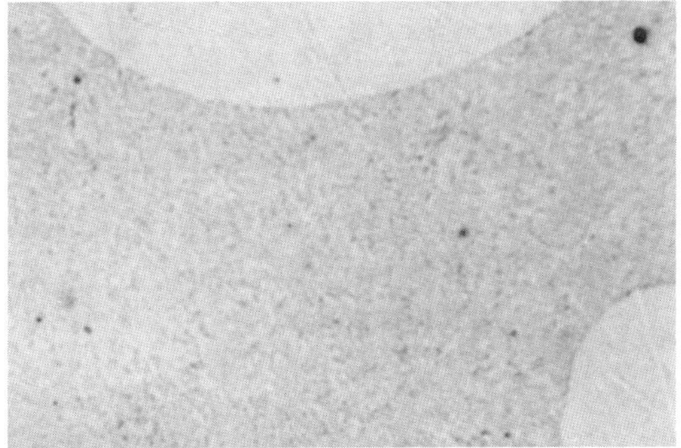

Class 6. Very Fine Structures

can, however, specify general trends so that the foundryman will know at least what direction to follow.

Type of Modifier

Both sodium and strontium are capable of producing the full range of microstructures, although sodium is the more powerful modifier in that it produces more uniformly modified structures at lower concentrations than does strontium. Antimony will yield only lamellar structures, class 2.

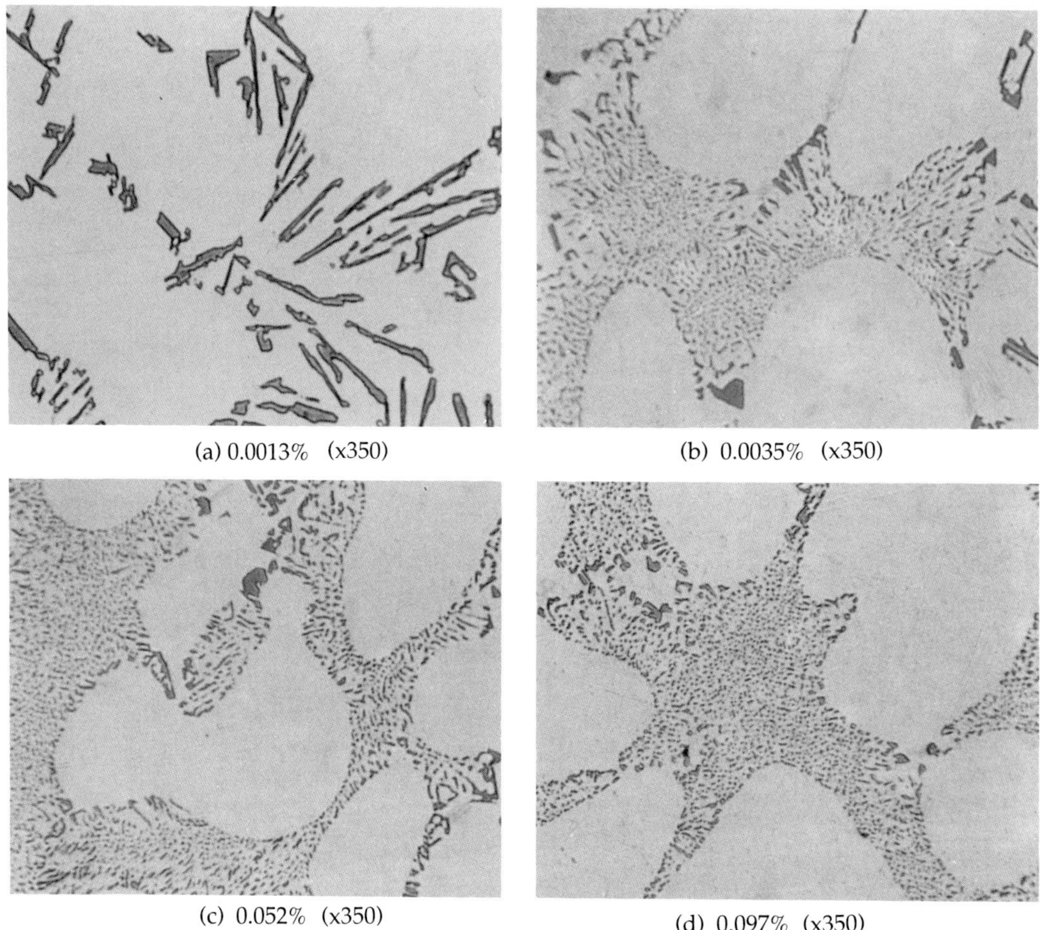

Figure 3.15. Change in microstructure of an A356 alloy with increasing strontium concentration [7].

Impurities Present in the Melt

This topic will be treated more fully in section 3.5 and Chapter 6. For the moment, it suffices to note that phosphorus in particular makes modification difficult, and that alloys which are easy to modify have a low phosphorus content. Antimony interacts with both sodium and strontium in a negative fashion, and antimony containing melts require exceptionally high levels of either modifier to produce structures of a class higher than 2. At the moment, the role of all other elements found in aluminum casting alloys is not clear, and is certainly a subject deserving of research. For example, magnesium has been reported by some to make modification easier and by others to make it more difficult.

Amount of Modifier Used

For a given set of casting conditions and alloy composition, there is a critical modifier level required to produce a given microstructure. In general, a higher concentration of modifier will produce a higher microstructure class, at least up to class 5. Too high a level is undesirable, as overmodification can occur (see 3.3). Figure 3.15 shows the development of a class 5 microstructure from a class 1 by the use of an increased strontium level.

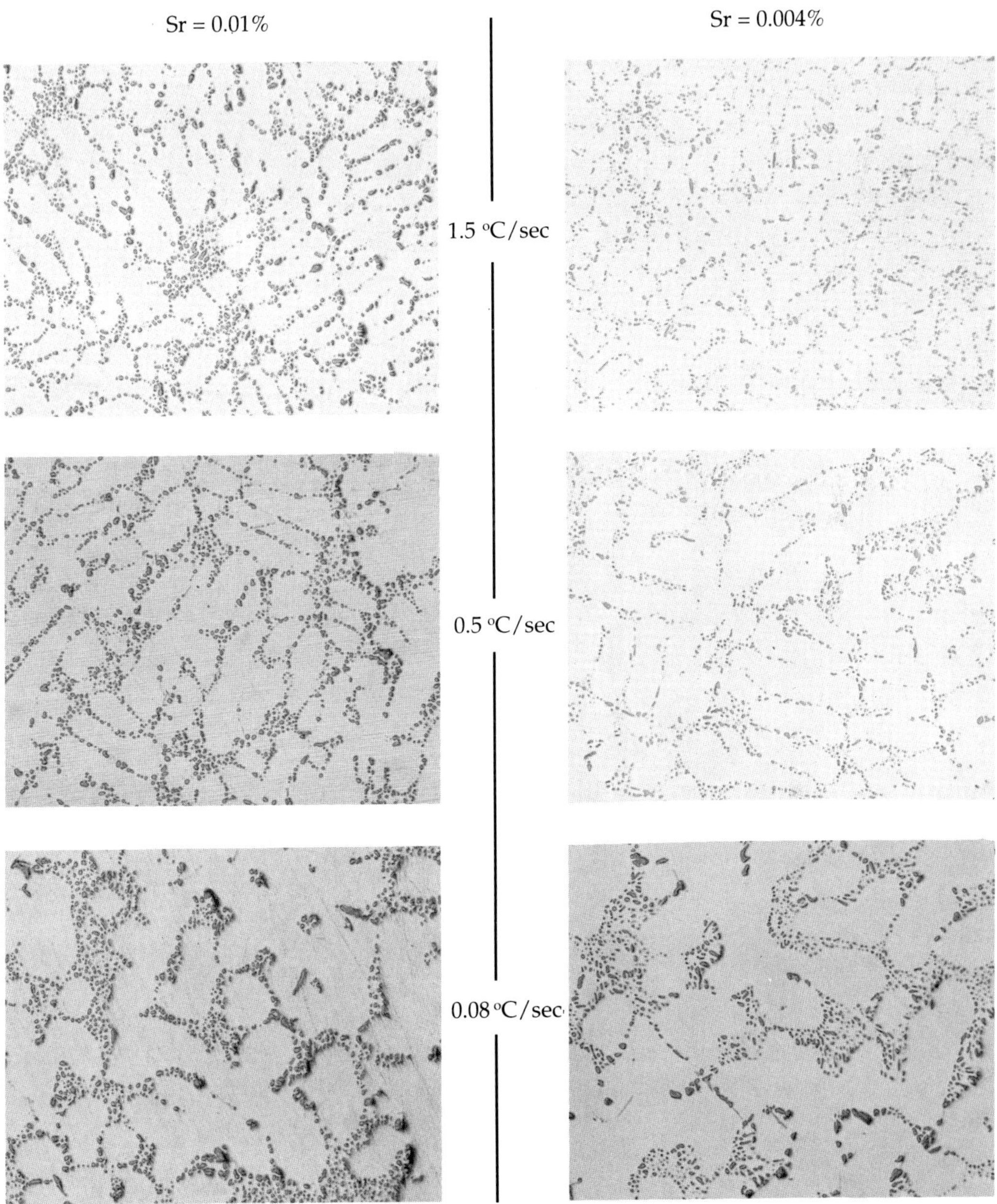

Figure 3.16. Effect of cooling rate through the solidification range and strontium level on the eutectic structure of an A356 alloy (x70) [7].

Freezing Rate

Higher solidification rates assist the modification process, and so lower modifier levels are required in permanent mold castings than in heavy-section sand castings. Modification has never been pursued in die casting because it is often said that die castings freeze so quickly as to produce a structure which is fine enough. However, some recent experiments on the use of strontium in die casting have indicated a potential beneficial effect of modification. Additions of 0.02% to 0.03% strontium to 380.0 alloy do lead to a noticeably finer microstructure which could result in improved machining properties. The microstructures of Figure 3.16 illustrate the interrelationship between freezing rate and modifier (strontium) level. At the fastest cooling rate used (1.5 °C/sec.), a four-fold increase in strontium concentration has virtually no effect on the microstructure. On the other hand, as the casting cools more slowly the structure coarsens, and a class 5 structure is obtained only if 0.01%Sr is used at a cooling rate of 0.08 °C/sec.

The lamellar structure produced by antimony treatment is particularly sensitive to freezing rate. Antimony treatment is not recommended for sand castings, as these solidify too slowly to ensure a uniform lamellar microstructure. Antimony treatment is therefore usually restricted to permanent mold applications.

Silicon Content

Higher silicon concentrations require larger amounts of modifier to produce complete modification. An increase of up to 50% in the amount of strontium needed is observed when the silicon level is changed from 7% to 11%.

3.3   Overmodification

Sodium or strontium levels higher than needed to produce a fully modified class 5 microstructure exert a deleterious effect on the properties of the alloy. Specific examples can be found in 5.1. Here we will examine only the effect of this overmodification on the microstructure.

Overmodification with sodium can take place if the sodium concentration exceeds 0.018% to 0.020%. Then, a coarsening of the silicon occurs associated with bands of primary

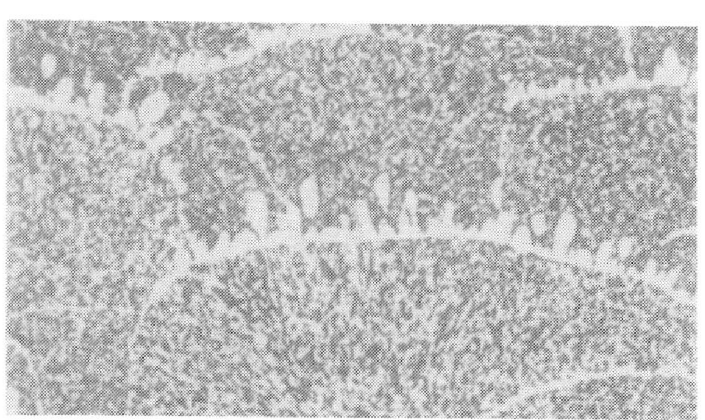

Figure 3.17. Overmodified structure in sodium modified eutectic alloy (x100) [8].

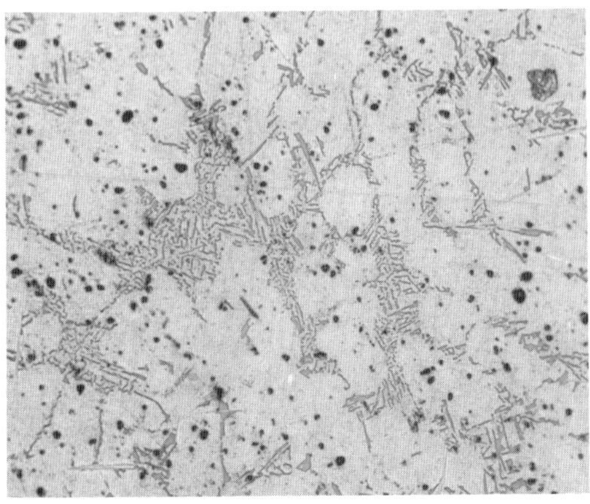

(a) optical micrograph (x400)

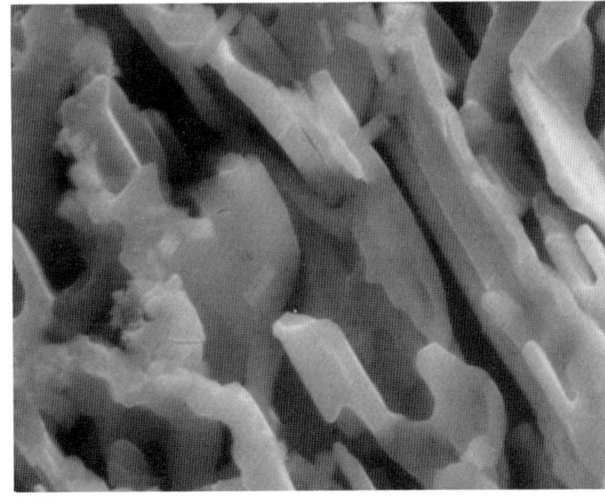

(b) scanning electron micrograph (x2000)

Figure 3.18. Coarse silicon caused by overmodification with 0.09% Sr in an A356 alloy **[9]**.

aluminum. These overmodification bands can be seen clearly in the photomicrograph of Figure 3.17. What appears to happen at these high sodium levels is a rejection of sodium in front of the solidifying interface. The compound AlSiNa can form, and this serves to nucleate the coarse silicon particles. After some growth of silicon occurs, the adjacent liquid is aluminum rich, and a sudden nucleation and growth of aluminum then takes place which envelopes the coarse silicon and leads to the appearance of an overmodification band in the final cast product.

Two distinct phenomena are associated with strontium overmodification. One of these is a coarsening of the silicon structure and the reversion of the fine fibrous silicon to an interconnected plate form. Figure 3.18 shows an optical micrograph of a 356 alloy treated with 0.09%Sr. The plate-like nature of the silicon is evident in both the optical micrograph and the SEM image of the same structure. This effect has been little studied and the reasons for its occurrence are unknown. A second evidence of strontium overmodification is the appearance of strontium containing intermetallic phases in the microstructure, such as the Al_4SrSi_2 particles seen in Figure 3.19. It is clear that both of these effects will reduce the properties of the alloy causing them to revert to values more typical of untreated material. Interestingly, the very limited knowledge available on this subject indicates that these two effects need not occur simultaneously,

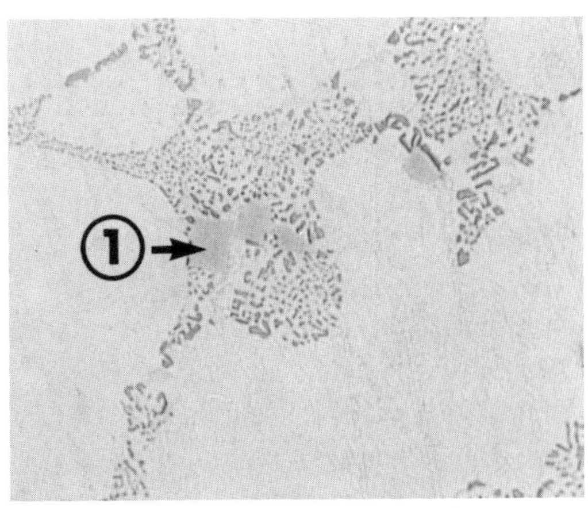

Figure 3.19. Al_4SrSi_2 phase ① caused by overmodification of 356 alloy (x270) **[10]**.

i.e., Al_4SrSi_2 can occur without appreciable silicon coarsening and vice versa.

3.4 Modifier Fading

Of the three commercially important modifiers, Na, Sr, Sb, two are subject to fading. Depending on the circumstances, the levels of Na or Sr present in the melt may decrease with time. Antimony is very stable chemically and is not known to undergo any natural fading effect. Two types of chemical reaction in the melt may cause modifier fading. The modifier may vaporize due to a high vapor pressure at melt temperatures, or it may oxidize due to an excessive chemical affinity for oxygen. In the latter case, the modifier will still remain in the melt, but in a chemically combined form. Such chemically combined elements are ineffective as modifiers; only free atoms in the liquid alloy can cause modification to occur according to the mechanisms outlined in 3.1.

Sodium Fading

Of the two, Na or Sr, sodium is by far the worst from the point of view of fading. This, in fact, is one of the major disadvantages of sodium usage: control of precise quantities in the melt is very difficult to achieve. With a vapor pressure of 0.2 atmospheres at typical foundry melt temperatures, sodium readily boils out of the melt, causing large losses. On the other hand, sodium once dissolved does not readily oxidize, and hence, this is not an important fading mechanism, particularly when it is first added.

The data presented in Figure 3.20(a) is typical of sodium fade. If metallic sodium is used, the rate of loss during addition is such that recoveries of only about 25% are possible immediately after addition. There is a subsequent loss over a period of 1-2 hours down to levels of 0.002%-0.005%. These values represent the limit of detectability of most analytical techniques for sodium, and hence precise values of residual sodium are difficult to specify. It is evident from the data of Figure 3.20(a), that if 0.005% to 0.01% is required for modification, then casting must be done within 30-40 minutes of melt treatment. Since sand castings cool more slowly than permanent mold castings, they will require a higher melt sodium, and so the time delay for pouring is more critical in sand than permanent mold casting.

Exact fading rates depend strongly on the circumstances of addition. Large melts are less prone to fading than small ones as they usually have a lower ratio of melt surface area to melt volume. Stirring will increase fading sharply, and hence degassing, even with an inert gas, is not recommended after sodium treatment.

Because the mechanistics of sodium treatment with fluxes and with pure metal are different, there is a marked difference in fading behavior as

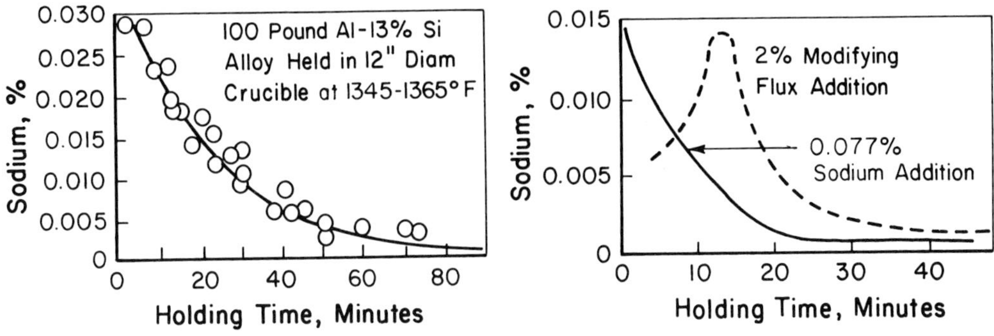

(a) loss of sodium during holding in a crucible

(b) comparison of sodium losses when additions are made as a pure metal and as a flux

Figure 3.20. Some typical sodium variations with time [12].

illustrated in Figure 3.20(b). Addition by fluxes requires a reaction at the flux-melt interface. This reaction is relatively slow compared to the addition of sodium in a pure form, and the maximum sodium concentration occurs significantly later. Sodium addition through flux treatment, therefore, offers a delay period of several minutes before fading begins. Since sodium loss is primarily by vaporization, the rate of fade is about the same in each case after the maximum concentration has been achieved.

It is possible to purchase ingot which is premodified. Remelting of such ingot will invariably entail a loss of modifier. In the case of sodium, it appears that a two-thirds loss can be expected under the best conditions.

It is appropriate at this point to mention melt treatments which will reduce the sodium content. In certain circumstances, it may be desirable to lower the sodium level to some predetermined value, or it may be required to remove sodium altogether. Of course, it is possible to allow natural fading, perhaps encouraged by stirring, to lower the sodium concentration, but the process can be accelerated and virtually all residual sodium eliminated by chlorination. Chlorine has a high affinity for sodium and will rapidly reduce its concentration through the formation of NaCl. The chlorine may be introduced by bubbling a nitrogen-freon mixture, by chlorine-inert gas fluxing, or by using hexachloroethane tablets. It has been reported[1], that using an amount of hexachloroethane equivalent to 0.375% of the charge weight will reduce the sodium level from 0.016% to 0.004%. Some loss of other elements, particularly magnesium, can also be expected through the formation of $MgCl_2$.

Strontium Fading

In general, strontium fades considerably more slowly than sodium, so much more so that it is regarded as a semi-permanent modifier. The fading mechanism of strontium is quite different from that for sodium. The vapor

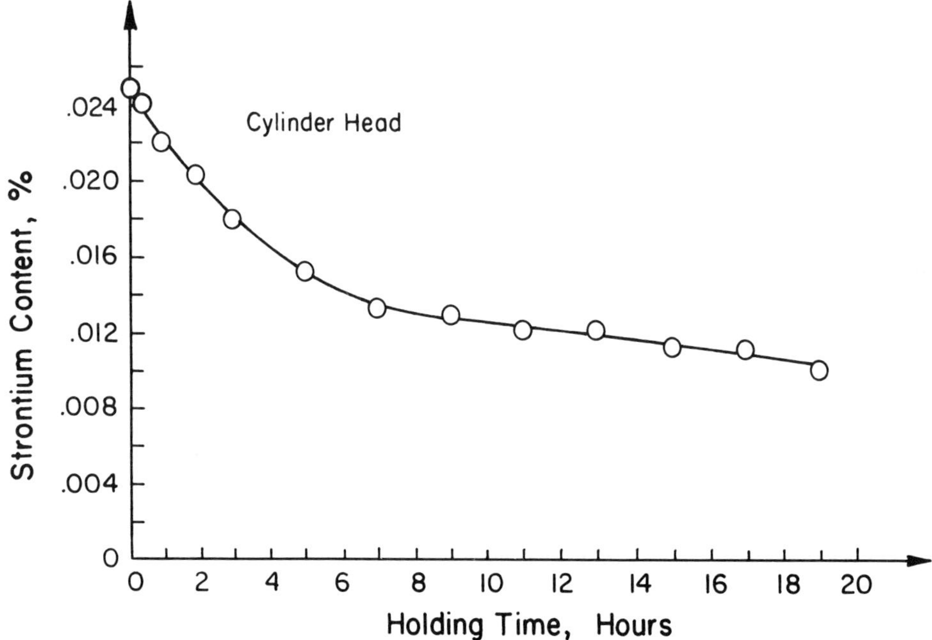

(a) for an initial concentration of 0.025% used to cast a cylinder head in 319.2 alloy

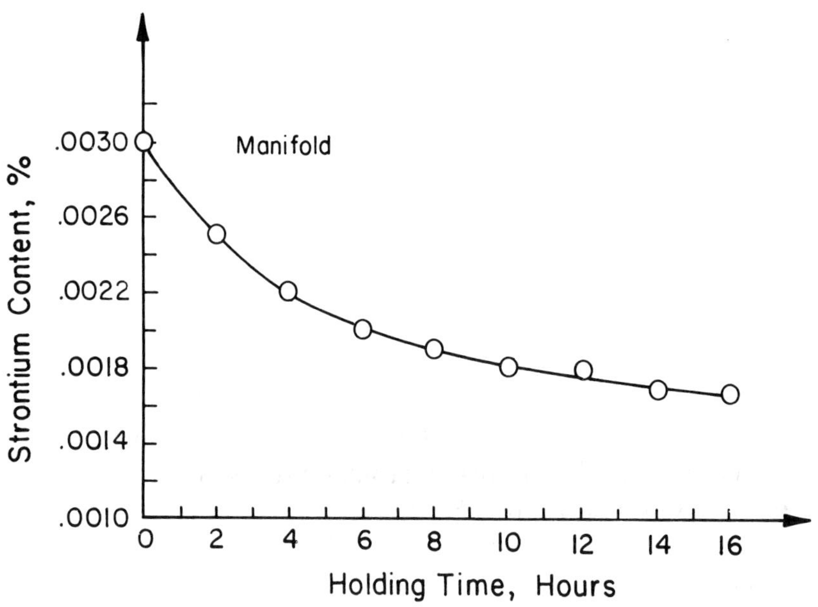

(b) for an initial concentration of 0.003% used to cast manifolds in 319.2 alloy

Figure 3.21. Some typical strontium losses during furnace holding [13].

pressure of strontium at 730C (1350F), is only 10^{-3} atmospheres, or 200 times smaller than that of sodium. Thus, vaporization of strontium is not an important factor. It appears that strontium losses from the melt are primarily by oxidation, since the oxide of strontium is somewhat more stable than that of either aluminum or silicon.

    The rate of strontium fading, like that of sodium, will depend highly on the melt circumstances, but there is ample evidence to demonstrate the much longer lifetime of strontium. For example, Figure 3.21 presents fading curves in 319.2 alloy at two initial concentrations. In this case, the alloy for casting the thinner section manifold solidifies more quickly, and hence requires less strontium for modification than does the alloy used to produce the thicker cylinder heads. When the starting concentration is 0.025%, it requires 9 hours for the

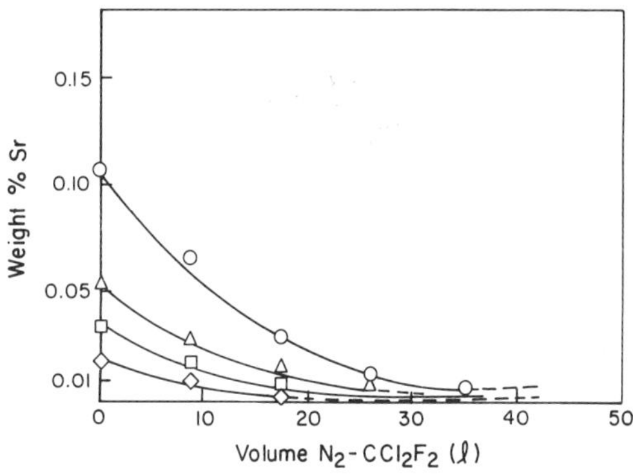

Figure 3.22. Strontium removal from an A356 melt by bubbling nitrogen-5% freon mixture. Four different starting concentrations are shown [7].

strontium to fade to one-half of its original value; when the initial concentration is 0.003%, this time is lengthened to something in excess of 16 hours. These are very long times, indeed, compared to the 20-30 minutes required for sodium to fade by a similar amount.

It is generally accepted that premodified ingots will lose strontium on remelting. Hess and Blackmun[14], in their classic paper on strontium modification, report melt losses of 30%-50% on remelting. However, unpublished work in the authors' laboratory on 10 kg melts indicate no significant losses on remelting. In the absence of fluxes, oxidation is the primary mechanism for strontium loss; any compositional changes on remelting must be highly dependant on such factors as turbulence and furnace atmosphere. Consequently, remelting losses, like fading, are probably very specific to the situation, and can likely be minimized by care during the remelting procedure. Losses in the presence of fluxes are discussed in Chapters 12 and 14.

Should it be desired to remove strontium from the liquid alloy, this is easily accomplished, since strontium, like sodium, reacts readily with chlorine to produce $SrCl_2$. Thus, bubbling of nitrogen-freon, chlorine-inert gas mixtures, or use of hexachloroethane tablets, will easily clean a melt of residual strontium. Some values are shown in Figure 3.22 for an A356 melt treated with nitrogen-5% freon mixture. At a flow rate of 1.75 l/min. the melt is cleared in about 20 minutes. The concentration of strontium reached after passing a given volume of gas can be calculated from the relation:

$$C = Co \; exp^{(-KV)}$$

where C = strontium concentration when a volume, V, of nitrogen-5% freon
         gas is used

    K = 0.085 l^{-1}

    Co = initial strontium concentration.

Alternatively, this equation could be used to determine the volume of nitrogen-freon necessary to lower the strontium level to any desired value.

As is the case with sodium removal, some loss of magnesium may be incurred during a melt treatment to remove strontium.

3.5 The Effect of Phosphorus

Aluminum foundry alloys may contain a dozen or more elements in addition to the major ones of aluminum, silicon and magnesium. Some of these elements are added intentionally as alloying agents, while others are present as impurities carried over from refractories, tools, etc. Since modification is essentially a surface effect (3.1), very small quantities of certain elements may either aid or hinder it. Our present state of knowledge of the influence of virtually all of these elements on the modification process is practically nonexistent, and it is only with the effect of phosphorus that we have any understanding at all.

Phosphorus enters aluminum casting alloys through contact with tools, refractories, refractory cements or crucible glazes. It is also contained in various amounts in alloying and other additions. Most of the work on phosphorus effects has been done in France by researchers at Pechiney who have developed a set of alloys with controlled phosphorus concentrations. In general terms, phosphorus interferes with modification by either Na, Sr or Sb, and alloys with higher phosphorus levels require larger retained modifier concentrations in order to produce an acceptable cast structure. If the phosphorus concentration can be reduced to less than 1 ppm, an Al-Si casting alloy will freeze with a lamellar (class 2) structure without the addition of any modifying agent. Thus, we may regard the "natural" structure of these alloys as lamellar. It is only the higher phosphorus values associated with commercial alloys which cause the eutectic silicon to solidify in an acicular form. Phosphorus is sometimes deliberately added to eutectic alloys for piston castings. Such additions cause a few primary silicon particles to nucleate and eliminate any traces of modification. These effects are generally felt to increase wear resistance and piston life.

The reasons for this coarsening effect of phosphorus on the eutectic structure are not known. In hypereutectic alloys, phosphorus is deliberately added because it reacts with the aluminum to form AlP particles which nucleate primary silicon (see Chapter 7). This results in a fine dispersion of primary particles, an effect which is quite the opposite to that in hypoeutectic alloys where the silicon is coarsened by phosphorus addition. One is left to conclude that the mechanisms operative in primary and eutectic crystallization of silicon are probably quite different.

Nevertheless, it remains a fact that the amount of modifier required *does* depend on the phosphorus content. Modifier-phosphorus interactions have been determined experimentally to some extent and these are illustrated in Figures 3.23-3.25 taken from reference 15. A lesser quantity of each modifier is required as the phosphorus level decreases, and although not much data exists on the effect of freezing rate, the permissible phosphorus concentration should increase as the freezing rate increases. This effect is seen in Figures 3.23 and 3.25 for sodium and antimony modification.

If the melt contains more phosphorus, more modifier is required, which also has implications for the allowable holding time after sodium treatment. This holding time will be less for a high phosphorus melt because the required sodium level to produce modification will be more. Thus, there is a shorter available fade time. Conversely, if the phosphorus concentration is low, a lower sodium level is needed, and the melt can be held for a correspondingly longer period. Figure 3.26 illustrates these considerations on a typical sodium fading curve. The same reasoning can be applied to strontium modification. However, the fading rate is so much slower than with sodium that differences due to phosphorus level are of much less practical concern.

Higher modifier concentrations required of higher phosphorus levels imply a reaction of some sort between the modifier and phosphorus. We are only now beginning to appreciate the existence of several complex chemical reactions which take place within liquid foundry alloys. Some of these are discussed in

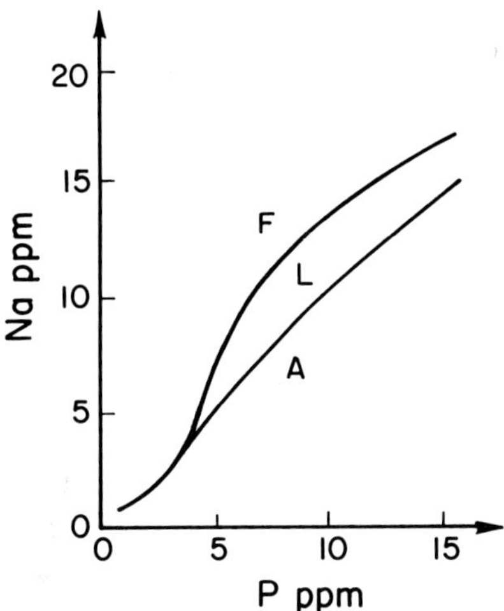

(a) in A356 alloy, solidification time = 13 sec

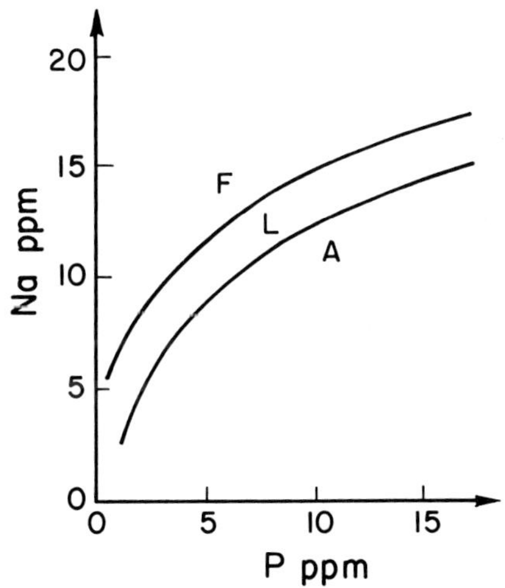

(b) in A356 alloy, solidification time = 60 sec

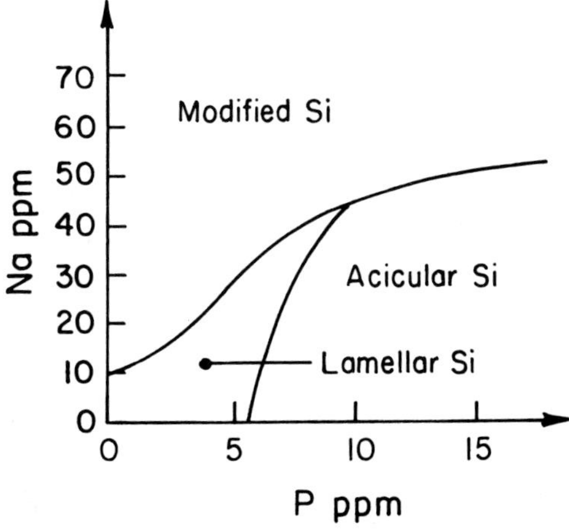

(c) in 413 alloy, solidification time = 13 sec

Figure 3.23. Sodium-phosphorus interactions[15].
(F = fibrous; L = lamellar; A = acicular).

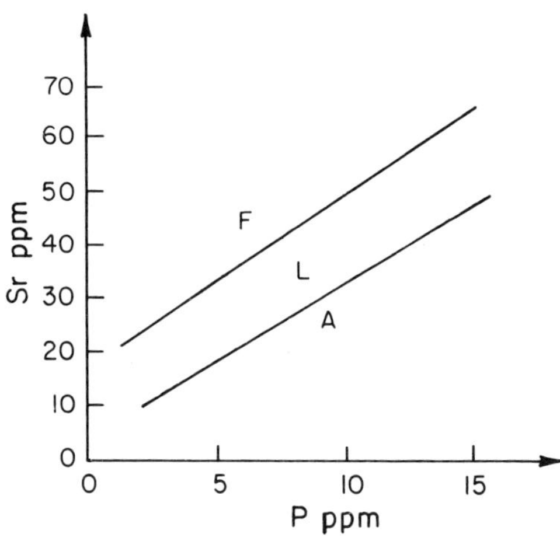

Figure 3.24. Strontium-Phosphorus interaction in A356 alloy [15]; solidification time = 60 sec (F = fibrous; L = lamellar; A = acicular).

Chapter 6, but in the case of modifier-phosphorus interactions there are at least two possibilities. Garat and Scalliet[15] have shown the presence of intermetallic compounds in modified alloy which contain measurable quantities of phosphorus. These compounds are presumed to dissolve the phosphorus as a solid solution, and hence to remove it from the liquid phase. In the case of sodium modification, it is $NaSi_2$ which forms. For strontium treatment, an unidentified AlSiSr compound has been found, and with antimony, Mg_3Sb_2 forms and dissolves phosphorus. A second possibility is that the modifiers react directly with phosphorus to form modifier-phosphorus compounds. Unfortunately, nothing is known about the formation of such phosphides in a liquid aluminum melt, and so this possibility must be left to speculation.

Control of phosphorus is clearly a major key in reproducible modification treatment. Pechiney has enjoyed considerable success with its controlled phosphorus ingot. Experience with North American suppliers has been that there are significant batch-to-batch differences in how their alloys respond to modification. Better control and lower levels of phosphorus could improve the situation significantly. In the foundry, phosphorus measurement was, until recently, virtually impossible. However, it is now possible to purchase emission spectrometers which can routinely analyze for this element. Of course, even the use of low phosphorus starting materials will not ensure a good control of modification if phosphorus is picked up during the melting process, as for example, by reaction with refractories. If this does occur, the ability to analyze for phosphorus will at least allow the problem to be identified, and its source located.

Incubation Period

An effect, probably related to phosphorus, is the existence of an incubation period with strontium modification. During the first 1-2 hours after a strontium addition is made, the degree of modification is observed to improve. A typical incubation effect is seen in Figure 3.27 where the modification ratings for samples which were sodium or strontium treated are plotted as a function of time after the treatment. As sodium fades, the microstructure deteriorates. With strontium an incubation period is noted. Ten minutes after the addition was made, some modification is evident (M.R. = 2.5); however, after one hour the structure is almost completely fibrous and the M.R. is 4.

At first this was thought to be due to slow dissolution of the strontium, but recently several studies of dissolution have shown that a maximum of 30 minutes is required to achieve the maximum melt strontium concentrations, even under the most unfavorable conditions. Significantly shorter times are necessary if more optimum dissolution conditions are used. Although virtually no detailed research has been conducted on this subject, the incubation is likely related to a reaction in the melt between strontium and phosphorus. For example, a slow strontium phosphide forming reaction would lead to such an effect. With sodium, an incubation period may or may not exist. It certainly is not detectable because it is masked by the much more pronounced phenomenon of fading. In any event, incubation with strontium is an advantage as it runs counter to any chemical fading, and is compatible with treatment in holding furnaces where long melt holding times are involved. This is particularly advantageous in the production of premodified ingots by primary or secondary producers.

It is not known if an incubation period exists with antimony treatment. Antimony is virtually always used as premodified ingots containing 0.1-0.15% Sb. It is not added directly at the foundry level, and hence it is almost impossible to observe incubation effects.

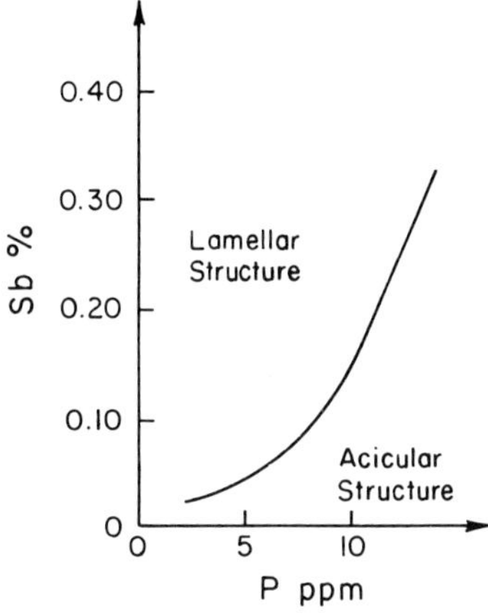

(a) in A356 alloy, solidification time = 13 sec

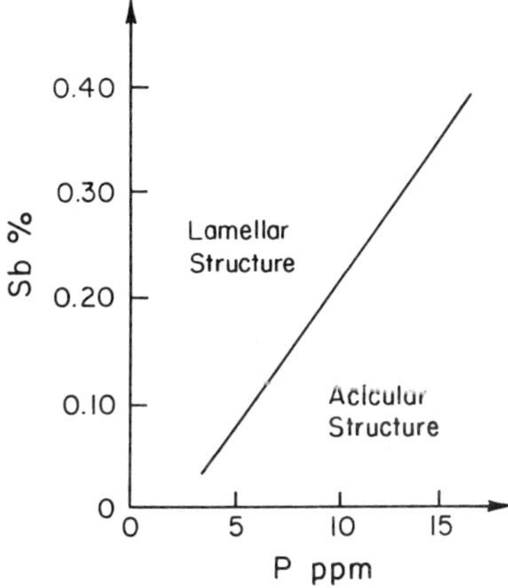

(b) in A356 alloy, solidification time = 60 sec

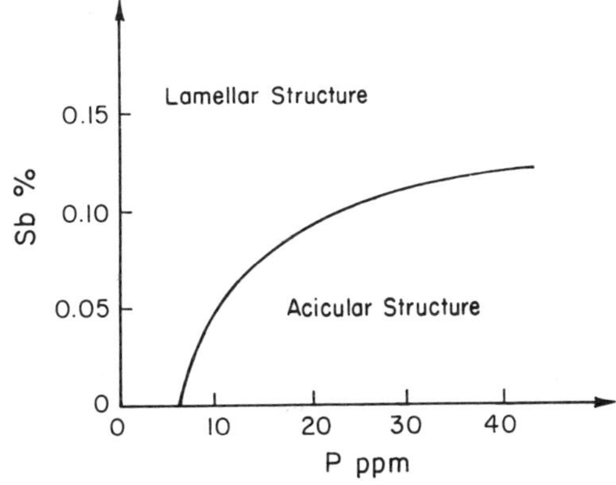

(c) in 413 alloy, solidification time = 13 sec

Figure 3.25. Antimony-phosphorus interaction[16].

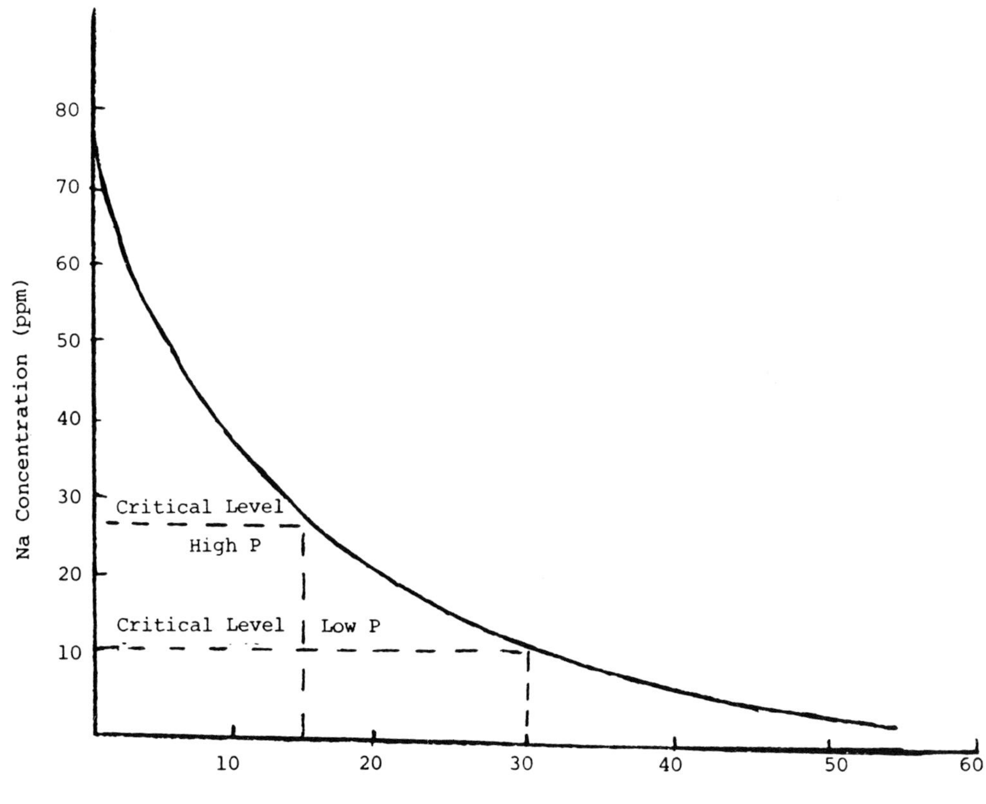

Melt holding time (min.)

Figure 3.26. The effect of phosphorus level on the allowable melt holding time with sodium modification. If there is a high phosphorus concentration, only 15 minutes of holding is possible; at a low concentration, the melt can be held for 30 minutes.

References

1. Pacz, Aladar. United States Patent 1,387,900 (Aug. 16, 1921).
2. Lu, Shu-Zu and A. Hellawell. "The Mechanism of Silicon Modification in Aluminum-Silicon Alloys: Impurity Induced Twinning," *Met. Trans. A.*, 18A, (1987) pp. 1721-33.
3. Closset, B., H. Dugas, M.O. Pekguleryuz, and J.E. Gruzleski. "The Aluminum-Strontium Phase Diagram," *Met. Trans. A.*, 17A (1986) pp.1250-53.
4. Pekguleryuz, M.O. and J.E. Gruzleski. "Dissolution of Reactive Strontium Containing Alloys in Liquid Aluminum and A356 Melts," *Met. Trans. A.*, 20B (1989) pp. 815-31.
5. ———. "Dissolution of Non-Reactive Strontium Containing Master Alloys in Liquid Aluminum and A356 Melts," *Canadian Metallurgical Quarterly* 28 (1989) pp. 55-65.
6. Apelian, D., G.K. Sigworth, and K.R. Whaler. "Assessment of Grain Refining and Modification of Al-Si Foundry Alloys by Thermal Analysis," *AFS Transactions*, 92 (1984) pp. 297-307.
7. Closset, B. and J.E. Gruzleski. "Utilisation du Strontium Metallique dans la Modification des Alliages Al-Si-Mg," *Fonderie-Fondeur d'Aujourd'Hui*, no. 16 (1982) pp. 41-46.
8. Friedriksson, H., M. Hillert, and N. Lange. "The Modification of Aluminum-Silicon Alloys by Sodium", *J. Inst. Metals*, 101 (1973) pp. 285-99.
9. Dasgupta, R., C.G. Brown, and S. Marek. "Analysis of Overmodified 356 Aluminum

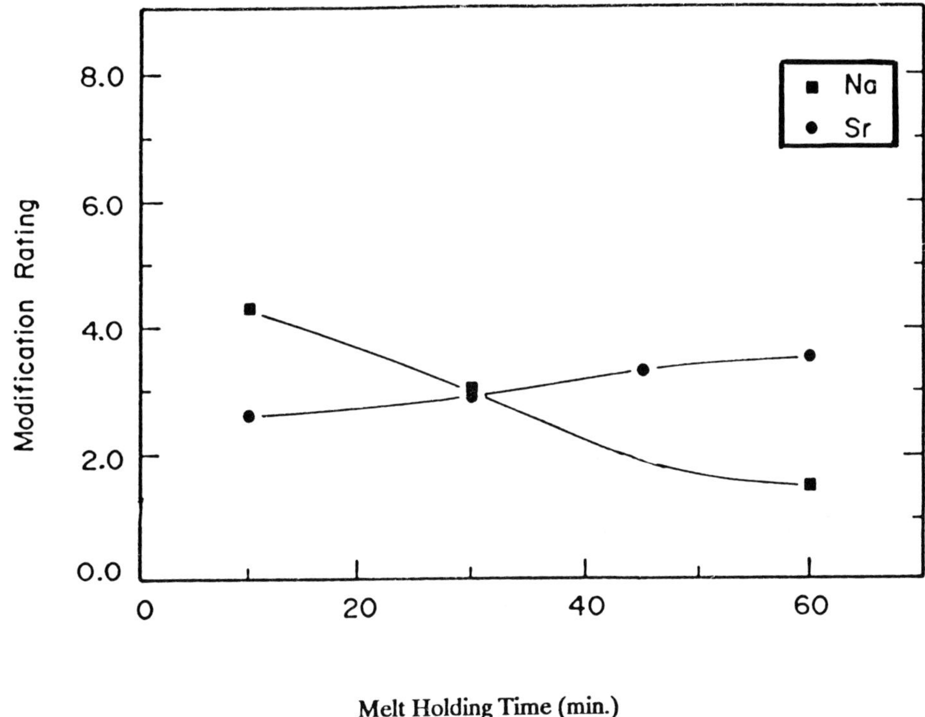

Figure 3.27. The evolution of microstructure with time as measured by modification rating for Na- and Sr-modified A356 alloy.

Alloy," *AFS Transactions*, 96 (1988) pp. 297-310.

10. Closset, B. and J.E. Gruzleski. "Structure and Properties of Hypoeutectic Al-Si-Mg Alloys Modified with Pure Strontium," *Met. Trans. A.*, 13A (1982) pp. 945-51.

11. Morrell, D.H. "The Effect of Degassing Treatment on the Sodium Content of Aluminum-Silicon Alloys," *Foseco Foundry Developments*, No. 6 (April 1960) pp. 20-25.

12. Kissling, R.J. and J.F. Wallace. "Refinement of Aluminum-Silicon Alloys," Crucible Melting Information Bureau, Crucible Manufacturers Association, Cleveland. (This data has been redrawn from a paper by L. Grand, "The Modification of Aluminum-Silicon Alloys," *Revue de l'Aluminum*, 29, [1952], pp. 5-15.)

13. Latkowski, A. "Strontium as a Modifier for Aluminum-Silicon Alloys," *Archiwum Hutnictwam*, 28 (1983) pp. 187-91.

14. Hess, P.D. and E.V. Blackmun. "Strontium as a Modifying Agent for Hypoeutectic Aluminum-Silicon Alloys," *AFS Transactions*, 84 (1975) pp. 87-90.

15. Garat, M. and R. Scalliet. "A Review of Recent French Casting Alloy Developments," *AFS Transactions*, 87 (1978) pp. 549-62.

16. Nagel, G. and R. Portalier. "Structural Modification of Aluminum Silicon Alloys by Antimony Treatment," *International Cast Metals Research Journal*, 5 (1980) pp. 2-6.

Chapter 4

Modification and Porosity

4.0   Introduction

Most foundrymen associate the process of modification with an increased tendency to produce more porous castings. Although little researched until very recently, the relationship between porosity and modification is frequently mentioned, in passing, in the scientific and engineering literature. It is often referred to by metalcasters, and there are some who will not use modification for fear of generating unacceptable levels of porosity. There is little doubt that some relation exists between modification of Al-Si foundry alloys and the soundness of castings produced from the alloys.

In this chapter we explore this relationship, and what is known of its causes. We will also present a few ideas on how to control porosity in the presence of modification. The subject matter necessitates frequent reference to hydrogen gassing, degassing, and the factors influencing pore formation in cast metals. For the reader who is not familiar with these topics as they relate to aluminum alloys, it is recommended that Chapters 9 and 10 be read in parallel with the present chapter.

Hydrogen is the only gas with any significant solubility in liquid aluminum, and since hydrogen is one of the main causes of porosity in castings, it is natural to associate the two. For this reason, it is often assumed that modifiers "gas" the melt, i.e., that somehow the addition of a modifier either adds hydrogen directly to the liquid alloy, or causes the molten material to absorb hydrogen at a faster rate than normal.

A careful reading of the literature, however, indicates that there is simply no convincing evidence to support the contention that modified melts contain more dissolved hydrogen. As to the relative behavior of Na, Sr and Sb in this regard, it is agreed that antimony does not increase porosity, but rather decreases it slightly. The influence of sodium and strontium is quite confused, with some claiming that strontium causes more porosity than sodium, while others claim the reverse.

One of the most quoted studies into the relationship between porosity and modifier content is that by Jacob[1]. He solidified sand cast bars, one end of which was chilled while the other end contained the riser. When either sodium or strontium was used as the modifier an increased porosity, as evidenced by a density decrease, was found at the riser end of the bars. Unfortunately, the hydrogen content of the melt was not measured in these experiments, and in the case of strontium, levels five to ten times higher than normally required were used. Porosity formation depends on a complex inter-relationship between hydrogen content, feeding, solidification range and liquid-solid surface energies, and changes in any one of these during modification, could lead to the behavior observed by Jacob.

The approach which we will take here is to first examine the relationship between modifiers and dissolved hydrogen concentration. We will then look at how the porosity distribution in a fixed geometry casting is affected by modification at a constant hydrogen level. On this basis, we will attempt to explain the foundryman's observation that sodium or strontium modified castings are more porous than unmodified castings.

4.1   Modification and Melt Hydrogen

A modification treatment with either sodium or strontium could increase the hydrogen content of a liquid alloy by one or a combination of the following three mechanisms:

- the direct addition of hydrogen to the melt along with the modifier;
- an increase in the rate of hydrogen pickup by a modified melt;
- an increase in the hydrogen solubility of a modified melt.

This last mechanism is not considered to be important, since the hydrogen level in aluminum alloy melts almost never attains the solubility limit. Thus, in practical terms, a modification treatment could add hydrogen either directly, or by influencing the subsequent rate of hydrogen absorption.

A review of the techniques available for measuring the hydrogen content of liquid aluminum is given in Chapter 9. Of these methods, the re-circulating gas techniques are by far the best if a direct, *in situ*, measurement in the liquid phase is desired. In what follows, all hydrogen concentrations were determined by this method. Since sodium and strontium behave somewhat differently with respect to their influence on melt hydrogen, they will be discussed separately.

Strontium and Melt Hydrogen

Strontium modification has often been stated to cause an increase in the hydrogen concentration of the melt. Careful measurements of the melt hydrogen level both before and after strontium addition have clearly shown that no such increase can be attributed to the utilization of strontium. For example, the experimental results illustrated in Figure 4.1 show the hydrogen concentration of a non-degassed A356 melt before and after addition of a 90%Sr-10%Al master alloy. It is evident that no increase in dissolved hydrogen occurs.

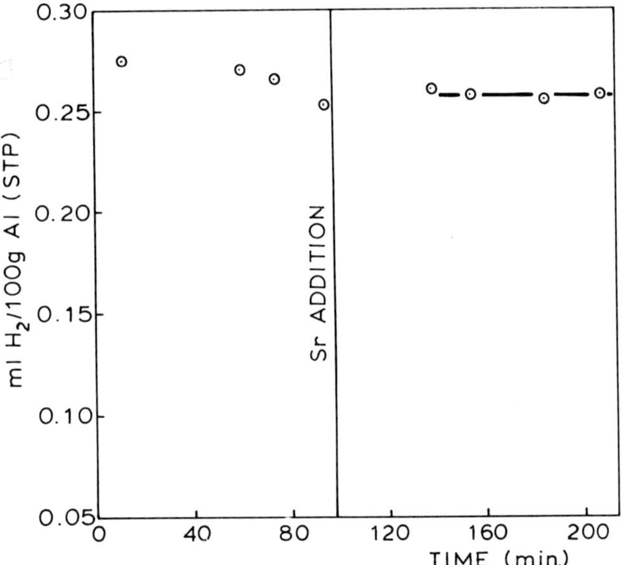

Figure 4.1. Melt hydrogen measured after the addition of 0.03% Sr to a non-degassed A356 melt at 710C. The strontium was added as a 90%Sr-10%Al master alloy [2].

In Table 4.1 are summarized the results of Telegas hydrogen measurements taken before and after strontium addition by a variety of master alloys. The melts to which these additions were made were all previously degassed. In no case is the increase in hydrogen greater than what would be expected due to normal regassing during the dissolution period of the modifier.

Melt degassing usually follows strontium treatment, and hence the question of regassing rate is important. An accelerated rate of regassing in the presence of strontium could certainly lead to enhanced hydrogen concentrations and increased levels of porosity. Several recent studies at McGill University[3-5], under highly controlled conditions, have demonstrated that regassing is not affected by strontium modification.

Figure 4.2 shows the results of a typical experiment in which strontium was added to a melt, the melt was degassed and then allowed to regas under a controlled moist atmosphere. Mass transfer coefficients for regassing were calculated from these results and are summarized in Table 4.2 for a variety of melting conditions. These coefficients are quantitative measures of the regassing rate, with significant differences being of at least a factor of 10. Of perhaps more interest to the foundry engineer is the last column of this table which gives the hydrogen levels in the melt after 1 hour of regassing from an initial level of 0.10 ml H$_2$/100 g Al. These values are applicable only to the conditions for which the appropriate mass transfer coefficients were measured, and hence different melting conditions will almost certainly yield different values. It is clear, however, that the presence of strontium has absolutely no effect on the rate of regassing. This is true over the temperature range 710C-780C (1280F to 1400F) and for both stirred and naturally convecting melts.

Table 4.1. Change in Melt Hydrogen Concentration During the Addition of Various Strontium Master Alloys

| Alloy | Melt Temp. (°C) | Modifier Type | Strontium (%) | | Hydrogen Level (ml H_2/100 g Al) | | | Hydrogen Increase upon Sr * Addition |
			Before	After	Before Degassing	After Degassing	After Sr Addition	
A356.0	700	A	0	0.021	0.28	0.17	0.20	0.03
	750	A	0	0.019	0.29	0.21	0.26	0.03
	700	B	0	0.012	0.25	0.18	0.20	0.02
	750	B	0	0.017	0.32	0.20	0.22	0.02
	700	C	0	0.021	0.26	0.16	0.21	0.05
	750	C	0	0	0.30	0.21	0.22	0.01
413.0	700	A	0	0.021	n/a	0.17	0.17	0
	750	A	0	0.016	0.21	0.15	0.20	0.05
	700	B	0	0.006	n/a	0.14	0.14	0
	750	B	0	0.022	0.23	0.16	0.19	0.03
	650	C	0	0.016	0.16	0.05	0.08	0.03
	700	C	0	0	n/a	0.11	0.13	0.02

* measured by Telegas 30 minutes after addition.
A: Al-3.5%Sr
B: Al-10%Sr
C: 90%Sr-10%Al

The effect of stirring the bath is to moderately increase the regassing rate. Melts prepared in an electric resistance furnace regassed faster if they were stirred than if they were allowed to simply convect naturally. Melts prepared in a gas fired furnace regassed faster than those unstirred in an electric furnace, but at about the same rate as those stirred during electric melting. The high temperature gradients generated in an open flame gas furnace are capable of creating as much mixing as in mechanical stirring. Since hydrogen is not added directly to the liquid when a strontium treatment is made, and since it does not lead to an increased rate of pickup after the treatment, the increased porosity observed in strontium modified castings cannot be due to hydrogen. The explanation must lie elsewhere and will be explored more fully in section 4.3.

Sodium and Melt Hydrogen

Our knowledge of the hydrogen concentration of sodium treated melts is less

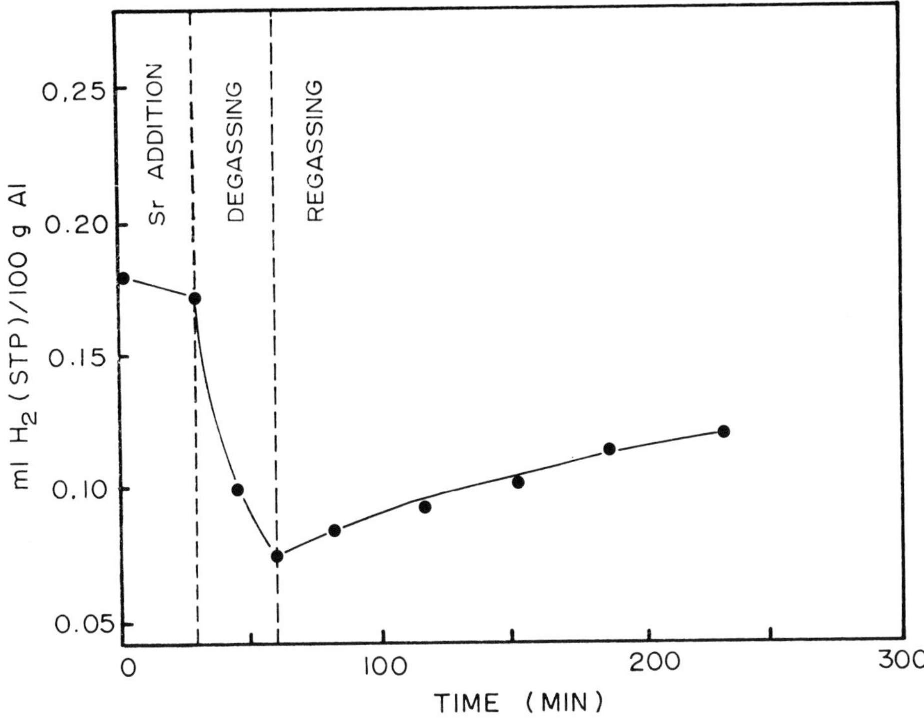

Figure 4.2. Changes in hydrogen level during the degassing and regassing of an A356 melt to which a 0.01% Sr addition was made [3].

perfect than that of strontium modified alloys, but it does appear that there can be a significant difference between the two. Sodium addition, at least in pure metallic form, is often accompanied by a violent reaction which leads to considerable fuming and agitation of the melt surface. In the presence of this reaction, there is a sharp increase in melt hydrogen (Figure 4.3), while if the reaction does not occur, no hydrogen uptake is found, and sodium behaves in an almost identical manner to strontium (Figure 4.3). It is not understood why a violent reaction occurs only on certain occasions. The gassing associated with the reaction may be due to melt agitation, or it may be due to the formation of NaOH on the surface of the metallic sodium. The formation of only 4% NaOH on the melt surface following a sodium addition will produce sufficient hydrogen to yield the observed increases.

While degassing after sodium addition is not a common practice, it is possible to retain sufficient sodium for modification provided a large enough excess was used initially. In such a case, melts will regas in a manner similar to that occurring with strontium modification. Figure 4.3 illustrates the behavior. Measured mass transfer coefficients for regassing of sodium treated A356 melts at 750C (1350F) are 5.2×10^{-6} m sec^{-1}. This value is almost identical to those reported in Table 4.2 for strontium containing melts. Thus neither sodium nor strontium affect regassing rates, but a sodium treatment can certainly lead to a direct hydrogen pickup, if it is accompanied by a violent addition reaction.

Table 4.2. Mass Transfer Coefficients for Hydrogen Regassing of A356 Alloy

Sr Level (%)	Melt Temperature (°C)	Furnace Type	Melt Stirring	Atmosphere Over Melt	Mass Transfer Coefficient (m sec^{-1})	Hydrogen Level after 1 hour* (ml H_2/100 g Al)
nil	710	electric resistance	natural convection	H_2O saturated at 40C (72F)	12×10^{-6}	0.12
0.02	710	"	"	"	12×10^{-6}	0.12
nil	750	"	"	"	5.0×10^{-6}	0.11
0.01	750	"	"	"	5.1×10^{-6}	0.12
0.02	750	"	"	"	5.2×10^{-6}	0.12
nil	780	"	"	"	8.4×10^{-6}	0.13
0.02	780	"	"	"	5.5×10^{-6}	0.12
nil	750	"	stirred	"	20×10^{-6}	0.16
0.02	750	"	stirred	"	27×10^{-6}	0.16
nil	750	gas	natural convection	normal furnace atmosphere	17×10^{-6}	0.15
0.01	750	"	"	"	19×10^{-6}	0.15

*Assumes regassing under a nitrogen atmosphere saturated with water at 40C (72F). The hydrogen concentration at the beginning of the hour is 0.10 ml H_2/100 g Al.

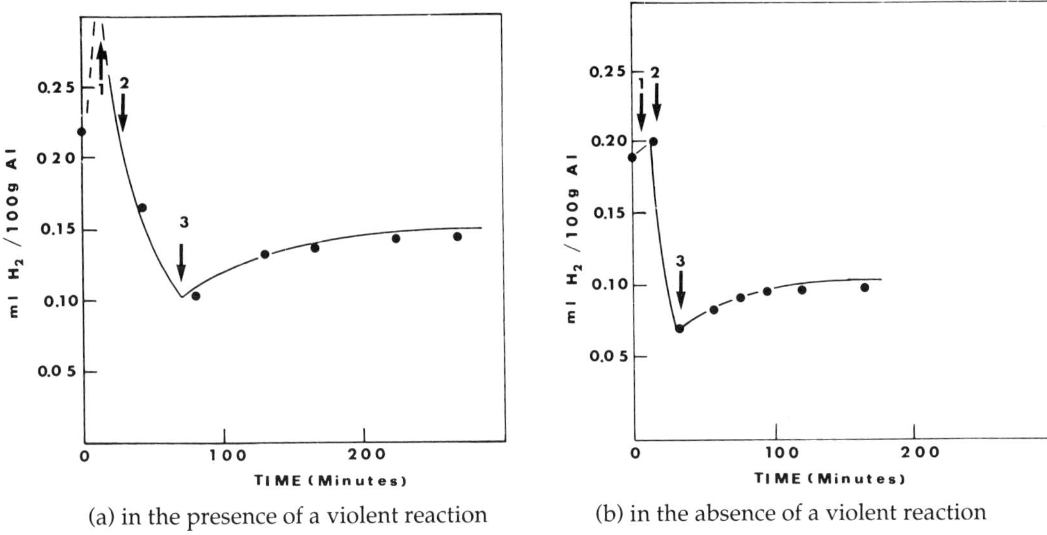

(a) in the presence of a violent reaction   (b) in the absence of a violent reaction

Figure 4.3. Some changes in dissolved hydrogen level on addition of metallic sodium to an A356 melt [5]:
1. sodium addition
2. start of degassing
3. end of degassing.

4.2   Modification and the Porosity Distribution

While dissolved hydrogen is definitely a very important factor in the generation of porosity in aluminum castings, it is by no means the sole cause. Changes in alloy freezing range, for example, or in surface energies could also account for increased porosity when modification is used. These considerations become all the more important with the demonstration that strontium treatment at least does not change the dissolved gas level of the alloy.

A careful review of the early literature on modification coming out of France and Germany leads one to suspect that modification may cause basic changes in the shrinkage/porosity patterns of cast alloys. For example, Mascre and Lefebvre[6] in 1959 published a set of photographs of sodium modified ingots which show clearly that an increase in sodium concentration changes the shrinkage from one large macropore to many finely dispersed pores of much smaller size. Arbenz[7] in 1962 published the photograph shown in Figure 4.4 of a casting in both the unmodified and sodium modified conditions. When unmodified, the casting section shows large macropores both internally and on the surface. These disappear with sodium modification to be replaced by a much finer dispersed porosity. Unfortunately, the dissolved hydrogen concentration of the melt was not determined in either of these examples. While the effects could be due to gassing, they are graphic enough to lead one to suspect that the process of modification may have, as one of its basic features, a rather dramatic change in the porosity distribution.

These early observations have recently been clarified in an investiga-

(a) unmodified. This casting contains large shrinkage cavities both internally and on the surface.

(b) sodium modified. The shrinkage cavities have disappeared to be replaced by much finer dispersed porosity

Figure 4.4. Variation in shrinkage behavior with modification [7].

tion by Argo and Gruzleski[8] using the Tatur Test at a measured and controlled hydrogen level of 0.20 ml H_2/100 g Al. The Tatur test utilizes a permanent mold of fixed geometry shown in Figure 4.5. By using simple techniques such as density measurement and water displacement, it is possible to quantify the volume of microshrinkage, macroshrinkage (pipe) and slumping and contraction. In order to draw reliable conclusions, it is necessary to perform a large number (about 20) of tests and to evaluate them statistically. When this is done, at a fixed hydrogen level, for unmodified, strontium modified, and sodium modified alloy, the results given in Table 4.3 are obtained.

Some very interesting features of modification are brought out by this work which goes a long way to explaining past observations. Total shrinkage, which is an alloy property, is not at all affected by modification. However, the way in which this shrinkage is distributed between macro-piping and microshrinkage depends strongly on whether or not the alloy is modified. Both sodium and strontium cause a significant diminution of the primary pipe and an increase in the amount of microporosity. In other words, shrinkage is redistributed when modification occurs, exactly as is indicated by references 6 and 7.

This realization allows us to explain how the addition of a modifier which does not change the dissolved hydrogen level of a melt can result in more porous castings. A proper casting design allows for formation of the pipe, or macroshrinkage, in the riser or running sections of the casting. *With modification, some of the piping is reduced and the amount of this reduction appears as distributed microporosity within the casting itself. Thus, the casting is more porous, not due to more dissolved gas, but rather due to what appears to be a fundamental feature of the modification process. Porosity is redistributed from primary piping to microporosity.* The diminished pipe associated with modification is very evident in the sectioned Tatur castings shown in Figure 4.6.

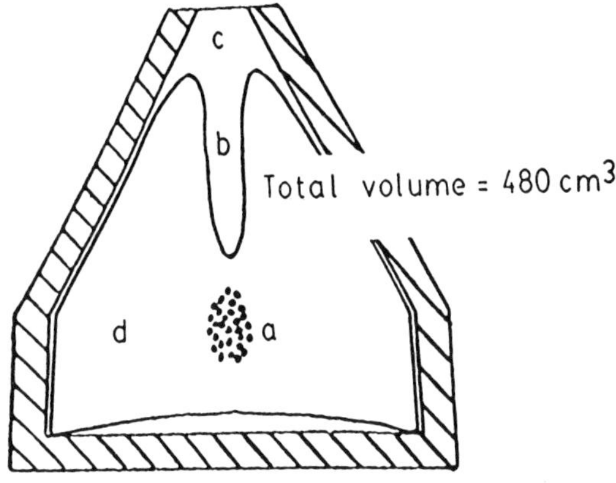

Figure 4.5. The Tatur Test showing mold geometry and the volume of the various shrinkage features.

The effects summarized in Table 4.3 take place at very low levels of modifiers. In Figure 4.7 we have reproduced the original data relating pipe volume to strontium concentration. Only very small amounts of strontium are necessary to cause a decrease in pipe volume. The levels required are, interestingly, much below those needed for full modification, and further increases cause no further change in piping. This same general behavior is observed for the other parameters listed in Table 4.3. We are thus dealing with an effect which takes place on the addition of only minute quantities of either Na or Sr, and at the moment, the most obvious explanation is that the modifier acts to decrease the surface tension of the liquid alloy, thus facilitating nucleation of micropores.

A graphic example of the effects of a sodium or strontium addition on the porosity distribution within a casting can be seen in the radiographs reproduced in Figures 4.8 and 4.9. These sand cast bars of A356 alloy were poured with a chill at the right side and an ingate on the left. In the complete absence of a modifier, the casting design is such that the central hexagonal section contains a region of concentrated porosity. When the same casting is poured from liquid which contains the same hydrogen level, but to which a modifier, either sodium or strontium has been added, we see that the concentrated porosity becomes dispersed throughout the piece. The major

Table 4.3. Tatur Test Measurements on A356 Alloy at Constant Hydrogen Concentration

	Unmodified	Sr-Modified	Na-Modified	Significant Difference from Unmodified[1]	
				Sr	Na
Apparent density (g cm^{-3})	2.660	2.655	2.645	yes	yes
Microshrinkage (cm^{-3})	2.9	4.0	5.75	yes	yes
Pipe (cm^{-3})	10.55	5.95	5.25	yes	yes
Slumping and contraction (cm^{-3})	26.55	31.2	28.4	yes	no
Total shrinkage (cm^{-3})	40.0	41.15	39.4	no	no
Microshrinkage as a % of total shrinkage	7.3	9.7	14.6	yes	yes

[1]Statistically significant by Student "t" Test.

shrink thus disappears, but at the expense of finer porosity in parts of the casting which were previously sound. This behavior shown here on a simple bar shape, is the same as that observed by Arbenz[7] on the more complicated casting in Figure 4.4.

While both modifiers are capable of causing this phenomenon, the available evidence is that sodium is more effective at porosity redistribution than is strontium. Sodium modified castings should therefore contain more microporosity. It is impossible to be very quantitative at this point, but examination of Table 4.3 indicates, for example, that almost 15% of the total shrinkage appears as microshrinkage when sodium is used, but only about 10% does so if strontium is the modifier.

Further confirmation of this can be obtained by measuring and plotting the density along sand cast bars. This is done in Figure 4.10 *for bars of the same hydrogen content* which are unmodified, sodium modified and strontium modified. Quantitative assessments of the amount of porosity can be obtained by measuring the areas over those curves using the density at the chill end as a base line. When this is done, the results shown in Table 4.4 are obtained. The greatest amount of porosity is found in the unmodified condition due to the concentration of macroshrinkage. With modification, the porosity is dispersed and the sodium modified bar contains more than the strontium treated bar.

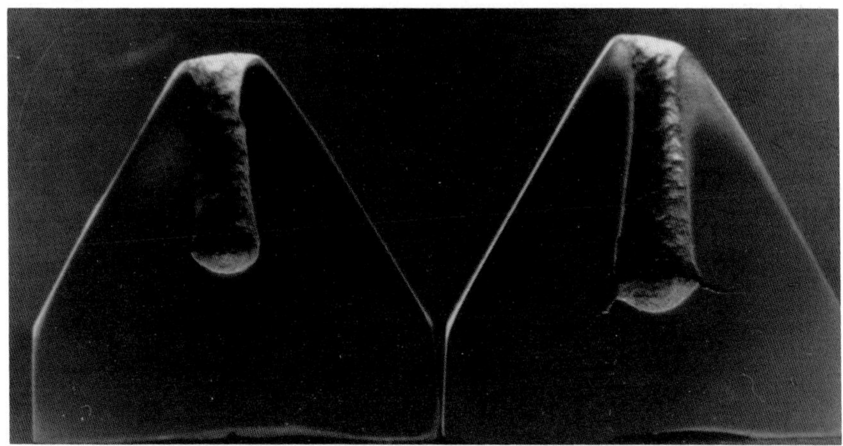

Figure 4.6. Sectioned Tatur castings showing the diminished pipe volume of modified castings (left) compared to unmodified castings (right) [8].

Table 4.4. Porosity Levels Calculated from Figure 4.10

Modifier	Profile Area (g cm^{-2})	Density (g cm^{-3})	Porosity cm^3
None	0.44	2.658	10.25
Strontium	0.22	2.669	5.15
Sodium	0.39	2.661	9.30

Pore Size

It was suggested earlier in this section that the effects described here are due to a reduction of surface energy by the modifier. This hypothesis is supported by the observation that micropores are larger in modified than in unmodified alloys. The castings shown in Figure 4.11, poured with the same hydrogen concentrations, show this effect. The porosity in modified alloys is therefore both more widely distributed and coarser than in unmodified alloys.

Slumping and Contraction

The one piece of information obtainable from the Tatur Test which we have not yet mentioned, is the slumping and contraction listed in Table 4.3. This relates to the deformation of the solid (or semi-solid) shell of the casting during solidification. Alloys which exhibit a large amount of slumping and

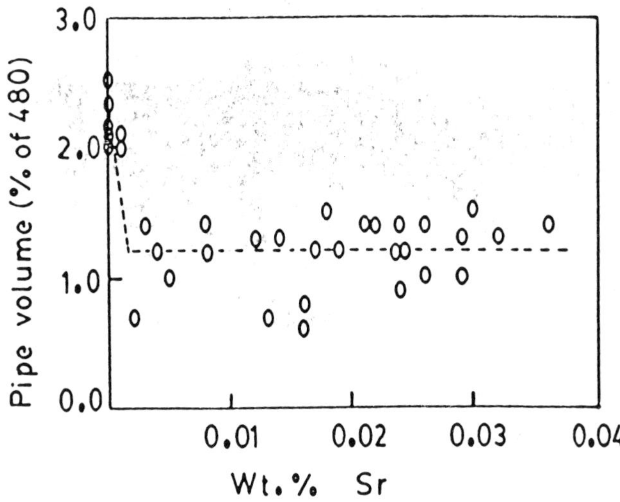

Figure 4.7. Variation of pipe volume with strontium level expressed as a percentage of the Tatur volume of 480 cm³ [8].

contraction are able to relieve internal stresses caused by shrinkage of the solid during and immediately after freezing. They thus possess a reduced tendency to hot tear. Strontium modified A356 has a significantly greater slumping and contraction than do either unmodified or sodium modified alloy, and hence should exhibit an enhanced hot tearing resistance. This is evident in the two sectioned Tatur castings of Figure 4.12. The one on the left, which is unmodified, shows pronounced hot tears at the base of the pipe, while the strontium modified sample on the left is free of this defect.

4.3   Avoiding Porosity in Modified Castings

The central theme of this chapter has been that a fundamental feature of modification is a change in the porosity distribution. Large areas of macroporosity are replaced by a widely distributed, but much finer porosity, so that the casting as a whole may appear to be less sound than in the unmodified condition. What, if anything, can the foundryman do about this, and is it ever to his advantage?

Let us first look at a case in which this effect is advantageous. Internal or surface shrinkage cavities may be eliminated by modification, e.g., as in Figure 4.4, provided the dispersed porosity which results does not create other serious problems such as a degradation of required mechanical properties. There are several foundries that add modifiers simply because they help to produce castings which are more macroscopically sound.

Modified castings very often do contain fine internal porosity. There may be cases where it is better to eliminate modification altogether, to produce very sound castings, and to attempt to develop the mechanical properties through heat treatment. We believe this to be an extreme case. In most situations, modification and sound castings are compatible, if some basic foundry rules are followed, but perhaps more rigorously than would otherwise be the case. Three aspects of melt treatment and molding need to be highly controlled and perhaps changed somewhat. These are: hydrogen gas control in the liquid, chilling and directional solidification in the mold, and gating and riser design.

Hydrogen Gas Control

Melts which contain higher hydrogen levels lead to more porous castings. The actual level of "acceptable" hydrogen will vary with each specific casting geometry and end-user application.

Modified castings should be poured from the most hydrogen-free liquid alloy which it is possible to produce. The effect of high quality degassing on porosity in a strontium treated casting is shown clearly in the photographs of Figure 4.12. When poured from non-degassed metal (0.2 mlH$_2$/100 g Al), the alloy contains easily visible overall porosity. On degassing to values in the range 0.02

Figure 4.8. Radiographs of sand cast bars, some of which have been treated with strontium. Porosity appears dark [8]. The strontium levels from top to bottom are: zero, zero, 0.003%, 0.009%, 0.012%, 0.018%.

mlH$_2$/100 g Al, measured by the Severn Science Technique[1], porosity becomes concentrated in the top region of this sand cast sample (Figure 4.12b). When degassing and modification are combined (Figure 4.12a), the low level of porosity is widely distributed and a high quality casting is produced containing only a few evenly dispersed pores. If strontium is the modifier, this is readily achieved. Degassing techniques are available which do not remove large quantities of strontium (Chapter 10), and as we have shown here, these melts will not regas any faster than non-modified melts.

Thus, strontium treated melts should be degassed after treatment using the best degassing technique possible to achieve the lowest hydrogen levels. Castings should be poured as soon as possible after degassing to

---

[1]The reader is referred to Chapter 9 for a description of measuring hydrogen gas.

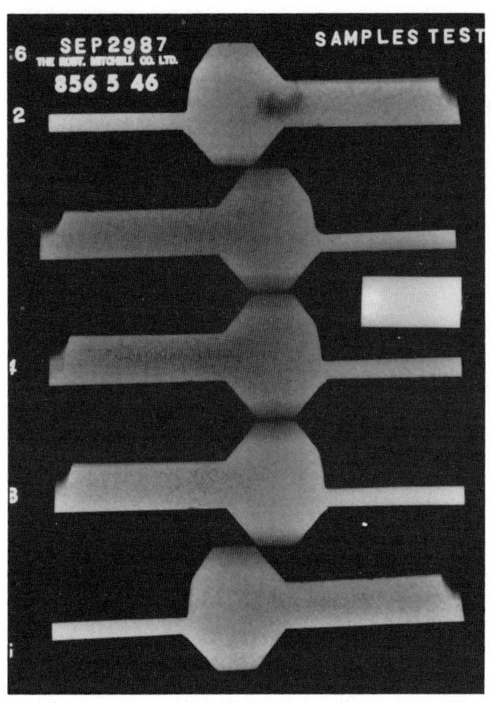

Figure 4.9. Radiographs of sand cast bars, some of which have been treated with sodium. Porosity appears dark[9]. The sodium levels from top to bottom are: zero, 0.002%, 0.003%, trace, 0.012%. (Previously published in the *Proceedings of the International Symposium on Production and Casting of Aluminum*, C. Bickert, [1988]. Reprinted with permission of the Canadian Institute of Mining and Metallurgy.)

minimize regassing, and, if necessary, more than one degassing operation can be carried out.

With sodium modification, the production of low porosity castings will not be so easy. Generally, sodium losses during degassing are so high as to preclude a degassing operation after modification treatment. It is, therefore, necessary to add sodium to a previously degassed melt and to cast as soon as possible. If metallic sodium is employed, and if a violent addition reaction occurs, some hydrogen pickup seems inevitable.

Chilling and Directional Solidification

The tendency of modified castings to freeze with dispersed porosity implies that feeding must be maximized in order to reduce this porosity to a minimum. Directional solidification, promoted through the use of chills and casting design, is the well accepted way to accomplish this. It is quite possible that different running and/or gating systems from those used with unmodified castings may provide improved feeding and sounder castings. Some ideas are presented schematically in Figure 4.13. The casting on the left (Figure 4.13a) is initially well designed and is sound when non-modified. With modification, some of the porosity initially present in the riser is dispersed into the riser end of the casting (Figure 4.13b). This can be moved back into the riser to once again produce a sound casting through use of chills (Figure 4.13c) or a larger riser (Figure 4.13d).

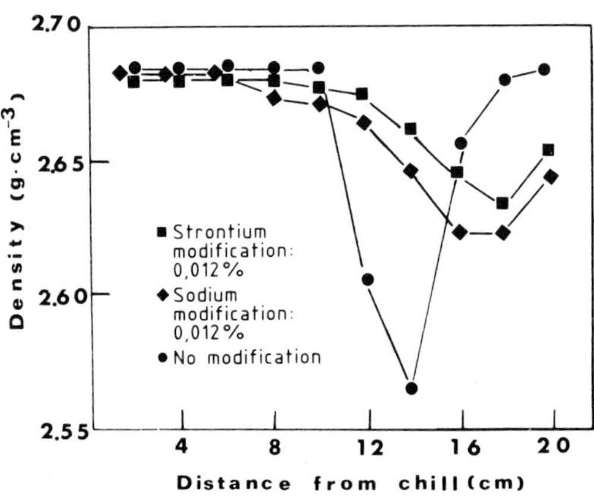

Figure 4.10. Variation of density versus distance from graphite chill for sand cast bars.

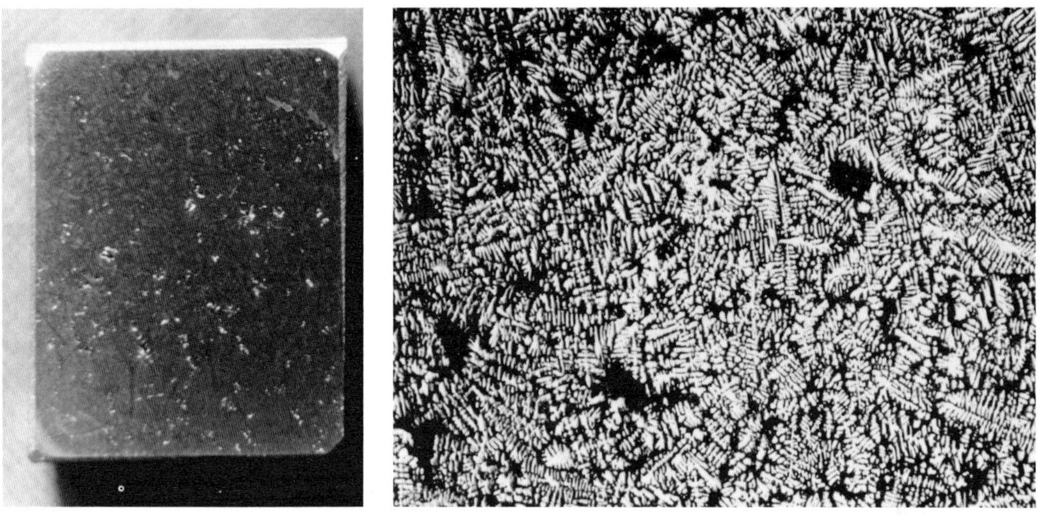

(a) strontium modified: macrograph (left); micrograph (right) (x10)

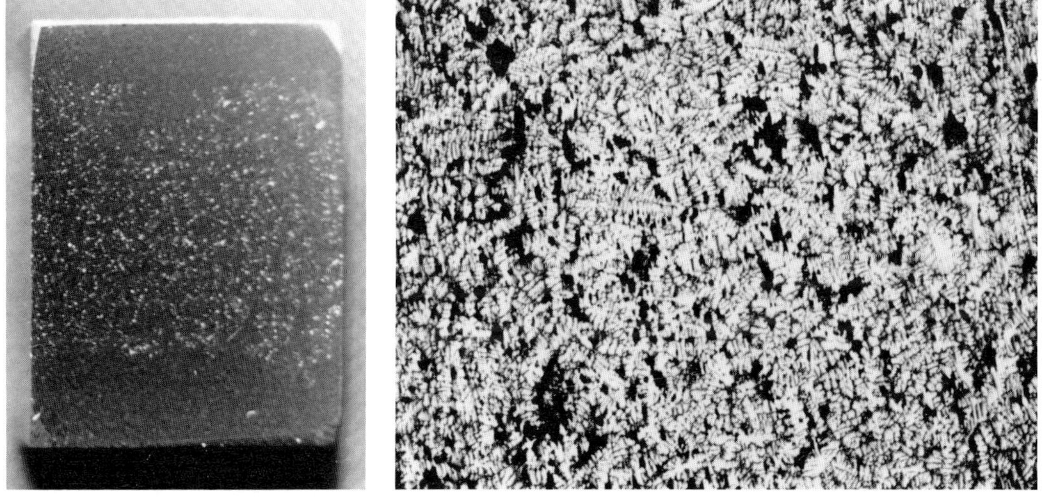

(b) unmodified: macrograph (left); micrograph (right) (x10)

Figure 4.11. Comparison of pore sizes in cast A356 of the same hydrogen level (0.20 ml H_2/ 100 g Al).

The casting on the right is not so well designed, and a shrink occurs in the casting close to the riser. (Figure 4.13e). When modified, this is dispersed throughout much of the piece (Figure 4.13f), but once again a sound part can result through proper use of chills (Figure 4.13g) or better riser design (Figure 4.13h). Special consideration should be given to the extra use of chills in critical locations if the casting is to be modified. Finally, it is suggested that a decision to use or not to use modification be taken as early in the design process as possible in order that everything feasible can be done in the mold design to maximize feeding and hence to minimize porosity.

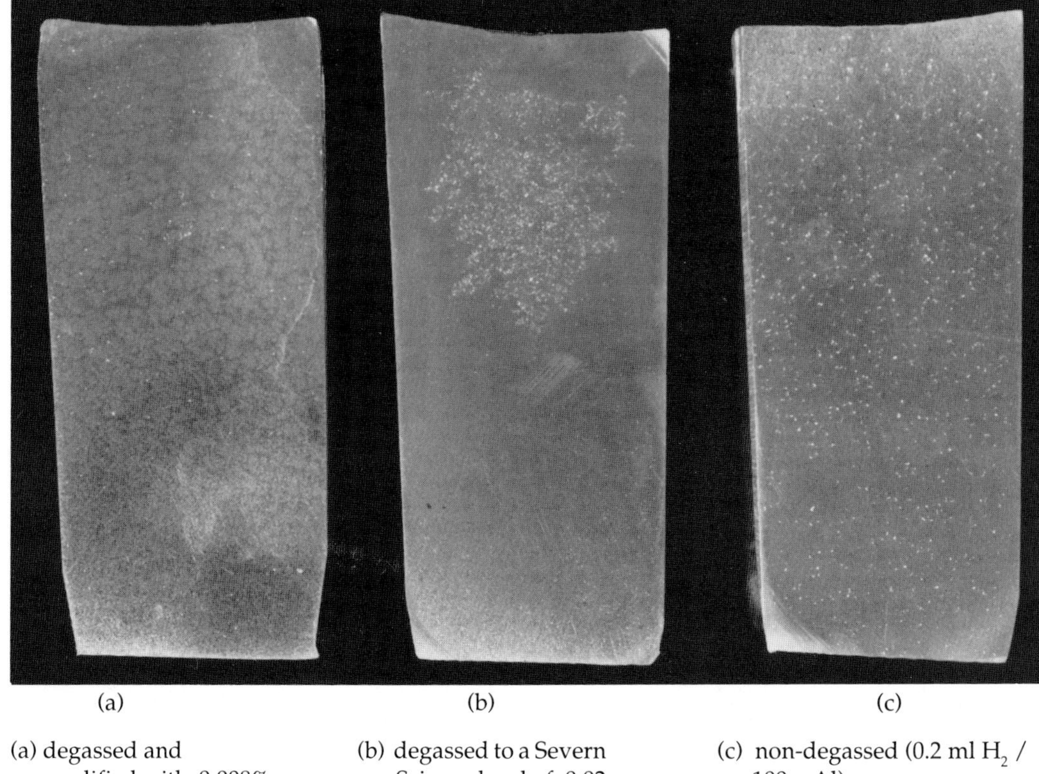

(a)   (b)   (c)

(a) degassed and modified with 0.008% Sr

(b) degassed to a Severn Science level of 0.02 ml H_2 /100 g Al

(c) non-degassed (0.2 ml H_2 / 100 g Al)

Figure 4.12. Porosity in a sand cast sample.

References

1. Jacob, S. "Modification de l'A-S7G06 par le Sodium, L'Antimoine et le Strontium," *Fonderie*, No. 363 (1977) pp. 13-25.
2. Gruzleski, J.E. N. Handiak, H. Campbell and B. Closset. "Hydrogen Measurement by Telegas in Strontium Treated A356 Melts," *AFS Transactions*, 94 (1986) pp. 147-54.
3. Dimayuga, F., N. Handiak and J.E. Gruzleski. "The Degassing and Regassing of Strontium Modified A356 Melts," *AFS Transactions*, 96 (1988) pp. 83-88.
4. Mulazimoglu M.H., N. Handiak and J.E. Gruzleski. "Some Observations on the Reduced Pressure Test and the Hydrogen Concentration of Modified A356 Alloy," *AFS Transactions*, 97, (1989) pp. 225-32
5. Closset, B. and J.E. Gruzleski. "Modification, Porosity and Hydrogen Content in Al-Si Casting Alloys," *USA Technical Communication*, 56th World Foundry Congress (May 1989) Dusseldorf.

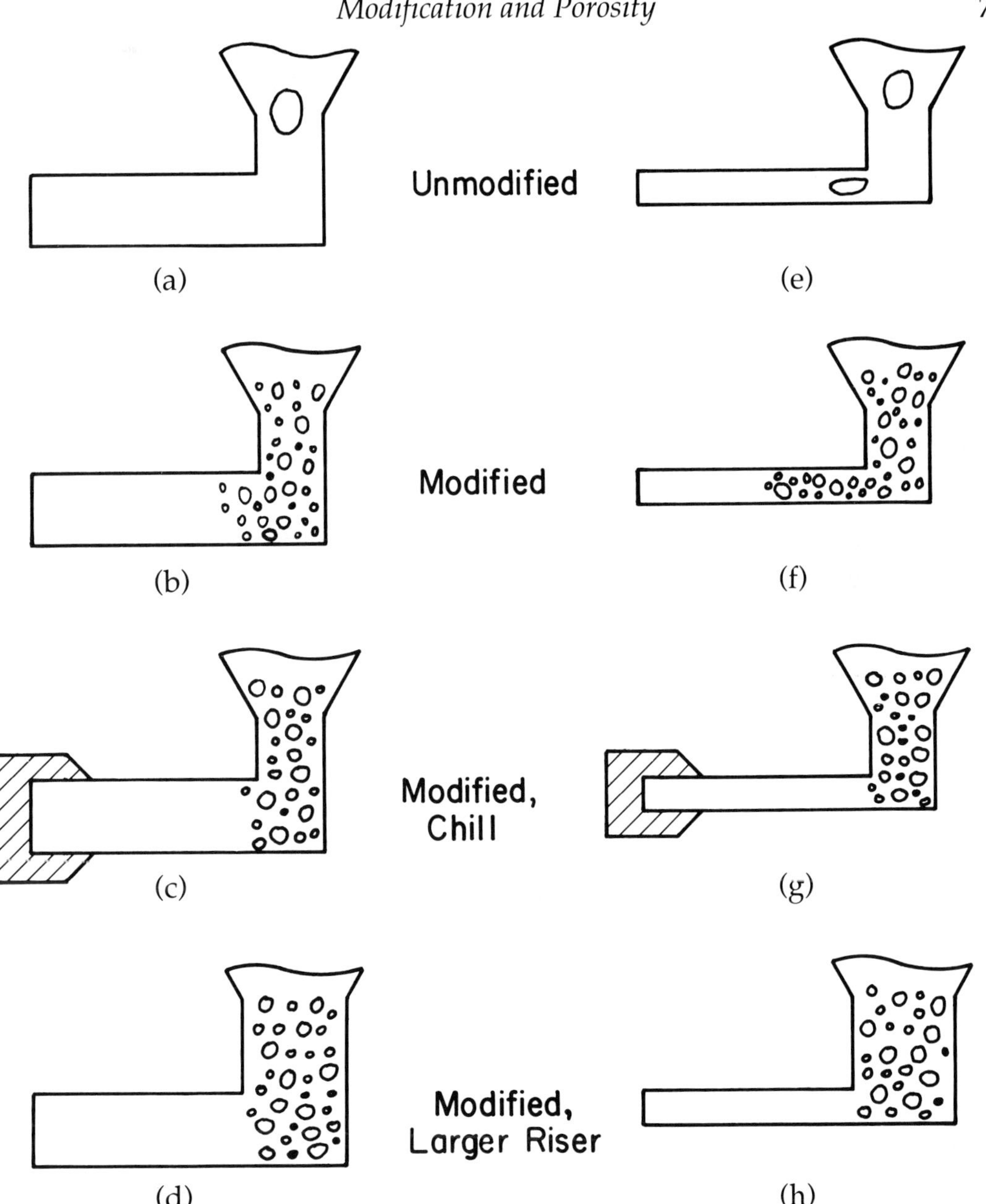

Figure 4.13. Porosity control in a modified casting.

6. Mascre, C. and M. Lefebvre, "La Solidification de l'A-S13 (Alpex)," *Fonderie*, No. 166, (1959) pp. 484-97.

7. Arbenz, H. "Lunkerneigung von Aluminum-Guslegierungen," *Giesserei*, 49 (1962) pp. 105-110.

8. Argo, D. and J.E. Gruzleski, "Porosity in Modified Aluminum Alloy Castings," *AFS Transactions*, 96 (1988) pp. 65-74.

9. ———. "Porosity in Aluminum Foundry Alloys—The Effect of Modification," *Proceedings of the International Symposium on Production and Casting of Aluminum*, ed. C. Bickert, Montreal (1988) Pergamon Press, pp. 263-82.

Chapter 5

The Properties of Modified Alloys

In this chapter we examine what is known of the properties of modified aluminum alloys. Many of the properties of these alloys in their unmodified form are already well documented in several handbooks, and therefore the emphasis here will be on the changes brought about by the process of modification.

5.1   Tensile Properties

There is more data available on the influence of modification on the tensile properties of Al-Si alloys than on any other property. Unfortunately, this data is often difficult to interpret and is sometimes contradictory. In this section, we will try to present a simple and concise view of how modification affects tensile properties.

Much of the confusion related to the tensile properties of cast aluminum alloys arises from the fact that these depend on several, often interrelated, variables such as:

- solidification rate;
- casting soundness;
- heat treatment (solutionizing, quenching, aging);
- eutectic modification;
- magnesium content.

In any effort to determine the effect of say, sodium modification on properties, all of these several variables must be controlled. If they are not the exact effect of the modification may be obscured by the influence of one or several uncontrolled factors. This can be seen in the technical literature in any one of several reports which contend that sodium modification has no effect on mechanical properties. Such a contention is directly opposite to the well established view that modification exerts at least a moderate influence on improving properties. The problem here is that sodium treatment is often associated with gassing, and modified alloys in general are more prone to microporosity for all of the reasons discussed in Chapter 4. Hence, it is

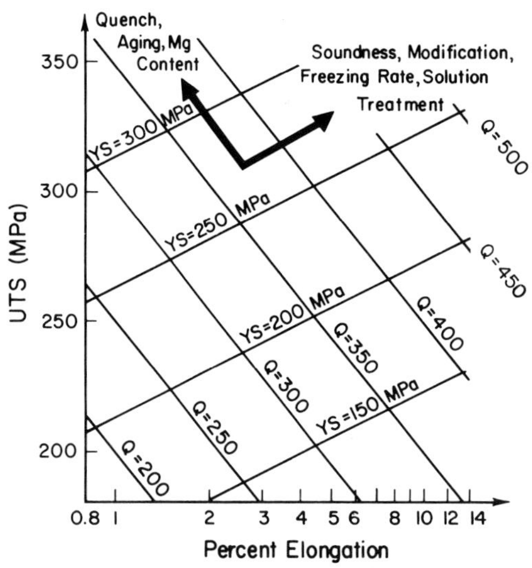

Figure 5.1. Relationship between Ultimate Tensile Strength, Elongation, Yield Strength and Quality Index for an Al-7% Si-Mg alloy [1].

difficult to separate modification from soundness, and in some cases the effect of decreased soundness can overcome the beneficial effects of modification.

The erroneous conclusion which is then made is that modification has no beneficial effects. Of course in a casting, the situation is complicated by the fact that it is often extremely difficult if not impossible to control all of the variables, and they will usually change greatly from one location to another within the casting. It is little wonder, therefore, that some founders praise modification, while others are not so convinced of its value.

The Quality Index

The influence of all of these many variables on tensile properties has been simplified somewhat by the introduction in the late 1970s of the concept of a Quality Index, Q[1]. This concept arose from a consideration of the relationship between ultimate tensile strength, elongation, and yield strength of Al-Si-Mg alloys. These parameters are plotted in Figure 5.1 for a 356 type alloy. It can be seen that an alloy having a yield strength of say 200 MPa can have either a low or a high quality. The high quality alloy could have an elongation of 10% and a UTS of 273 MPa. The low quality version of the same alloy might have an elongation of only 1% and a UTS of 212 MPa. One could imagine alloys of equal quality as having combinations of UTS and elongation which lie along lines more or less perpendicular to the lines which describe the yield strength. This concept gives rise to the Quality Index defined as:

$$Q = UTS + (K)\log \text{ elongation.}$$

It is necessary to choose the coefficient, K, such that Q is independent of yield strength (or hardness), i.e., so that the lines of equal Q are at right angles to the lines of equal yield strength in Figure 5.1. For heat treated 356 alloys, this results in a Quality Index of:

$$Q = UTS + (150)\log \text{ elongation.}$$

This Quality Index has units of stress (MPa or ksi). The usefulness of this parameter in describing the effects of modification is that the Quality Index depends on the fineness and compaction of the microstructure. It is precisely these features which are of concern when modification is performed. Since the Quality Index combines both strength and ductility, it is much more descriptive of the true tensile properties of a casting than either the tensile strength or the elongation alone.

Some parameters change the Quality Index while others do not affect it. The direction of influence of several casting/heat treatment variables is indicated by the arrows at the top of Figure 5.1. Magnesium concentration does not influence the Index, but soundness, for example, can change it. Sounder castings will exhibit higher Index values. The variables that do not change the Index can be manipulated to alter the tensile properties while maintaining the same Quality Index. For example as shown in Figure 5.1, an alloy for which Q = 400 MPa might have a UTS equal to 290 MPa and an elongation of 5.5%. If an elongation of 8% is desired while maintaining the same Quality Index, then the aging time could be changed to accomplish this. In this case, the UTS. would drop to 265 MPa and the yield strength would be slightly less than 200 MPa.

Variables which change the Quality Index can compensate for each other. Thus, a rapidly solidified but poorly modified structure may possess the same Quality Index rating as one which is more slowly solidified but better modified. We will return to these considerations later in this chapter.

How Modification Changes Tensile Properties

We have seen that the main effect of chemical modification is on the microstructure, and it is these microstructural changes which influence directly the mechanical properties. The coarse silicon plates of the unmodified acicular silicon structure act as internal stress raisers in the microstructure and provide easy paths for fracture. With modification, the structure becomes finer and the silicon more rounded, both of which contribute to somewhat higher values of ultimate tensile strength and greatly increased values of ductility.

The improvements which are observed in tensile properties depend on the structural differences between the modified alloy and the unmodified alloy. For example, a non-modified sand cast hypoeutectic alloy will possess coarse acicular silicon; the same alloy modified and cast in a permanent mold will contain fine fibrous silicon. The structural differences are great and large differences in tensile properties can be expected. On the other hand, both chill cast modified and unmodified alloys have somewhat similar structures due to their rapid solidification rates, and here the effect of modification on properties is minimized.

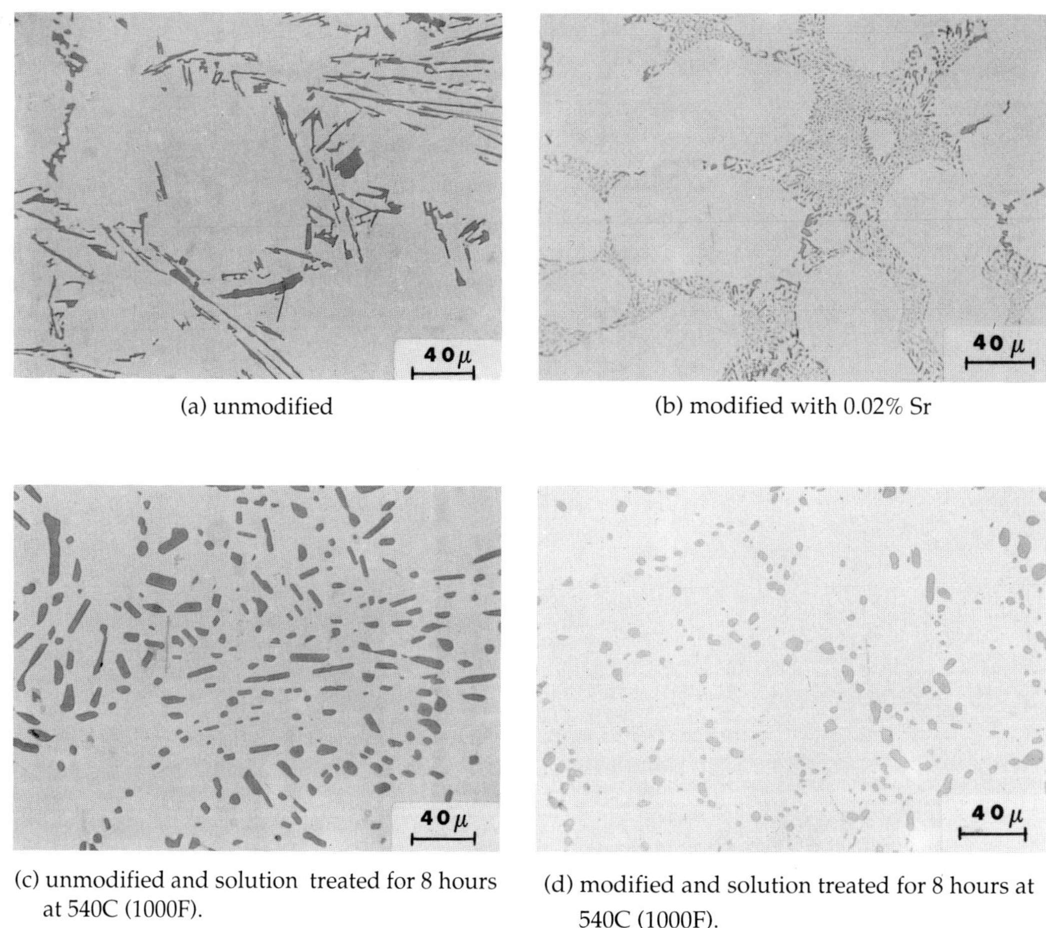

(a) unmodified   (b) modified with 0.02% Sr

(c) unmodified and solution treated for 8 hours   (d) modified and solution treated for 8 hours at
   at 540C (1000F).   540C (1000F).

Figure 5.2. Microstructures of A356 alloy.

Most Al-Si alloy castings are heat treated in a standard age hardening process. During the solution treatment, unmodified silicon becomes more spherical, and modified silicon particles coarsen. These changes are illustrated in the photomicrographs of Figure 5.2. It is evident that the microstructural differences between modified and unmodified eutectic silicon are somewhat diminished by the heat treatment, and we observe also that the differences in tensile properties are also less, although a significant difference is still found.

These effects are reflected in the typical values of tensile properties given in Tables 5.1 and 5.2. The properties quoted are taken from data for various French alloys. They are typical values for the chemistries stated and should not be regarded as either maximum or minimum obtainable properties.

Table 5.1. Properties of Some Heat Treated Alloys

Alloy Chemistry	Al-7%Si-0.3%Mg (A356)						Al-7%Si-0.6%Mg (A357)						Al-9%Si-0.3%Mg*			
	Permanent Mold			Sand			Permanent Mold			Sand			Permanent Mold		Sand	
Silicon Structure	UTS**	E†	Q**	UTS	E	Q	UTS	E	Q	UTS	E	Q	UTS	E	UTS	E
Acicular	290	12	452	275	2.5	335	330	10	480	290	2	335	300	11	280	2
Lamellar	290	17	475	280	4	370	330	13	497	285	2	330	300	14	280	2.5
Fibrous	290	17	475	280	6	397	330	13	497	285	2.5	345	300	14	280	4

*the data permitting the determination of Q for this alloy chemistry has not been developed
**MPa
†%

Table 5.2. Some Properties of Non-Heat Treated Alloys[*]

Silicon Structure	Al-7%Si-0.3%Mg		Al-11%Si	
	UTS (MPa)	E (%)	UTS (MPa)	E (%)
Acicular	180	7	150	6
Lamellar	200	12-16[**]	170	14-18[**]
Fibrous	200	16	170	18

[*] cast in permanent mold
[**] according to the fineness of the lamellar eutectic

Effect of Amount of Modifier, Cooling Rate and Fading

Regardless of the modifier used, the as-cast or the heat treated microstructure is a function of modifier level. For a given freezing rate there is an optimum concentration of modifying element; less than this level creates undermodification; amounts greater cause over-modification. Since microstructure varies with modifier content, so do the tensile properties. These effects are illustrated in Figure 5.3 for strontium modified A356 alloy cast at three different cooling rates. The scatter in data points for each casting condition is typical of cast alloys, and illustrates the difficulty in specifying properties on the basis of only a few samples. The properties change more slowly if too much strontium is added than if too little is used. Hence, a slight overmodification is less serious than an undermodification. The tensile properties decrease as the cooling rate of the casting decreases. This is a reflection of two aspects of the cast structure: the difficulty of obtaining a fine modified structure at very slow freezing rates; and the increased tendency for castings to be less sound if they freeze slowly. Nevertheless,

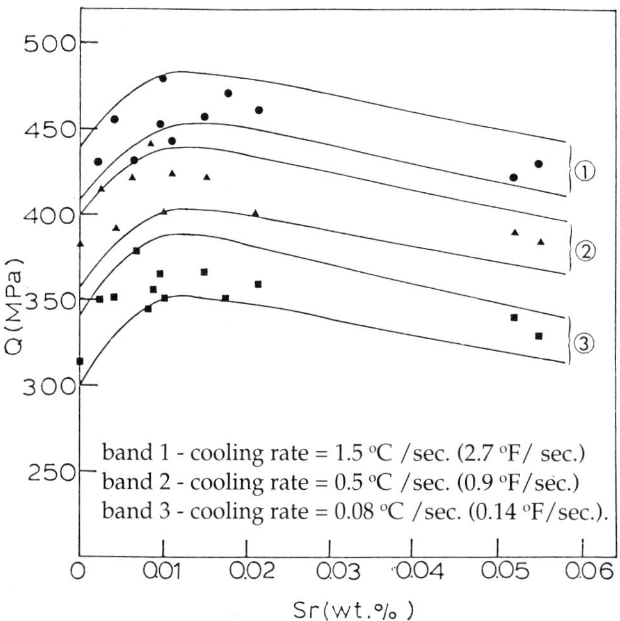

Figure 5.3. Variation of the Quality Index, Q, with strontium concentration in A356 alloy cooled at three different rates[2].

in all cases, the use of modification does result in improved properties.

Some effects of sodium modification on the Quality Index of A356 are illustrated in Figure 5.4. The range of properties obtained in castings chilled various amounts is indicated for three heat treatments by parallelograms. The solid line parallelograms represent sodium modified alloy while the dotted line parallelogram is for non-modified alloy. If the heat treatment entails a 140F (60C) quench followed by aging at 325F (163C) modification

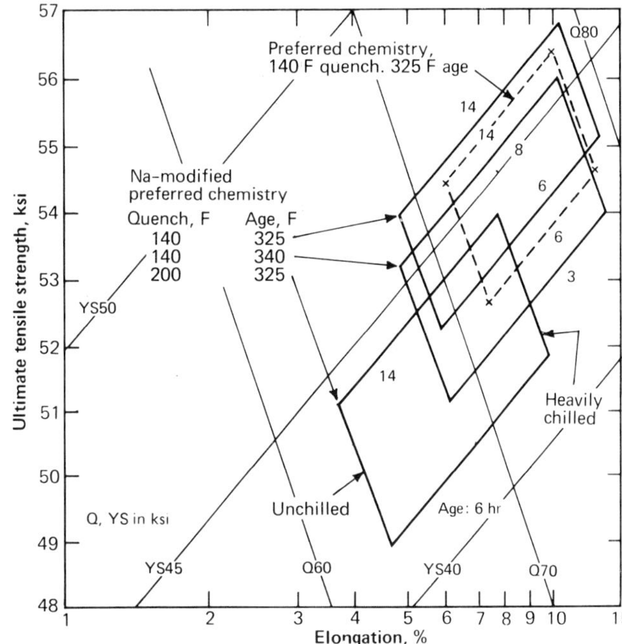

Figure 5.4. *The effect of sodium modification on the Quality Index of A356 alloy[3].*

improves the properties of heavily chilled castings. On the other hand, there is a deterioration of properties in unchilled castings due to a lack of soundness caused by the increased tendency for porosity in modified alloys. (see Chapter 4).

If antimony is used for modification, the lamellar structure is the only one which will form. This structure is more sensitive to freezing rate than the fibrous structures produced by strontium or sodium. It is recommended that antimony treated alloys only be used if the solidification time is less than 20-40 seconds, depending on the alloy.

With modifier fading, there will be a coarsening of the microstructure, and of course, a decrease in the tensile properties. Sodium is very much worse than either strontium or antimony in this regard as the data in Table 5.3 taken from reference 4 shows. One hour of holding at 760C (1400F) causes the properties to revert to their unmodified values. Both strontium and antimony, on the other hand, are stable in the melt, and their modifying effect resists reversion. Hence, the tensile properties are stable with time and are only changed slightly by remelting.

Modifier Type and Tensile Properties

Since tensile properties depend so greatly on the silicon morphology, it would be expected that similar microstructures, no matter how produced, would yield about the same properties. A well modified structure produced through sodium treatment should exhibit more or less the same tensile properties as a similar structure produced by strontium modification. Furthermore, since the differences between unmodified silicon and either a fine lamellar or a fine fibrous structure are so large, antimony treatment should produce about the same improvement in properties as is achieved with either sodium or strontium. Indeed this seems to be the case as indicated by

Table 5.3. Effects of Fading and Remelting on Tensile Properties of Al-9% Si Alloy[4]

Liquid treatment	Microstructure	UTS (MPa)	E (%)
no modification treatment	coarse,acicular Si	180	5.5
held 60 min at 760C after Na treatment	poorly modified	175	6.0
held 60 min at 760C after Sb treatment	fine lamellar	195	11.5
held 60 min. at 760C after Sr treatment	well modified, fibrous	210	12.0
Sb treated, cast and remelted at 760C	fine lamellar	185	11.5
Sr treated, cast and remelted at 760C	well modified	200	10.0

Table 5.4. Mechanical Properties of A356 Alloy Treated with Different Modifiers[5]

Modifier	Structure	As-Cast		Heat Treated[*]		
		UTS (MPa)	E (%)	UTS (MPa)	E (%)	Q (MPa)
None	acicular	180	6.8	304	11.8	465
Sodium	fibrous	195	16.4	292	15.1	469
Strontium	fibrous	196	15.9	301	14.4	475
Antimony	lamellar	201	11.9	293	16.5	475

[*]solution treated at 540C (1000F) for 10 hours, quenched, aged 6 hours at 160C (320F)

the values in Table 5.4 for A356 alloy modified in different ways. Choice of a modifier is not really dictated by its effects on properties, but by other considerations such as ease of dissolution, cost, resistance to fading, or resistance to gassing.

Property Variations Within a Casting

The tensile properties obtained on coupons cut from castings are often inferior to those obtained from separately cast test bars. The latter are generally the source of data found in handbooks, but we frequently observe that the mechanical properties of actual castings do not achieve what might be regarded as typical values for the alloy used. The reason for this discrepancy, of course, is that castings are often less sound than cast test bars used to determine alloy properties, and even within a casting the more highly chilled regions will possess better mechanical properties than areas that cooled more slowly.

Modification can be used to overcome, at least partially, the negative effect of porosity on properties in slowly solidified regions of a casting. An interesting series of tensile property values obtained from a test casting modified with sodium and containing regions having very different solidification rates is given in Table 5.5[6]. Test bars were cut from four zones in the casting which cooled at the following rates: zone 1, 230°C/min; zone 2, 110°C/min; zone 3, 22°C/min; zone 4, 17°C/min. These zones were radiographed and classed according to the ASTM reference album which describes 8 classes of radiographs. An increasing class number corresponds to an increasing severity of microporosity.

At all freezing rates, except perhaps the fastest, sodium modification causes increased porosity for all of the reasons discussed in Chapter 4. Furthermore, the severity of microporosity increases when the freezing rate decreases, as is evidenced by the higher ASTM radiographic classes found with the low freezing rate zones. Associated with this is a decrease in UTS and an increase in ductility which combine to increase the Quality Index of the casting. This is true in all but zone 4, and even here the Quality Index is relatively unchanged from the non-modified case, even though a substantial deterioration in casting soundness is evident.

The reasons for this seemingly contradictory behavior are that the beneficial effects of modification on tensile properties can outweigh the negative effects caused by decreased casting soundness. Hence, in a casting, a slight increase in porosity associated with modification does not necessarily mean that the tensile properties of that casting will be lower than if it were unmodified. Indeed, in many cases real benefits of modification can be found even in the presence of increased porosity. It is quite likely that modification

Table 5.5. Tensile Properties and Radiographic Ratings for A356 Alloy Modified with Sodium[6]

Na conc. (ppm)	Zone 1				Zone 2				Zone 3				Zone 4			
	UTS (MPa)	E (%)	Q (MPa)	ASTM Class	UTS (MPa)	E (%)	Q (MPa)	ASTM Class	UTS (MPa)	E (%)	Q (MPa)	ASTM Class	UTS (MPa)	E (%)	Q (MPa)	ASTM Class
0	292	5.7	405	0-1	263	2.5	322	1	247	1.6	277	1	242	1.4	263	1
24	280	10.3	431	1	263	5.1	369	2	228	1.7	262	2	213	1.5	239	4-5
40	281	10.1	431	1	260	5.5	371	3	240	3.1	313	3	216	2.0	261	5-6
120	271	11.1	427	1	251	5.7	364	3	235	3.7	320	3	213	3.4	292	6

could be used to advantage in castings where porosity is a problem. The use of modification might result in significant improvements in tensile properties, and in only slightly more unsoundness than is normally experienced.

Typical Tensile Properties of Some Modified Alloys

In Table 5.6 we have summarized the typical tensile properties of several modified alloys. These properties were determined on separately cast test bars, in either a sand or a permanent mold, by workers at Pechiney. The values reported were thus all determined in the same manner and are comparable from alloy to alloy. Where applicable, the AA alloy designation has been used. In some cases the specific alloy composition is not frequently used in North America and the ISO designation is given along with an indication of the major alloying elements present. The reader should keep in mind that these are typical values only, and that they may well be less if measured on samples taken from actual castings. The modifying agent has been given in each case, but the same values should be applicable to other modifiers since it is the structure and not the specific modifying element which determines the properties.

5.2   Impact Properties and Fracture Toughness

Modification exerts a significant improvement on the impact strength of both as-cast and heat treated alloys. Little specific data is available for a wide range of metallurgical conditions, but some idea of the beneficial effects can be gained from Figure 5.5. Here, the variation in unnotched impact strength for both as-cast and heat treated strontium modified A356 and 413.0 alloy is given. Impact values depend strongly on the testing technique used, but it is obvious that modification, particularly if combined with heat treatment, can lead to increases of several hundred percent.

Impact strength is given to the material by the ductile aluminum matrix which separates the brittle silicon phase. Any process which reduces the size of the brittle phase particles or increases their separation will improve impact properties. Modification does the former, and the coarsening process, which takes place during the solution treatment, accomplishes the latter. The result is a much larger effect than might be supposed by studying improvements in tensile ductility alone.

All indications so far point to impact properties being much more sensitive to modification than simple tensile properties. For example, several mechanical properties of strontium treated A356 alloy are shown in Figure 5.6. Impact strength increases by about 150% when the alloy is modified,

Table 5.6. Representative Tensile Properties and Hardness of Modified Alloys [7]

Alloy	Type of Casting	Heat Treatment	Precipitation Treatment Temp. (°C)	UTS (MPa)	0.2% Yield Strength (MPa)	E (%)	Brinell Hardness
356 Sr modified	Permanent Mold	As-cast	-	200	90	14	55
		T6	160	290	200	16	90
		T6	170	310	250	12	100
	Sand	T6	160	260	200	3	90
		T6	170	275	250	2	100
356 Sb refined	Permanent Mold	As-cast	-	200	90	14	55
		T6	150	260	155	22	75
		T6	160	280	190	18	85
		T6	170	300	230	15	95
	Sand	As-cast	-	170	90	4.5	55
		T6	150	250	155	8	75
		T6	160	275	190	5	85
		T6	170	290	230	4	95
357 Na modified	Permanent Mold	As-cast	-	200	100	11	60
		T6	150	300	200	18	90
		T6	160	325	250	14	100
		T6	170	340	290	11	110
	Sand	As-cast	-	160	100	3	60
		T6	150	260	200	4.5	90
		T6	160	285	250	2	100
		T6	170	310	290	1.5	110

Table 5.6 (Con't)

Alloy	Type of Casting	Heat Treatment	Precipitation Treatment Temp. (°C)	UTS (MPa)	0.2% Yield Strength (MPa)	E (%)	Brinell Hardness
357 Sb refined	Permanent Mold	As-cast	-	200	100	10	60
		T6	150	300	200	18	90
		T6	160	325	250	14	100
		T6	170	340	290	10	110
	Sand	As-cast	-	160	90	4.5	55
		T6	150	260	200	4.5	90
		T6	160	285	250	2	100
		T6	170	310	290	1	110
Al-Si$_{11}$Mg (Sr modified) Fe-0.14	Permanent Mold	As-Cast	-	180	80	14	55
Cu-0.02 Mg-0.20		T6	160	260	175	14	80
	Sand	As-cast	-	150	80	4	55
Al-Si$_{12}$CuNiMg (Na modified) Cu-1.25	Permanent Mold	As-cast	-	240	200	1.5	85
Mg-1.35 Ni-1.00	Permanent Mold	T6	210	300	260	1.0	110
Al-Si$_5$Cu$_3$Mg (Sb refined)	Permanent Mold	As-cast	-	240	170	2	70
Mg-0.35	Permanent Mold	T6	170	415	345	4	125
	Permanent Mold	T4	5 days @ 20C	360	220	9	100

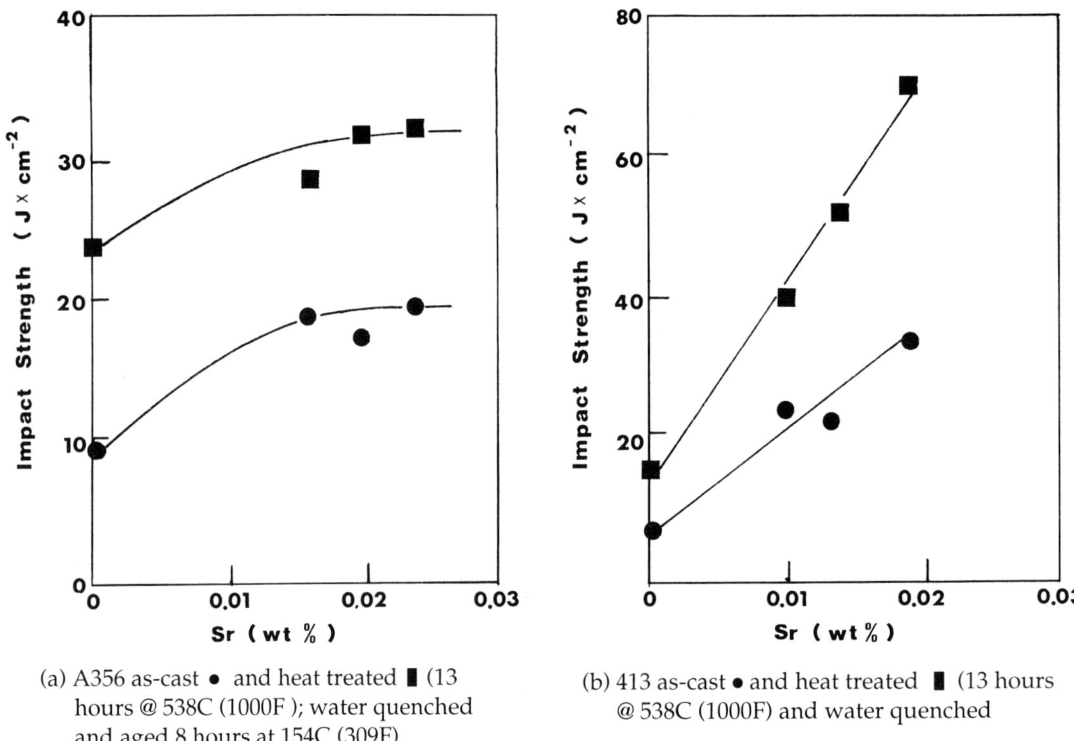

(a) A356 as-cast ● and heat treated ■ (13 hours @ 538C (1000F); water quenched and aged 8 hours at 154C (309F)

(b) 413 as-cast ● and heat treated ■ (13 hours @ 538C (1000F) and water quenched

Figure 5.5. Impact strengths of strontium treated alloys[8].

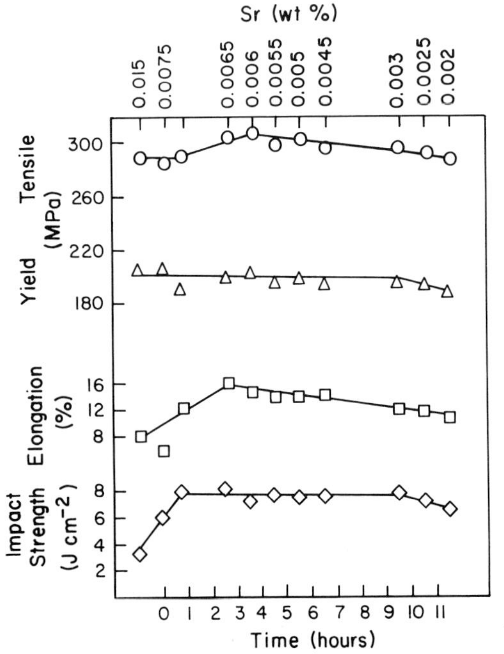

Figure 5.6. Mechanical properties of A356.0 alloy modified with strontium and held for times of up to 11 hours before casting.

while elongation improves by about 100%. Both the yield and tensile strength are basically unaffected. It is also worth noting that property values obtained from these permanent mold samples do not vary much with retained strontium level. Once the initial modification is made there is good retention of mechanical properties.

5.3   Fatigue Properties

As with fracture toughness, specific information on the effect of modification on fatigue strength is scanty. Some of the available data is summarized in Table 5.7 where fatigue strength was determined under conditions of rotating bending using a permanent mold cast Alkan specimen. The fatigue strength values are the mean value of applied stress which gives a 50% probability of failure at 10^7 cycles.

Table 5.7. Fatigue Strength of Some Modified and Unmodified Alloys[7]

Alloy	Melt Treatment	Heat Treatment	Fatigue Strength (MPa)
A356	non-modified	T6-6hrs. @ 160C	110
A356	Sb refined	T6-6hrs. @ 160C	105
A356	Sb refined	as-cast	90
A357	non-modified	T6-6hrs. @ 160C	105
A357	Sb refined	T6-6hrs. @ 160C	110
$Al\text{-}S_{11}$	Sb refined	as-cast	60
$Al\text{-}Si_{11}Mg_{0.2}$	Sr modified	as-cast	70

The dramatic changes in properties on modification which are seen in the tensile and impact properties are not evident in fatigue strength. Fatigue strength drops with increased amounts of silicon, but seems little affected by antimony treatment or even by strontium modification. This is perhaps not so surprising. Fatigue cracks propagate along the silicon-aluminum interfaces, and the stress required to cause decohesion at these interfaces should be little influenced by a modification treatment.

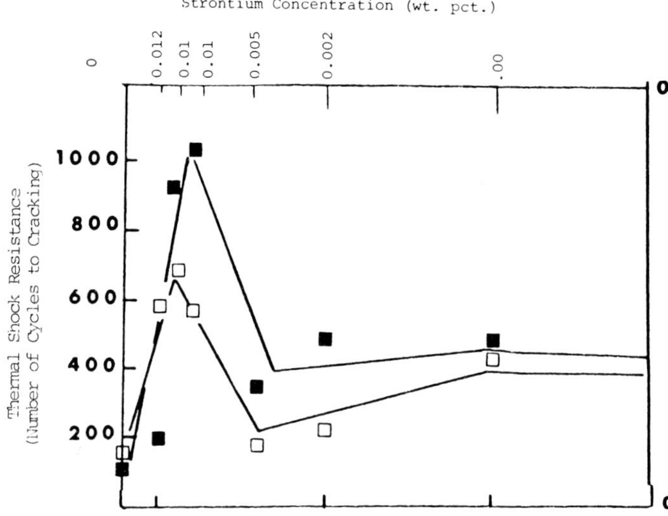

Figure 5.7. Thermal shock resistance of strontium treated 319.0 alloy. Samples were permanent mold cast into molds at two temperatures:
■ *mold at 250C*
□ *mold at 400C.*

5.4   Thermal Shock Properties

Modification exerts a dramatic effect on the resistance to thermal shock. Data is available for strontium treated 319.0 (Fig. 5.7) which indicates a five-fold increase in the number of thermal cycles required to cause cracking of a fully modified alloy. A clear peak is evident in the data and the thermal shock resistance appears very microstructurally sensitive, dropping off significantly as the degree of modification decreases.

5.5   Machinability

Intuitively, one would expect that the structural refinement and better distribution of silicon that is associated with modification would lead to an improved machinability. Machining expenses can account for a sizeable fraction of the cost of producing a finished casting, particularly for inexpensive, large volume parts produced by die casting. Despite this, there have been few attempts to quantify the effects of modifi-

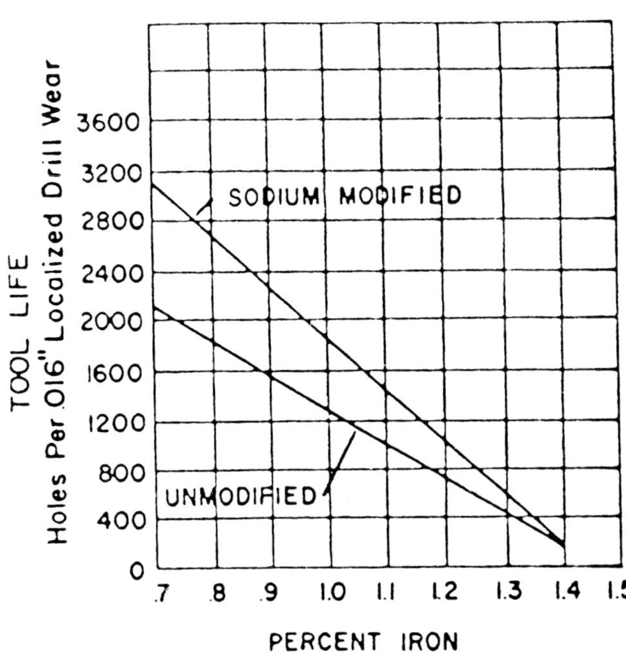

Figure 5.8. Tool life variation with iron content and modification in a 380 alloy[9].

cation on machinability. This is probably due to the difficulty and expense involved in doing realistic machining tests.

The indications are that substantial improvements in machinability are possible if alloys are modified. For example, in Figure 5.8 we show measured tool life in machining a 380 alloy, both unmodified and sodium modified. Tool life is a strong function of the iron content, but the benefits of modification are clear—a 50% increase in tool life at 0.7% iron.

Although rare earths are not used to any extent as commercial modifiers they are capable of bringing about at least partial modification. Both cerium and mischmetal additions have been shown to decrease the horsepower required to machine an Al-12%Si alloy, and to increase both the chip thickness ratio and the shear plane angle[10]. No studies of strontium or antimony treatment on machining properties appear to have been done, but it is expected that both of these elements would exert positive effects on machining properties.

5.6   Foundry Properties

The foundry properties which influence the ease of producing a casting and the quality of that casting are threefold: shrinkage properties, hot tearing tendency and mold filling ability or fluidity. In Chapter 4, we deal at length with the interrelationship between modification and shrinkage. Here, we will examine the effects of modification treatment on the remaining two foundry properties.

Hot Tearing Tendency

The slumping tendencies as assessed by the Tatur Test of strontium and sodium modified A356 alloys were discussed in Chapter 4. Strontium modified alloys show a greater tendency to slump compared to unmodified alloys. This implies that the solid shell which forms at the mold wall during solidification of the casting is less strong and is able to deform in order to relieve tensile stresses set up during freezing.

Although Al-Si foundry alloys are known for their intrinsic hot tearing resistance they can, in some circumstances, hot tear. The "H" shaped casting in Figure 5.9 is an example. Here, the very considerable restraint caused by the casting geometry has led to tears at the junction of the horizontal and vertical portions of the casting. When 0.02%Sr is added to modify the structure, the hot tears are eliminated completely due to the less rigid solid shell of the modified alloy. Modification, therefore, improves the already excellent hot tearing resistance of Al-Si alloys.

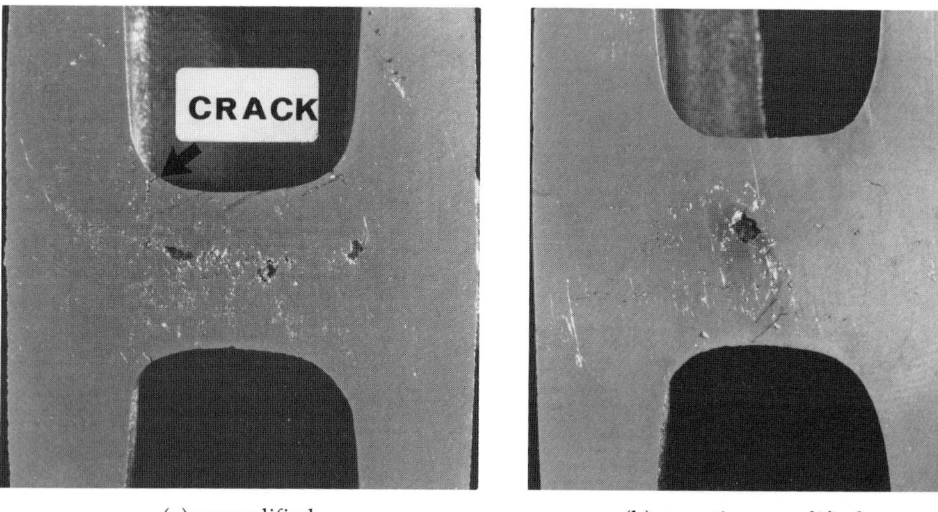

(a) unmodified   (b) strontium modified

Figure 5.9. The elimination of hot tearing through the use of modification[8].

Fluidity

The fluidity of an alloy is used as a measure of its mold-filling ability. It is determined by measuring how far the liquid alloy will flow until it stops. Units of fluidity are, therefore, units of length, inches or centimeters. Fluidity is very often confused with viscosity, although the viscosity of the liquid is only a small factor in determining the fluidity. A metal stream stops flowing because it freezes, or more specifically because the solidifying dendrite mesh becomes so thick as to block liquid flow through it. The fluidity is therefore determined primarily by the solidification characteristics of the alloy. Alloys with long freezing ranges exhibit poor fluidity, while short freezing range alloys, such as eutectics, have excellent fluidity.

Fluidity is usually measured by pouring a spiral, either in sand or a permanent mold, and measuring the length of the solid spiral. Vacuum suction techniques are also used, and in die casting a mold consisting of channels of different diameters is employed. Values obtained from one technique cannot be compared directly to those determined by another method.

The effect of modification on fluidity is far from clear. It is generally accepted that sodium treatment reduces fluidity somewhat and that antimony has no effect. Data on strontium is not plentiful. The reductions reported with sodium are in the range 10%-20%, but at the same time it is possible to find reports which indicate an increase of about the same magnitude. The experimental determination of fluidity is extremely sensitive to casting variables, so that a precision of ±10% on all reported measurements is probably reasonable. As this is the range of changes reported due to modification, it is most probable that any effect of modification is small, and is not more than 10%.

The sand cast fluidity of A356, A319 and A413 alloys is shown in Figure 5.10 over the temperature range 700C to 750C (1290F to 1380F). No significant differences are found between the unmodified alloy or the sodium or strontium modified alloys, the maximum differences being of the order of ±10%. It is quite possible that modification might influence the fluidity of alloys cast into other types of molds, e.g. permanent or investment molds. Here the surface roughness is very different from sand, and the conclusion cited above may not be valid.

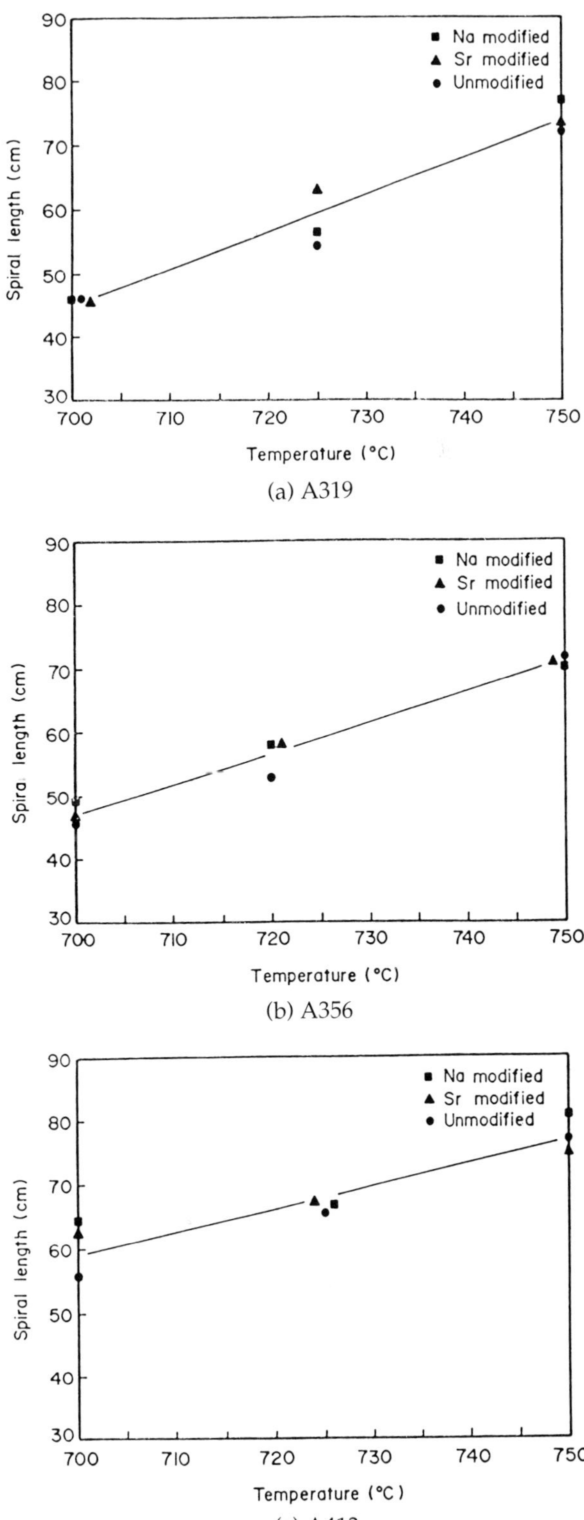

Figure 5.10. The temperature dependence of sand cast fluidity for three aluminum alloys, unmodified, sodium or strontium modified **[11]**.

References

1. Drouzy, M., S. Jacob, M. Richard. "Le Diagramme Charge de Rupture des Alliages d'Aluminium," *Fonderie*, No. 355 (1976) pp. 139-47.
2. Closset, B. and J.E. Gruzleski. "Mechanical Properties of A356.0 Alloys Modified with Pure Strontium," *AFS Transactions*, 90 (1982) pp. 453-64.
3. Vaccari, J.A. "Optimizing Al-Si-Mg Castings," *American Machinist* (August 1981) pp. 122-26.
4. Poniewierski, Z. "Effect du Type de Modification sur la Microstructure de l'Eutectique Aluminium-Silicium," *Rev. int. Htes. Temp. et Refract.*, 14 (1977) pp. 253-60.
5. Bercovici, S. "Controle des Structures de Solidification et des Proprietes des Alliages Al-Si," *Revue de l'Aluminium* (February 1979) pp. 85-99.
6. Jacob, S. and J. Boutault. "Aspect Radiographique et Caracteristiques Mecaniques des Pieces Moulees en AS7G03," *Fonderies-Fondeur d'Aujourd'hui*, 44 (1985) pp. 9-17.
7. "Aluminium, Alliages de Fonderie," Societe de Vente de l'Aluminium Pechiney. (Various technical brochures published in the period 1979-81.)
8. Closset, B. "Modification and Quality of Low Pressure Aluminum Castings," *AFS Transactions*, 96 (1988) pp. 249-60.
9. Queener, C.A. and W.L. Mitchell. "Effect of Iron Content and Sodium Modification on the Machinability of Aluminum Alloy Die Castings," *AFS Transactions* , 73 (1965) pp. 14-19.
10. Sharan, R. and N.P. Saksena. "Rare Earth Additions as Modifiers of Aluminum-Silicon Alloys", *AFS International Cast Metals Journal*, 3 (1987) pp. 29-33.
11. Argo, D. and J.E. Gruzlesk., "The Fluidity of Sodium and Strontium Modified Sand Cast Aluminum-Silicon Foundry Alloys," *Cast Metals* , Vol 2, (1989), pp. 109-12.

Chapter 6

Interactions of Modifiers

6.0   Introduction

Up to the early 1970s sodium was the only modifying element used. Since that time, strontium has become widely accepted, particularly in North America, while antimony finds extensive use in Europe and Japan. In recent years, one country or group of countries have tended to use more than one modifying agent. For example, strontium is being used to a greater extent in Japan alongside antimony. All three common modifiers are used in Europe, and there has been some indication that antimony may find its way into the United States foundry industry.

The possibility of mixing modifiers has, therefore, become very real. This could arise in several ways. Should a given foundry use more than one modifying element in its operation, it is quite possible that modifiers will become mixed when in-house scrap is remelted. Sodium contamination of antimony-refined melts can occur through the use of sodium containing fluxes.

A more general problem is a mixing of antimony, strontium and sodium in the scrap cycle of a nation or trading bloc. This would pose no particular difficulty if all three modifiers were compatible with each other or acted in a synergistic manner; indeed, it could even be advantageous. Unfortunately, this is not the case as it has been shown that antimony reacts in a negative way with both sodium and strontium. Furthermore, antimony is the only modifier which must be used in relatively large amounts (0.1% to 0.4%) in order to produce the desired microstructural effects. It is, therefore, present in larger quantities in scrap. It is also the most stable of the modifiers when mixed with liquid aluminum—it has a low vapor pressure and little tendency to oxidize from the melt. In addition, it poses health hazards. Thus, when antimony is mixed into the scrap cycle, it becomes very difficult to remove, and presents definite metallurgical and environmental problems for any foundry attempting to modify with either sodium or strontium.

In this chapter, we examine what is known of the nature of interactions between antimony and sodium or strontium and what can be done to overcome them. We will also examine the feasibility of using sodium and strontium in concert to achieve a better and more prolonged modification than is possible with either of them alone.

6.1 Strontium-Antimony Interactions

When antimony was first used to treat Al-Si casting alloys, it soon became evident that it was incompatible with strontium. In the presence of antimony, much larger quantities of strontium than usual are required to cause modification, and in the case of slow cooling rates modification becomes extremely difficult, and even impossible to achieve. The exact behavior will depend on the alloy used, the phosphorus content, and of course the antimony concentration.

Little work has been done to determine the structure-chemistry relationship, but two alloys have been at least partially investigated. The structure of an Al-13%Si (413 type) and A356 alloy, as a function of strontium and antimony concentrations, is given in Figure 6.1. When both strontium and antimony are present in the 13%Si alloy, it does not seem possible to obtain

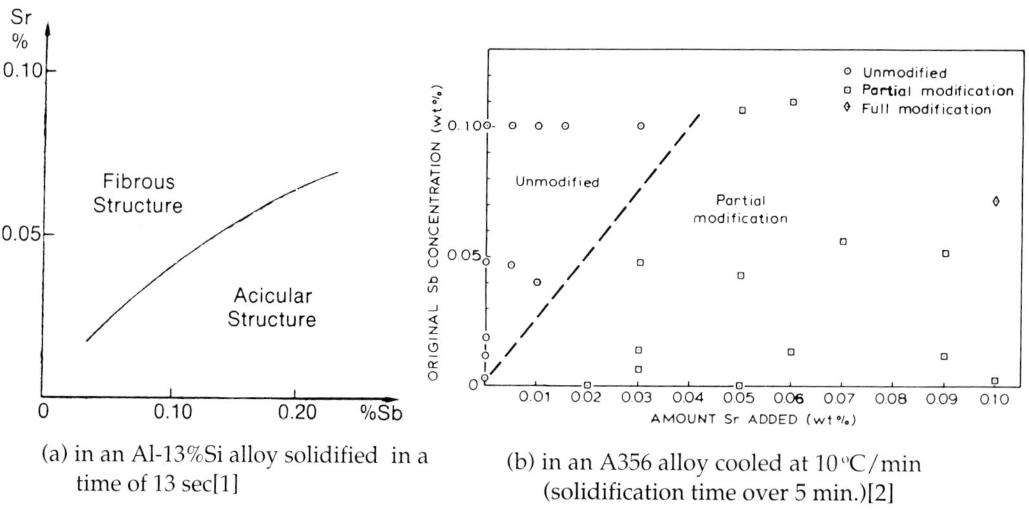

(a) in an Al-13%Si alloy solidified in a time of 13 sec[1]

(b) in an A356 alloy cooled at 10 °C/min (solidification time over 5 min.)[2]

Figure 6.1. Eutectic structures observed in the presence of both strontium and antimony.

a lamellar microstructure—the alloy is either modified or acicular.

In the A356 (Fig. 6.1b), a partial modification consisting of a mixture of lamellae and fibers is possible, but full modification is very difficult to achieve. The reader should bear in mind that these relationships are valid only for the specific solidification rates used. For example, the A356 alloy froze very slowly—a worse case situation. At a faster freezing rate, reasonably well modified structures can be obtained at certain strontium-antimony concentrations. Figure 6.2 presents the modification rating, as a function of strontium level, for 356 alloys which are either free of antimony or contain

0.01% or 0.04%Sb. This alloy cooled three times faster than that in Figure 6.1(b), and good modification is obtained if the strontium content is about 0.08% at 0.04%Sb, or 0.04% at 0.01%Sb. The usual strontium concentration required in a 356 alloy is 0.01%-0.02%. Very poor modification is achieved at these levels if even 0.01%Sb is present.

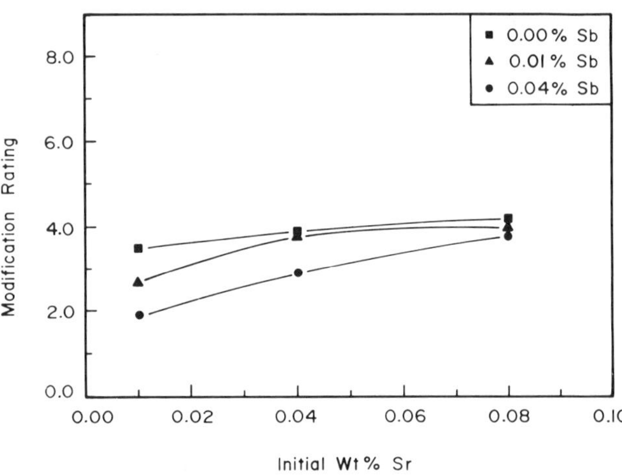

Figure 6.2. Modification rating of strontium modified A356 containing antimony. These samples were cast one hour after the strontium treatment and cooled at 0.5°C sec^{-1} through the solidification range[3].

The mechanism of this negative interaction involves the chemical reaction of both strontium and antimony in the bath to form solid intermetallic compounds. There are likely several types which can form depending on the conditions, and so far at least one has been identified. Mg_2Sb_2Sr (Fig. 6.3) has been found in A356 alloy containing both antimony and strontium. Formation of such a compound chemically binds strontium atoms, thus preventing them from acting as modifiers according to the mechanism outlined in section 3.1. Of course, if sufficient strontium is added to react with all of the antimony present, then any excess will be free to modify. Thus, much higher amounts of strontium are required when antimony is present than when it is absent.

The compound Mg_2Sb_2Sr has a density of 4.2 gm cm,$^{-3}$ and hence is approximately twice as dense as liquid Al-Si alloys. Upon formation, it rapidly sinks in the bath and concentrates at the bottom of the furnace or crucible. It acts in a similar manner to sludge in die casting alloys, and will form a thick mushy region at the bottom of the melt. Under normal conditions, strontium recoveries are in the 90% to 95% range. However, if antimony is present they can appear to drop dramatically due to the formation of Mg_2Sb_2Sr. The strontium is not lost, but becomes concentrated in the dense compounds at the bottom of the melt. Since most sampling techniques involve taking material from the upper half of a bath, lower than expected

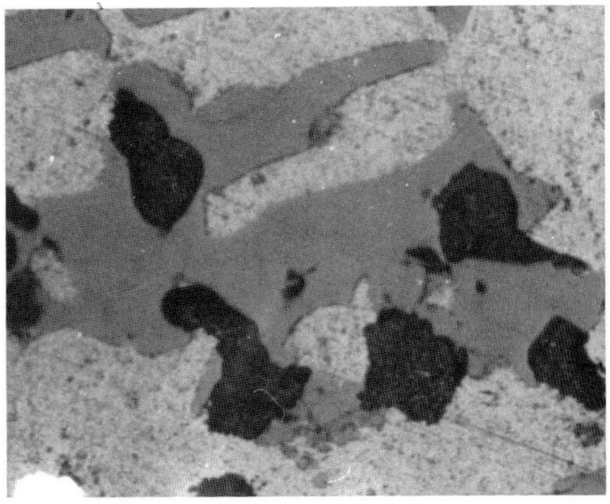

Figure 6.3. Mg_2Sb_2Sr compound (the black phase) formed during the strontium modification of A356 alloy containing antimony[2] (x1000).

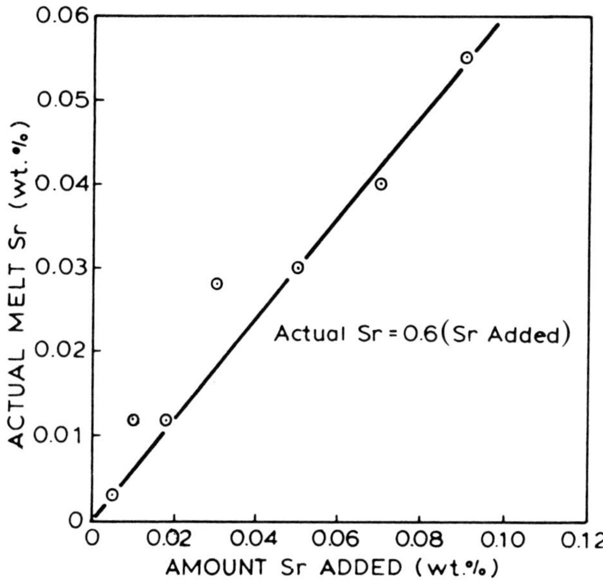

Figure 6.4. Plot of the amount of strontium added to an A356 melt containing 0.05% Sb versus the analyzed strontium in the upper half of the melt [2].

strontium analyses will result. This behavior is shown graphically in Figure 6.4 where a recovery of only 60% is recorded near the top of the melt.

Of course, the reaction to form compounds also removes antimony from the upper part of the melt, and dramatic drops in analyzed antimony levels are seen when strontium is added (Fig. 6.5). The most pronounced drop occurs in the first 20-30 minutes, which is the period required for strontium dissolution. The reaction to form Mg_2Sb_2Sr apparently occurs readily during this interval, and the antimony concentration then levels out at a new and lower level. Of course, if no strontium is added, the antimony concentration is completely stable with time, reflecting the lack of fading of this element.

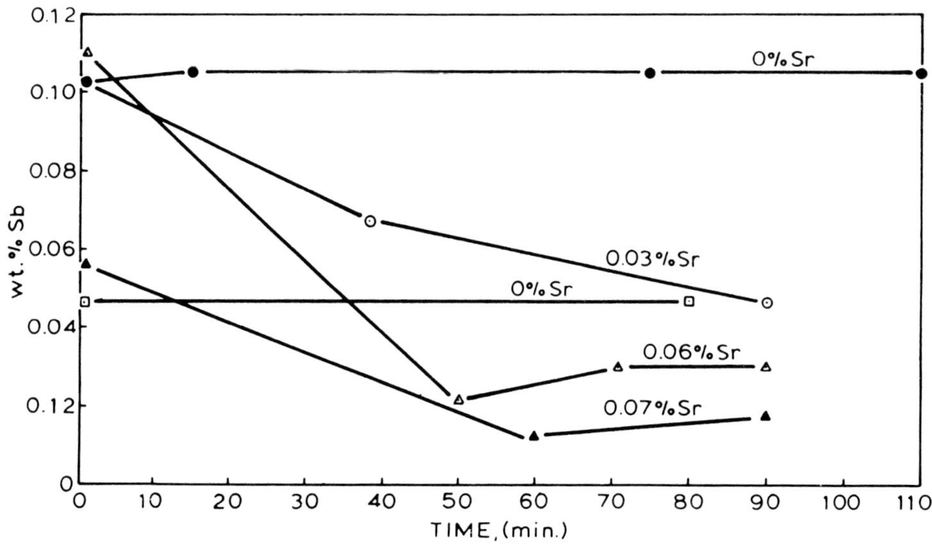

Figure 6.5. Antimony concentration near the top of melts to which various amounts of strontium are added [2].

6.2 Sodium-Antimony Interactions

Sodium also interacts with antimony in a negative manner requiring higher than normal levels for good modification if antimony is present. Microstructure-chemistry relationships for a 413 type alloy are presented in Figure 6.6. Here the effect of freezing rate on the microstructure is evident. At the slow cooling rates characteristic of sand castings, there is only a narrow compositional "window" through which well modified structures can be produced. High freezing rates facilitate modification even in the presence of antimony. In a permanent mold (Fig. 6.6), the allowable composition range over which well modified structures are possible is much larger. Nevertheless, significantly higher sodium levels are necessary to produce a modified structure.

The sodium-antimony interaction mechanism is similar to the strontium-antimony mechanism. Sodium reacts with antimony and other elements in the melt to form compounds. The sodium is thus combined chemically and unable to modify. Probably several types of compounds form. They are difficult to find and to analyze, but several with a stoichiometry of the type $NaMgSb_{2-5}$ have been identified (Fig. 6.7). The formation of such compounds creates compositional changes in the upper part of the bath.

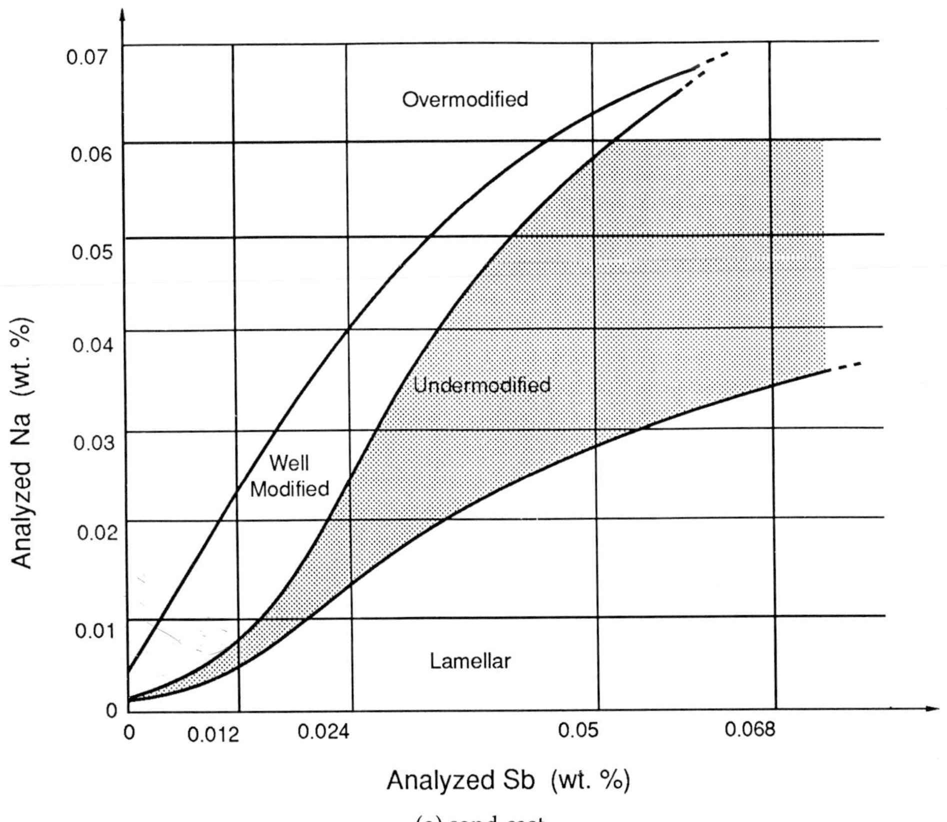

Figure 6.6. Structures obtained in an Al-13% Si alloy (413 type) sodium treated and containing various amounts of antimony [4].

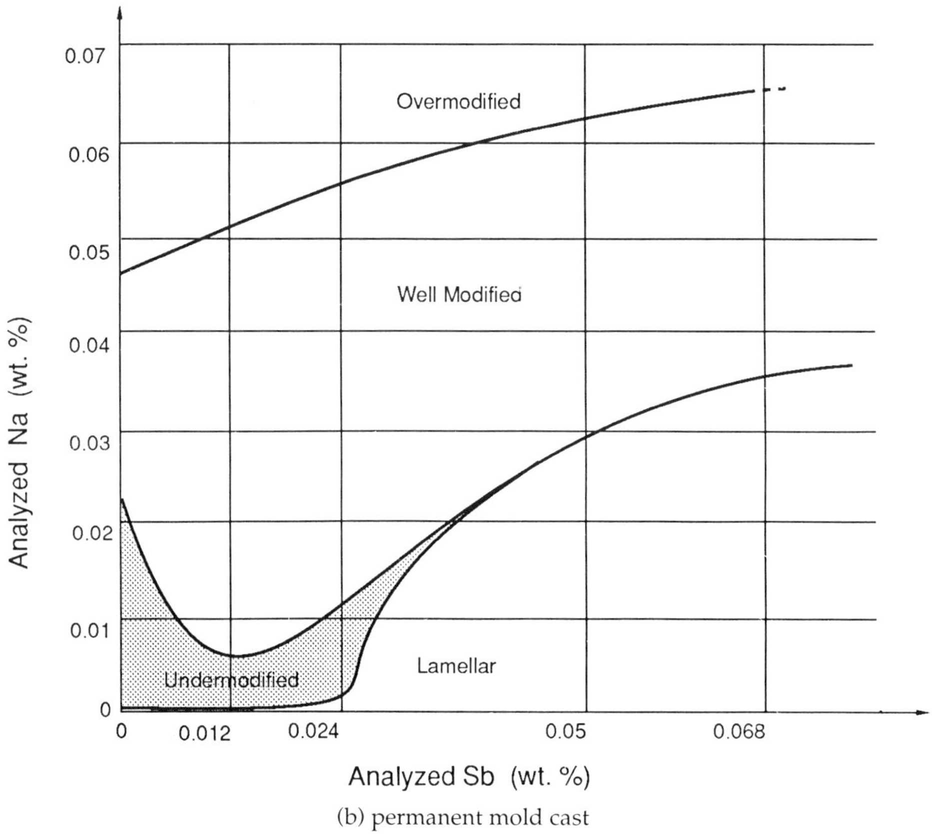

(b) permanent mold cast

Figure 6.6, cont'd.

As is the case with strontium, a decrease in antimony level is measured with sodium treatment, the change being greater at higher starting concentrations of antimony and with larger sodium additions. Figure 6.8 shows this change ($\Delta\%Sb$) in an A356 alloy sampled 15 minutes after sodium treatment.

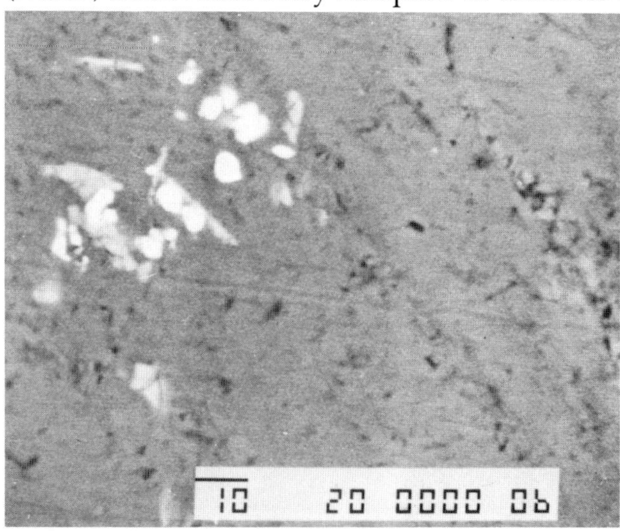

Figure 6.7. NaMgSb$_{2-5}$ compound (the white phase) found in sodium treated A356 alloy containing antimony (x700) [3].

If sufficient sodium is added to react with all of the antimony, any excess will be an effective modifier. In A356, as in 413 (see Figure 6.6(b), this amount is about 0.04%, as demonstrated by the data in Figure 6.9. Modification, which is almost as good as that obtained in the absence of antimony, can be achieved if the initial sodium level is 0.04% or higher. Again, the reader is reminded that these values pertain only to the freezing rate for which they were measured. Higher rates will probably require somewhat less sodium; slower rates, a bit more.

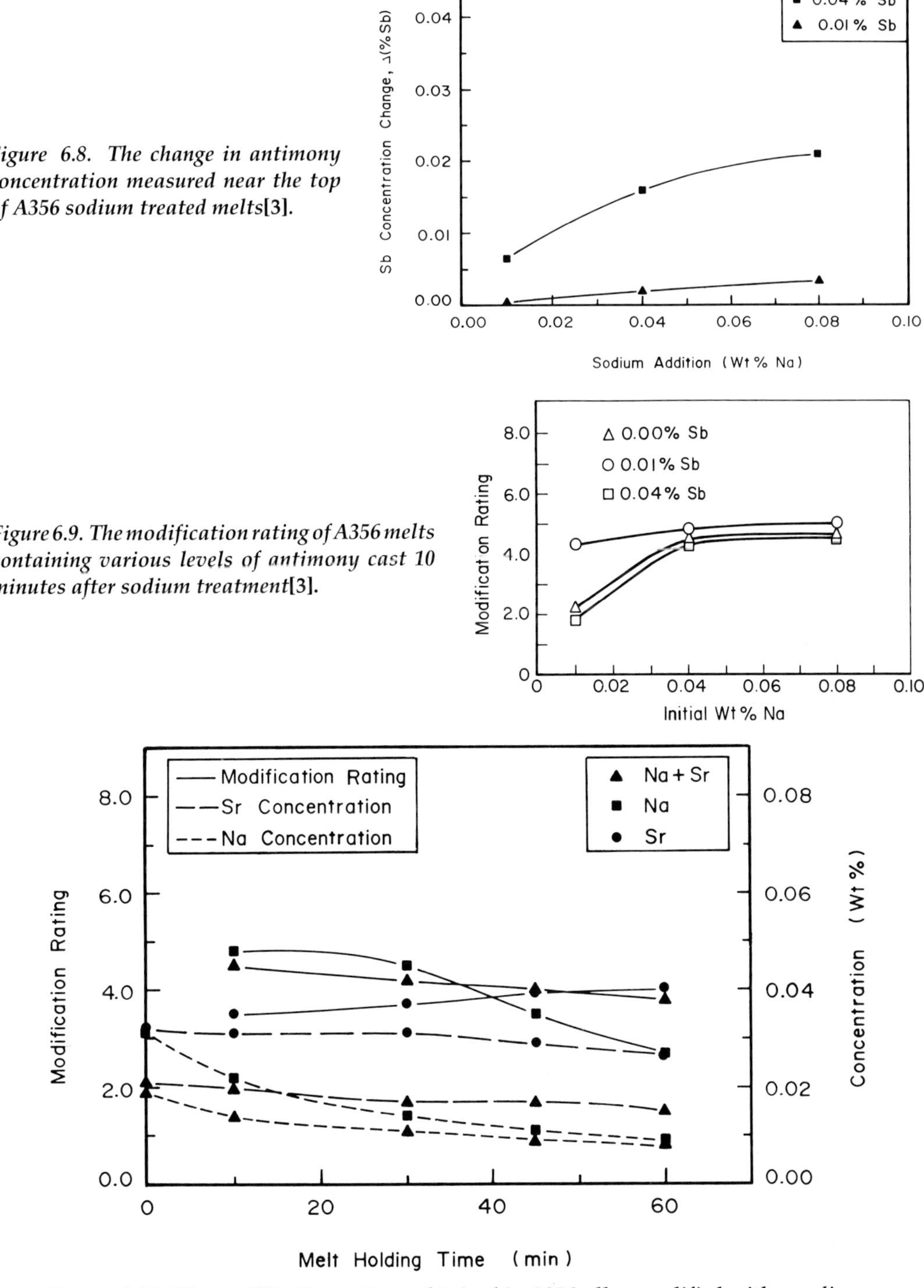

Figure 6.8. The change in antimony concentration measured near the top of A356 sodium treated melts[3].

Figure 6.9. The modification rating of A356 melts containing various levels of antimony cast 10 minutes after sodium treatment[3].

Figure 6.10. The modification ratings obtained in A356 alloy modified with a sodium-strontium combination compared to modification ratings produced by individual sodium or strontium treatment[3].

6.3   Sodium-Strontium Interactions

Some early reports in the literature indicated that sodium and strontium were not compatible. This has proven to be untrue, and, in fact, the two can work together to provide excellent long term modification.

Sodium is known to be highly effective at short holding times, and then to decrease in effectiveness due to fading. Strontium, because of its incubation period, improves the structure with time. There are no known Sr-Na compounds which form in aluminum alloys, and hence no interactions such as those experienced with antimony would be expected. Strontium-sodium mixtures should provide a stable, long-term modification, with sodium acting predominantly at short times, and strontium at longer times after treatment.

The effects of such a combined treatment can be seen in Figure 6.10. Modification ratings lying consistently in the 4-5 range are obtained with little change over a one-hour period. Under identical casting conditions, sodium alone yields similar structures up to 40 minutes, but these then degenerate due to fading. Strontium modification does not give such good microstructures initially, but these do improve with time. The mixed modifier, however, yields acceptable and constant structures over the entire time span.

No commercial modifiers containing both sodium and strontium are currently available. The few casting producers who use this technique make separate additions of each modifier.

6.4   Overcoming Negative Interactions

There are three approaches which can be taken to combat the negative effects of antimony contamination. These are to remove the antimony from the melt before subsequent modification, to dilute the antimony concentration, or to add an excess of modifier in order to precipitate the antimony as dense compounds. To our knowledge, none of these methods is used in common commercial practice. No satisfactory techniques have been developed to allow easy antimony removal from the melt in a foundry environment. This may not prove to be too difficult, but is does not seem to have been tried.

An example of how antimony dilution can be used to allow strontium modification is given in Table 6.1. During this in-plant test, an A356.0 alloy containing approximately 0.08% Sb was treated with 0.01% Sr, but this did not alter the original lamellar structure which is characteristic of antimony treated melts. Increasing the strontium level to 0.017% also did not allow modification. However, a fully modified structure was readily produced after the addition of sufficient virgin ingot to dilute the antimony level to approximately 0.03% and the subsequent addition of 0.02% Sr. This is shown by samples 8 and 9 in Table 6.1.

Table 6.1. Strontium Modification of An Antimony-Containing A356.0 Melt

Sample No.	Sb (%)	Sr (%)	Microstructure	Comments
1	0.075	0	lamellar	as received
2	0.08	0.008	lamellar	0.01% Sr added
3	0.065	0.009	lamellar	after degassing
4	0.069	0.017	lamellar	additional 0.01% Sr added
5	0.054	0.013	lamellar	after degassing
6	0.026	0.003	acicular	dilution with virgin ingot
7	0.033	0.004	acicular	after degassing
8	0.043	0.017	modified	0.02% Sr added
9	0.032	0.013	modified	after degassing

The other approach, that of adding an excess of modifier, is successful. While an excess of either sodium or strontium alone can be employed, there are definite advantages to using a combination of the two. The amounts required will depend on the antimony concentration. In the case of recycled scrap, this is likely to be less than 0.01%, and for this level, a modification treatment with 0.03%Sr and 0.01%Na yields excellent short- and long-term modification. Higher antimony concentrations require larger additions. For example, if 0.04%Sb is present, a treatment with 0.06%Sr and 0.02%Na is suggested.

Some experimental results on an A356 alloy under these conditions are given in Figures 6.11 and 6.12. Since the success of this technique depends on the reaction of antimony to form dense solid compounds, melts so treated should not be stirred and should probably be filtered. It will be necessary from time to time to remove the sludge from crucibles and furnaces in order to prevent its incorporation into castings.

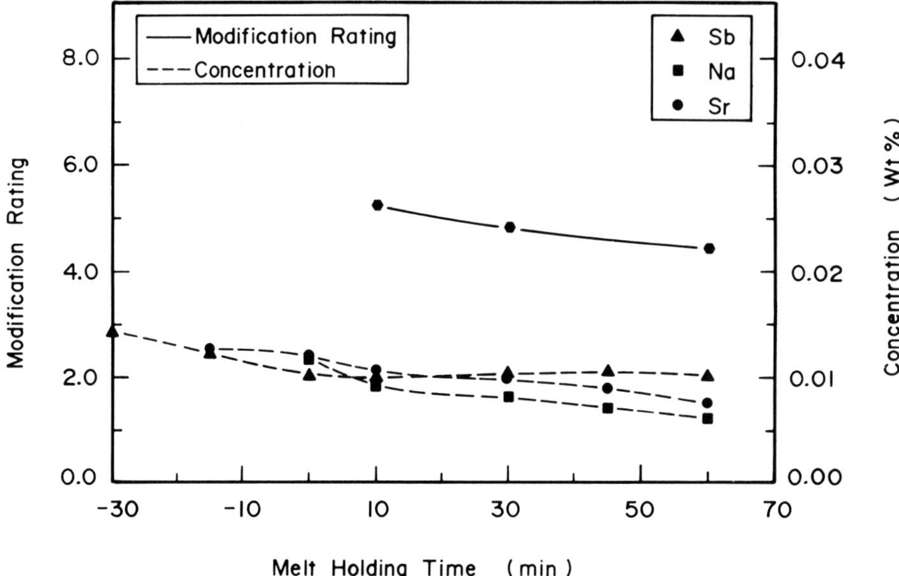

Figure 6.11. Modification ratings obtained in an A356 alloy containing 0.014% Sb by modifying with a combination of 0.03% Sr and 0.01% Na. Separate additions were made and the zero holding time represents the time of sodium treatment[3].

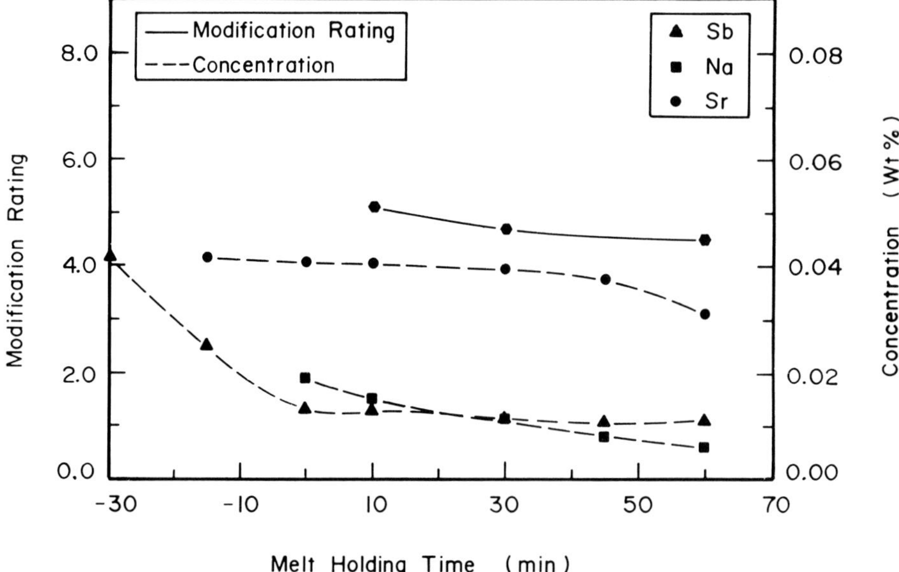

Figure 6.12. Modification ratings obtained in an A356 alloy containing 0.04% Sb by modifying with a combination of 0.06% Sr and 0.02% Na. Separate additions were made and the zero holding time represents the time of sodium treatment[3].

References

1. Nagel, G. and R. Portalier. "Structural Modification of Aluminum-Silicon Alloys by Antimony Treatment," *AFS International Cast Metals Journal*, 5 (1980) pp. 2-6.
2. Handiak, N., J.E. Gruzleski and D. Argo. "Sodium, Strontium and Antimony Interactions During Modification of AS7G03 (A356) Alloys," *AFS Transactions*, 95 (1987) pp. 31-38.
3. Wang, W and J.E. Gruzleski. "Sodium and Strontium Modification in the Presence of Antimony", *Materials Science and Technology*, 5 (1989) pp. 471-75.
4. "Incompatibilite du Sodium et de l'Antimoine dans le Traitement des Alliages Aluminium-Silicium," *Fondeur d'Aujourd'Hui*, 272 (1976) pp. 12-13.

Chapter 7

Refinement and Modification in Hypereutectic Alloys

7.1   Structure of Hypereutectic Alloys

The eutectic in the Al-Si alloy system occurs at 12%Si; therefore, one would expect that all alloys containing more than about 12%Si should exhibit a normal hypereutectic microstructure consisting of a primary silicon phase in a binary eutectic matrix. While this does occur, more complex microstructures are frequently observed, which are difficult to explain simply in terms of the equilibrium phase diagram.

Depending on the circumstance, three (seemingly odd) types of microstructures may occur in cast hypereutectic alloys:

- primary aluminum dendrites may occur in alloys which are just slightly hypereutectic (12% < Si < 14%);
- some hypereutectic alloys which contain up to 0.1% Sr exhibit a completely eutectic microstructure;
- hypereutectic alloys often contain primary aluminum dendrites in addition to primary silicon and eutectic.

The existence of these structures, which seem to defy the phase diagram, reflect the complexity of the solidification process of a casting. As they are often due to a melt treatment, or the casting solidification rate, it is appropriate to discuss the origin of each here.

Primary Aluminum Dendrites in Slightly Hypereutectic Alloys

These are found in alloys to which a modifier, such as strontium, has been added or in alloys which are rapidly solidified. Both eutectic modification and/or rapid freezing displace the eutectic concentration to higher silicon values, and depress the eutectic temperature. Typically, the eutectic composition may be raised from 12% to 14%Si and the eutectic temperature lowered by up to 10°C. Consequently, alloys in the 12% to 14%Si range can be hypereutectic under conditions of slow freezing and no modification treatment, but they will become hypoeutectic if treated with sodium or strontium, or if rapidly frozen. The microstructure of a eutectic alloy, and the same alloy to which 0.02% Sr was added is shown in Figure 7.1. The formation of primary aluminum dendrites in the strontium containing alloy is obvious. These dendrites form due to the displacement of the eutectic, illustrated on the phase diagram in Figure 7.1c.

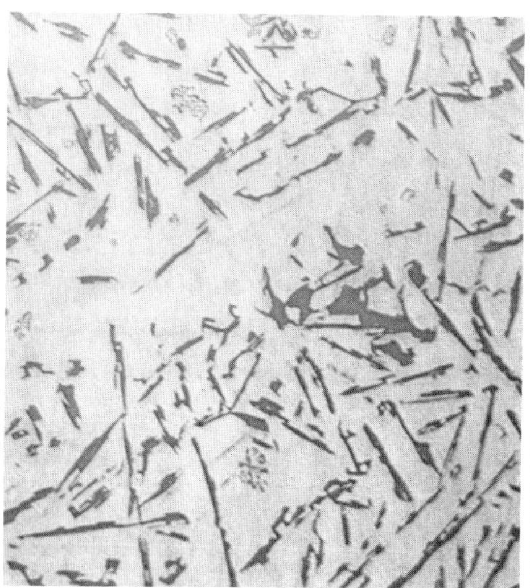

(a) eutectic alloy microstructure (x100)

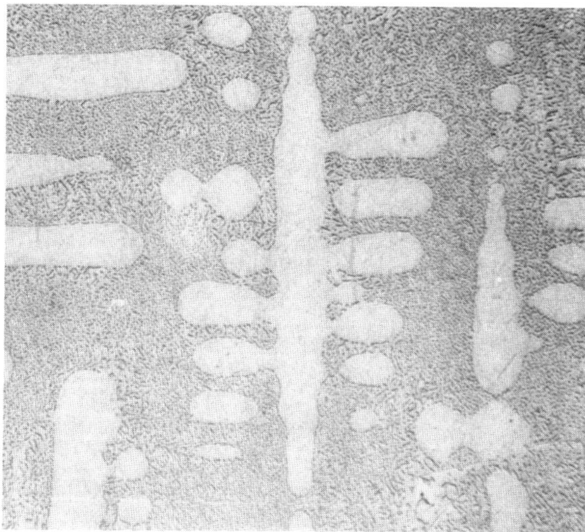

(b) eutectic alloy to which 0.02% Sr has been added (x100)

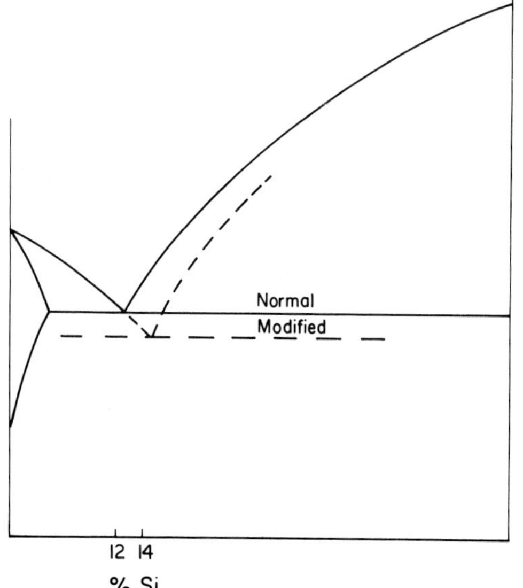

(c) schematic phase diagram illustrating the eutectic shift

Figure 7.1. Displacement of the eutectic point caused by modification or high freezing rates.

Hypereutectic Alloys with a Completely Eutectic Microstructure

The 3HA alloy which contains up to 15%Si, 0.1%Sr and substantial amounts of Cu, Ni, Mg and Fe solidifies with a completely eutectic microstructure, despite its relatively high silicon concentration. The origin of this structure is believed to be related to the eutectic shift associated with strontium, coupled with displacements due to the other alloying elements. Thus, the alloy composition corresponds to the eutectic in a complex multi-component alloy system.

Hypereutectic Alloys with Primary Aluminum, Primary Silicon and Eutectic

These structures are commonly found in alloys with more than about 15%Si, which freeze relatively quickly as in permanent mold or die casting (see Fig. 7.2). The microstructure arises from the fact that the freezing process is only quasi-equilibrium, and follows the phase diagram in only an approximate way.

Figure 7.2. Chill-cast A390 alloy with a microstructure consisting of primary silicon, aluminum dendrites and eutectic.

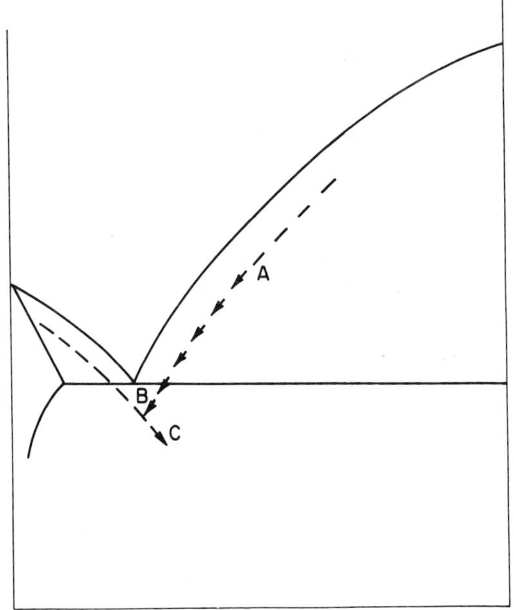

Figure 7.3. Solidification sequence of an alloy exhibiting the microstructure shown in Figure 7.2.

The solidification sequence is illustrated in Figure 7.3. Primary silicon crystals nucleate and begin to grow at a temperature below the equilibrium liquidus temperature (point A in Fig. 7.3). As the silicon grows, the liquid around it becomes highly enriched in aluminum, and follows the depressed liquidus line to point B where solid aluminum nucleates and freezes as a halo around the primary silicon. As this halo grows, it may become dendritic in form, or remain as a layer encapsulating the silicon. With growth of the aluminum phase, the neighboring liquid is enriched in silicon, and finally the eutectic is nucleated at B. When solidification of the eutectic is completed, the alloy is entirely solid and will contain two primary phases as well as the eutectic.

Primary Silicon Morphology

In hypereutectic alloys in which primary silicon does occur, it can appear in several different forms, and it is not uncommon for many of these to be found in the same casting. The silicon morphology is highly dependent on the solidification parameters such as freezing rate, temperature gradient in the liquid and local liquid composition. All of these change continuously during the freezing of a casting, and so it is not surprising that different silicon forms should occur almost side by side. Among the most common are:

- star-shaped primary silicon (Fig. 7.4a), which forms along five axes originating from a single nucleus (several variations of this shape can be found);
- polyhedral primary silicon (Fig. 7.4b), which appears hexagonal or octagonal on a polished section (sodium treatment results in a so-called spherical primary silicon, but this is simply a multi-faceted form of the basic polyhedral shape (see section 7.5);
- dendritic primary silicon (Fig. 7.4c) which resembles the dendritic form of simple metals, such as the primary aluminum dendrites in hypoeutectic alloys.

Eutectic Silicon Morphology

The eutectic silicon in hypereutectic alloys can occur in either an acicular (plate-like) or fibrous (modified) form exactly as in hypoeutectic alloys. In addition, local regions of a very geometric lamellar eutectic can sometimes be found (Fig. 7.5). This is known by a variety of names such as feathery, skeleton, web, etc., and is confined to hypereutectic alloys.

7.2   Phosphorous Refinement

The primary silicon phase of hypereutectic alloys is not readily nucleated by the usual impurities which are present in these alloys. As a result, primary silicon often does not begin to crystallize until a considerable undercooling is reached below the liquidus temperature. The paucity of nuclei allows only a relatively few silicon particles to form, with the result that the microstructure contains extremely large primary crystals. An example is shown in Figure 7.6(a). In addition, primary silicon is slightly lighter than the liquid alloy and tends to float. As a result, castings which solidify slowly, such as sand castings, can exhibit pronounced gravity segregation of silicon. In the example of Figure 7.7, the upper portions of the casting contain a high concentration of primary silicon while the lower part is of essentially eutectic composition.

Each of these rather unique features of the solidification of cast hypereutectic alloys are unfortunately detrimental. As outlined in Chapter 2, hypereutectic alloys are used mainly for their wear resistance, which is imparted by the hard particles of primary silicon. In any cast part, uniform and predictable wear resistance is essential, and therefore macrosegregation of the wear resistant phase is intolerable. The machining of hypereutectic alloy castings is more difficult than their hypoeutectic counterparts because of the presence of the hard primary phase. Large unevenly distributed particles, such as those in Figure 7.6(a) cause greater tool wear than does a smaller, more uniformly distributed primary phase. Thus, fine, evenly distributed primary silicon, is an essential feature of cast hypereutectic alloys. This microstructure, an example of which is seen in Figure 7.6(b), can be promoted by facilitating the nucleation of the primary silicon through the addition of an appropriate nucleating or refining agent.

There are several ways in which silicon nucleation can be promoted. Methods which have been reported to work

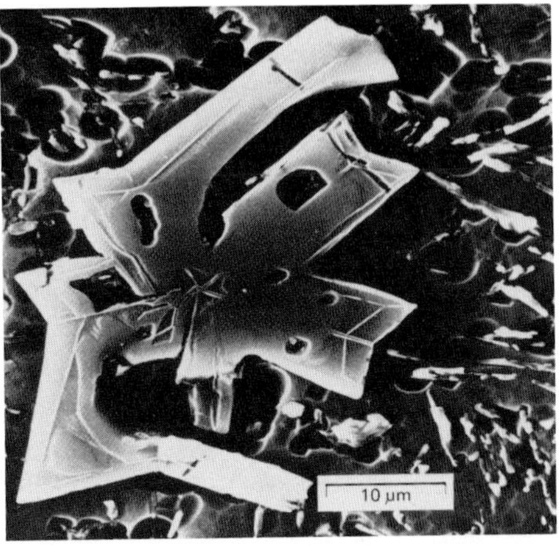

(a) star-shaped

(b) polyhedral

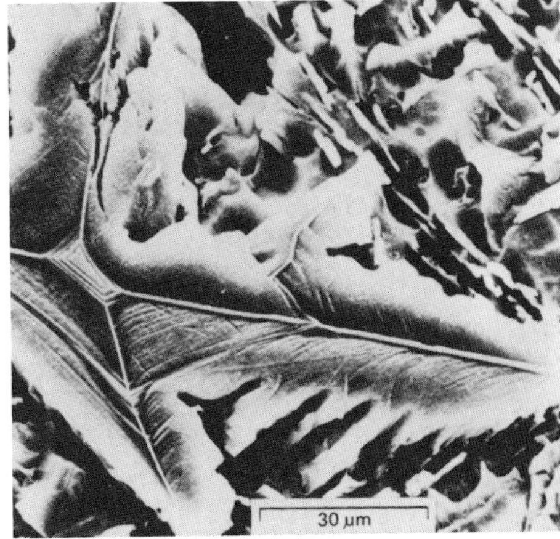

(c) dendritic

Figure 7.4. Some common morphologies of primary silicon (scanning electron micrographs).

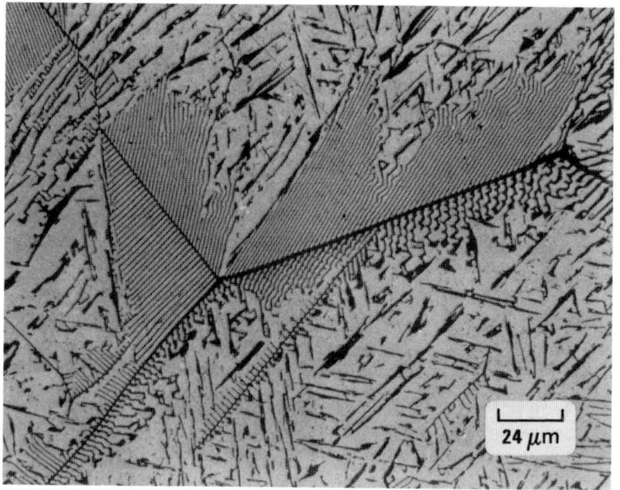

Figure 7.5. Feathery eutectic structure found in hypereutectic alloys.

but which have not found industrial application are dealt with in section 7.6. Here, we will discuss the only technique in common foundry use. Refinement of primary silicon is usually achieved by the addition of phosphorus to the melt. Phosphorus reacts with the liquid aluminum to form aluminum phosphide, AlP, which has a crystal structure very similar to that of silicon, and acts as an effective heterogeneous nucleant (see Chapter 8). We will now examine various features of phosphorus treatment.

(a) unrefined   (b) refined with 0.003%P

Figure 7.6. Typical primary silicon sizes. Al-17% Si alloy chill cast at a rate of 29°C sec⁻¹[1]

Types of Additives

It is not difficult to add phosphorus to hypereutectic alloys, and many different phosphorus salts and phosphorus containing compounds have been shown to be effective. Some are easier to work with than others, and two techniques have emerged as the most successful. A Cu-8%P alloy is very convenient to use and is available as a brazing alloy in rod or shot form. Of course, use of this alloy adds considerable copper to the melt. Many hypereutectic alloys do contain copper as a strengthener, and some of the required

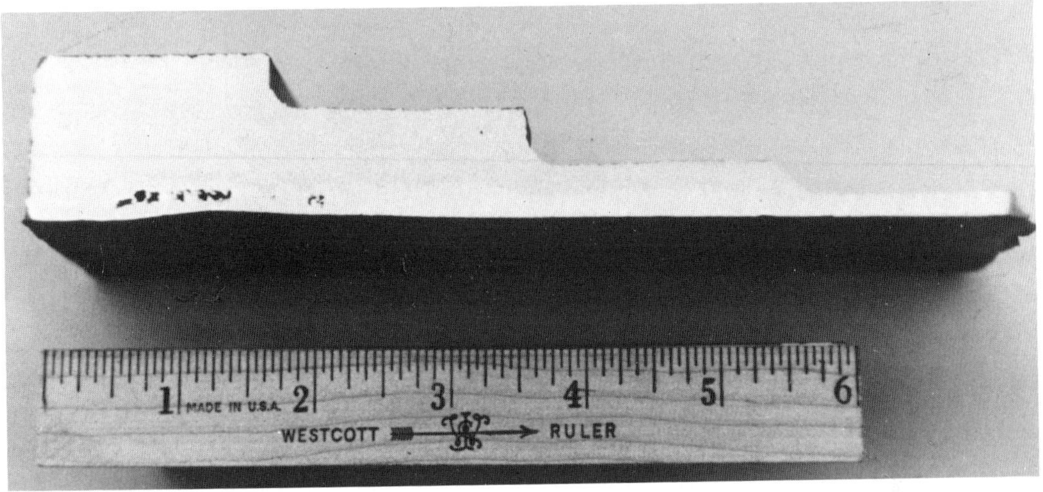

(a) general view of a section through the casting

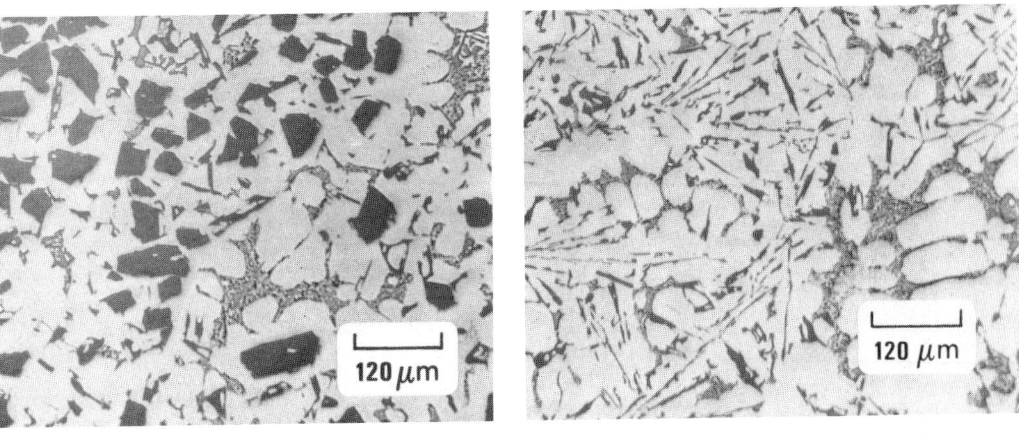

(b) micrograph of structure in the upper part of the casting

(c) micrograph of structure in the lower part of the casting

Figure 7.7. Macrosegregation in sand cast A390 alloy[1].

amount can be added during the phosphorus treatment. Copper is, however, detrimental to the corrosion properties of these alloys, and hypereutectic alloys used in salt water applications should ideally contain no copper.

In this case, phosphorus treatment can be done with one of a number of specially designed commercial nucleants, or through a flux. Most fluxes contain red phosphorus as the active agent and have additions of other salts to clean and degas the melt. In using any of these reagents, the manufacturer's instructions should be followed closely. Recoveries of phosphorus vary somewhat according to the type of additive used, and the exact addition technique, but they are always low, and range from 5% to 20%. An increasingly popular form of phosphorus addition is by means of Ni_3P. This offers excellent phosphorus recoveries.

Optimum Phosphorus Concentration

This will depend on many variables, the most important being the silicon concentration of the alloy and the metal solidification rate. It is difficult to obtain precise values for the best phosphorus level, because the treated alloys are rarely analyzed for phosphorus. However, it does lie in the range 0.003% to 0.015% depending on the casting conditions. A quick metallographic check which is often used is to look for star-shaped primary silicon (Fig.7.4a). This is usually found in alloys that are deficient in phosphorus.

As the silicon concentration of hypereutectic alloys increases, the primary particles have a natural tendency to be larger. For example, under identical freezing conditions, and in the absence of phosphorus, the silicon size can increase from 40 to 180 μm when the alloy silicon concentration changes from 12% to 20%. Larger amounts of phosphorus will, therefore, be required for higher silicon alloys, and it is expected that complete refinement of very high silicon alloys, e.g., 30%Si, will be difficult to achieve.

A rapid solidification rate can exert a significant refining effect even in the absence of phosphorus. A 390 alloy processed by conventional die casting does not require a phosphorus treatment due to the rapid cooling rates associated with die casting. On the other hand, hypereutectic alloys such as 390 are very difficult, if not impossible, to sand cast. Even with phosphorus treatment, solidification rates are so slow that unacceptably large primary phase particles form and float to the upper surfaces of the casting. The effect of cooling rate, through the liquid-solid range, on the primary silicon size in a phosphorus treated alloy is illustrated in Figure 7.8. A silicon particle size in the range 15-20 μm is optimum, and it is seen that a wide spectrum of cooling rates exists which is capable of yielding the optimum size. Some observed particle sizes obtained in a phosphorus-refined Al-22%Si alloy are summarized in Table 7.1.

Overrefinement caused by the addition of too much phosphorus is possible. This is manifested as a slight coarsening of the silicon particles and a broadening of their size distribution. The development of microstructure from the unrefined to the overrefined state is shown in Figure 7.9. The untreated alloy contains a broad spectrum of silicon

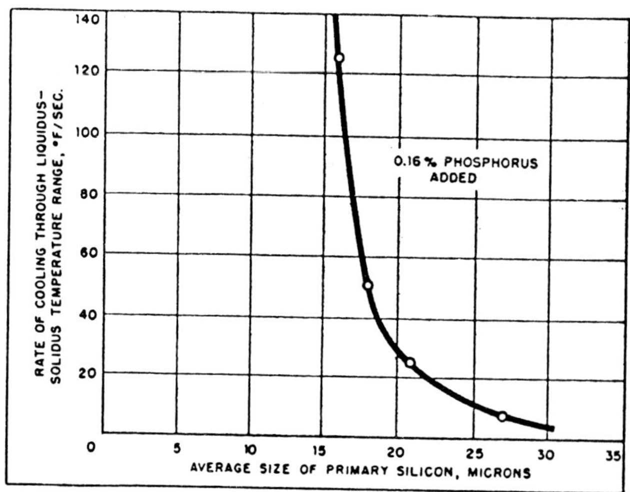

Figure 7.8. The effect of cooling rate through the solidification range on the size of primary silicon in an Al-23%Si alloy refined with phosphorus[2].

sizes (varying by a factor of 6) from less than 20 μm to about 120 μm. The best refinement occurs with 0.003-0.006%P, and this causes the silicon particle size distribution to become much narrower and to peak at about 20 μm. Overrefinement takes place if the phosphorus level is raised to 0.009%. A broadening of the size distribution occurs, and the most frequently occurring particle is now about 30 μm in size.

Table 7.1. Silicon Particle Sizes Found in Different Types of Castings Produced from an Al-22% Si Alloy Treated with 0.055% P[3]

Casting	Measured Particle Size (μm)
Conventional Pressure Die Casting	20
3/16 inch plate poured in a copper mold	15
1 inch diameter bar in cast iron mold	29
1-1/2 inch thick block in cast iron mold	48
1-1/2 inch thick block sand cast	78

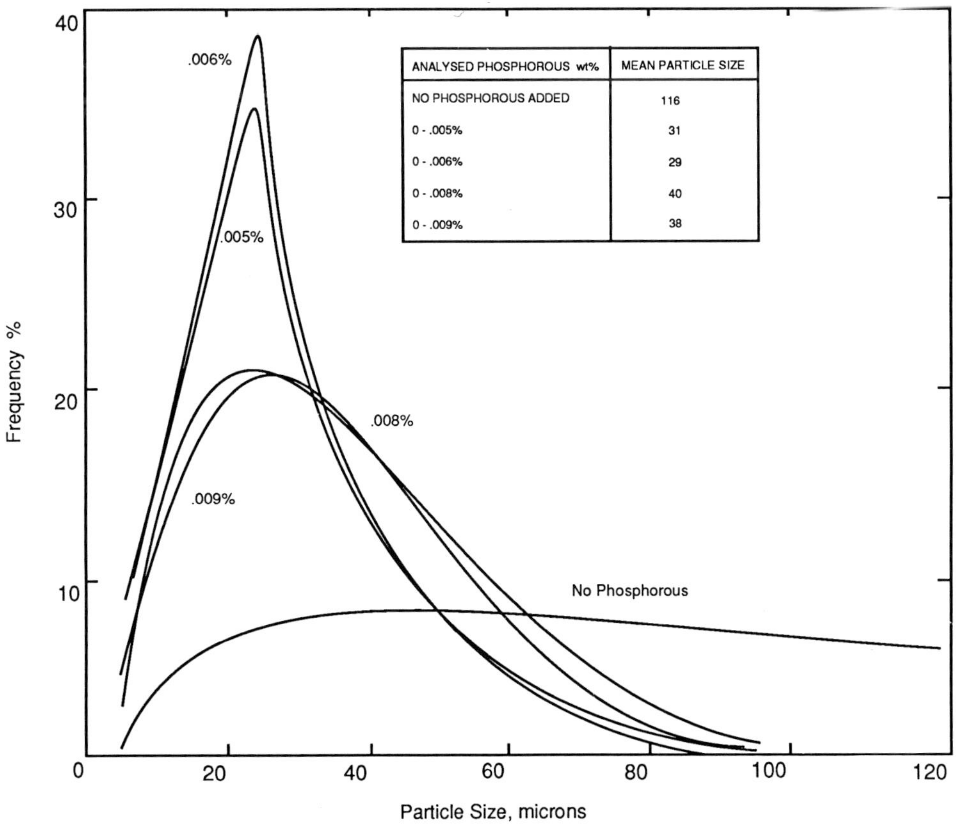

Figure 7.9. Particle size distribution curves from phosphorus refined hypereutectic alloys[3].

Holding Time and Remelting

Like many melt additives, phosphorus is subject to fading, although at a much slower rate than sodium, for example. Fading is evidenced by a gradual increase in the silicon particle size with time, and can be compensated for, somewhat, by the use of higher concentrations. The mechanism of phosphorus fading seems to be an agglomeration of the AlP nuclei. This reduces the number of possible nucleation sites and results in a coarsening of the silicon. The effect of phosphorus fade in an Al-23%Si alloy is illustrated in Figure 7.10.

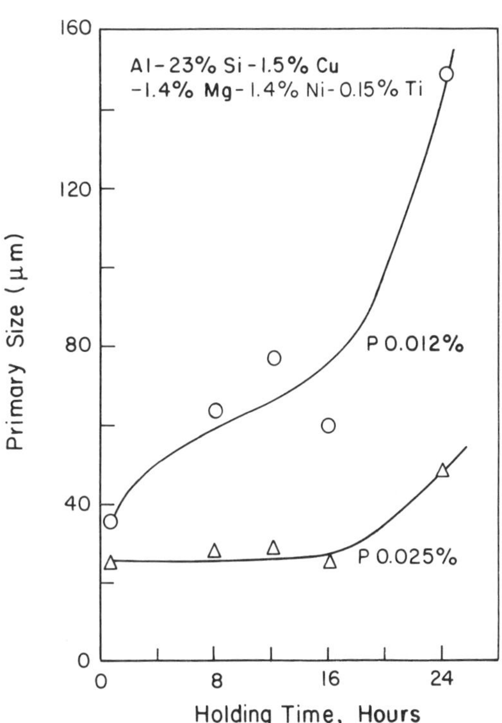

Figure 7.10. The effect of phosphorus fade on silicon size [4].

At the higher level of refiner, the microstructure is not affected by holding at 860C (1580F) for up to 16 hours, and even at the lower level, several hours are needed before serious fading occurs. It is generally considered that phosphorus treated melts can be held for 3 to 5 hours before fading becomes a problem and re-refinement becomes necessary.

The remelting of phosphorus treated ingot has only a small effect on the phosphorus refiner. Up to 5 remelts can be performed before the level of refinement becomes unacceptable. For example, an Al-22%Si alloy treated with 0.016%P was reported to have a primary silicon size of 33 μm. On one remelting, this increased to 45 μm and to 50 μm after the fourth remelt[3]. Phosphorus, therefore, poses relatively few problems in this regard, especially when compared to the eutectic modifiers, sodium or even strontium.

Effect of Degassing

Although silicon somewhat decreases the solubility of hydrogen in liquid aluminum, the higher melting and casting temperatures used with hypereutectic alloys encourage hydrogen dissolution. Degassing is, therefore, an important part of the technology of these alloys. Degassing agents may be various mixtures of nitrogen-chlorine or nitrogen-freon, pure nitrogen, pure argon or hexachloroethane. Some loss of refinement is to be expected with degassing and so the degassing treatment time should be kept to the minimum, consistent with achieving the desired hydrogen concentration. Figure 7.11 shows the loss of refinement experienced in one example of degassing at 800C. The rate of silicon coarsening during degassing should be compared to that experienced by simply holding the melt without degassing. The mechanisms of interaction between the degassing agent and the nuclei are not well understood, but it is believed that nuclei are floated out of the melt by attachment to the rising bubbles of gas.

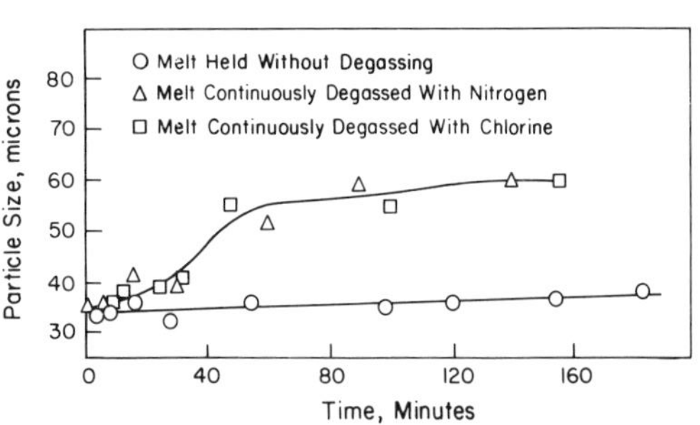

Figure 7.11. The effect of degassing on primary silicon size in an Al-22%Si alloy[3].

There are some reports that a brief degassing treatment exerts a beneficial effect on refinement. This probably has nothing to do with the degassing itself, but is the result of stirring of the melt to produce a more uniform distribution of AlP nuclei.

Treatment Temperature and Casting Temperature

While the details of the melt phase reactions to form AlP are not at all understood, it is apparent from experimental observations that more and better distributed nuclei are formed at higher rather than lower temperatures. Therefore, treatment and casting temperatures should be as high as are feasible. Tagami and Yo[5] have shown in an Al-20%Si alloy treated with 0.02%P that much finer silicon is obtained if the melt is treated and cast from 900C (1650F) than from 700C (1290F). Remelting a coarse material which was previously cast at a lower temperature will reverse the earlier structure, provided the second casting operation is done at a high temperature.

If a hypereutectic alloy is to be die cast and refined without the benefit of phosphorus, it is necessary that the alloy temperature in the holding furance not be allowed to fall below the liquidus temperature. If it does, large silicon crystals will grow and be injected into the die. The resultant microstructure will then contain a mixture of coarse and fine silicon.

Even if the alloy is phosphorus refined, it is important that the casting temperature be high enough to allow complete filling of the mold cavity before precipitation of the primary silicon begins. If this is not done silicon will form during the mold-filling process, and the forced convection which accompanies pouring will result in clustering of the primary silicon. Some areas of the microstructure will therefore appear extremely rich in silicon while others will be depleted. In the next section we examine the influence that phosphorus refinement has on various properties of hypereutectic alloys.

7.3   The Effect of Refinement on Properties

Fluidity and Feeding

Hypereutectic alloys possess excellent fluidity due to the high latent heat of fusion of the primary silicon phase. In the unrefined state, however, the large primary silicon crystals interlock during the final stages of solidification and act to impair feeding. Phosphorus refinement improves this situation somewhat by creating much finer silicon which has less tendency to interlock. Nevertheless, this family of alloys remains prone to feeding difficulties. It has the widest freezing range of any of the aluminum casting alloys, and should be used with generous risering and chilling.

Tensile Properties

Refinement has a beneficial effect on the tensile properties since it produces a finer and more even distribution of the brittle silicon phase. The improvement in tensile strength can vary from 10% to almost 100% depending on the silicon content of the alloy. Compositions near the eutectic yield relatively little primary phase. Refinement has a lesser effect than on alloys whose composition is far from the eutectic, and which contain very large quantities of primary silicon. Of course, the tensile strength generally decreases as the silicon concentration is raised. These effects are illustrated in Figure 7.12.

The tensile ductility of high silicon alloys is quite low due to the presence of the brittle phase. In the unrefined condition, values less than 0.05% are common. With refinement, these can be increased to close to 1%.

Machinability

Improvements in machinability are the chief reason for utilizing phosphorus refinement. When large silicon particles are present in the microstructure, machining may be almost impossible. Excessive tool wear is experienced, especially with non-carbide tools. In addition, uneven heating of the workpiece may occur result-

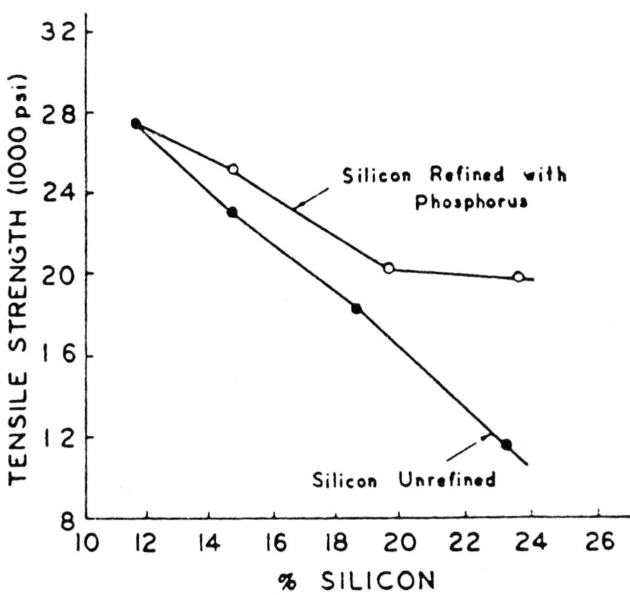

Figure 7.12. The tensile strength variation of unrefined and phosphorus refined hypereutectic alloys[1].

ing in poor dimensional control on machining.

The extent of improvement in tool life will depend on the type of casting process used. Machinability of sand castings is improved the most by refinement, since these contain the largest and most non-uniformly distributed silicon in the unrefined state. The improvements in die casting are less, due to the finer as-cast structure produced by this process. Since hypereutectic alloys are rarely sand cast and may not require refining if die cast, the greatest advantages to machinability are to be found in permanent mold castings where improvements in tool life of 50% can be realized.

Surface roughness can also be improved by refining as indicated by the data in Table 7.2 taken from reference 7. The surface roughness values quoted were measured by a Leitz surface roughness gauge. A lower value indicates a smoother machined surface.

Table 7.2. Surface Roughness of Machined Piston Alloys[7]

Alloy	Primary Silicon	Rough Turned	Fine Turned	Turned with Diamond-Tipped Tool
Al-17%Si	unrefined	5	3	2.5
	refined	3	2	1.2

Wear Properties

Hypereutectic alloys are most often used in applications which exploit their wear resistance. Surprisingly, it is not clear that refinement has any effect on wear properties. This is probably related to the difficulty of performing meaningful wear tests.

One school of thought says that it is the total amount of silicon rather than its morphology which determines wear properties. Thus, higher silicon alloys wear better than lower silicon alloys. Another point of view is that refinement does indeed maximize wear properties. Convincing experiments to support each hypothesis can be found. Perhaps the conflicting information simply illustrates that silicon shape and size is only a minor factor in determining wear, and that other variables such as lubrication and applied load are of much greater importance. Related to wear is the observation that in some cases, hardness is reduced by refinement, while in others it is increased. The differences are never very large, and the erratic behavior reflects the influence of other variables on the properties. In automotive applications, wear properties are enhanced by a special etching process which partially removes some of the eutectic matrix. The primary silicon particles then stand out on the surface to provide wear resistance.

7.4   Strontium Treatment

In section 3.5 we discussed the fact that phosphorus leads to a coarse acicular eutectic structure, and so its deliberate addition to hypereutectic alloys guarantees an unmodified eutectic silicon surrounding the refined primary phase.

The benefits of eutectic modification of hypoeutectic alloys were discussed at length in Chapter 5. It is only reasonable to expect that similar benefits could be found in hypereutectic alloys, if primary phase refinement and eutectic modification could be caused to occur simultaneously.

Unfortunately, this has proven to be an elusive goal for foundry metallurgists, since modifying elements are chemically incompatible with the phosphorus added to refine the primary silicon. Both strontium and sodium react in the melt to form either strontium phosphide or sodium phosphide. These compounds are more stable than aluminum phosphide and hence a coarsening of the primary silicon occurs. However, with sufficient additions, the eutectic can be completely modified, and morphological changes can be brought about in the primary silicon. In many cases the resultant tensile properties are as good as those in phosphorus refined alloys. Unfortunately, other properties have not yet been investigated. In this and the following section, we will briefly examine the effects of adding strontium or sodium to commercial hypereutectic alloys which are phosphorus pre-refined.

In Figure 7.13, we show the result of adding various amounts of strontium modifier to a chill cast A390 alloy. Levels typical of those required to cause modification in hypoeutectic alloys coarsen the primary silicon, but allow it to retain its blocky shape (Fig. 7.13b). Very much higher concentrations cause a change in the primary silicon morphology itself, resulting in a dendritic shape (Fig. 7.13c). In this example, sufficient strontium is present at even 0.03% to modify the eutectic. This modification has, however, been achieved at the expense of a three-fold increase in the primary silicon size. Such effects are very cooling rate dependent. For example, the same alloy sand cast solidifies with very coarse dendritic silicon and almost no modification of the eutectic (Fig. 7.14b).

Despite the coarsened primary structure, the addition of strontium to the A390 alloy does not change its tensile properties, and may even be regarded as giving them a marginal improvement. In the phosphorus modified condition, this alloy had a UTS of 198 MPa, and an elongation measured on 50 mm of 0.8%. With 0.16% strontium added, the UTS increased slightly to 212 MPa while the elongation remained unchanged. These property values illustrate the relative importance of eutectic modification and primary silicon refinement in these alloys.

In a binary Al-17%Si alloy, the equilibrium weight fraction of alloy which freezes as eutectic is approximately 95%. This is much greater than the approximately 5% which solidifies as primary silicon. As a result, modification of this alloy affects considerably more phase constituent than does refinement. At least as far as the tensile properties are concerned, modification of the eutectic can more than compensate for a considerable coarsening of the primary phase. Some typical tensile properties of a strontium treated, binary Al-17%Si alloy are summarized in Table 7.3. Significant increases in both the tensile strength and the elongation are found when the eutectic becomes modified.

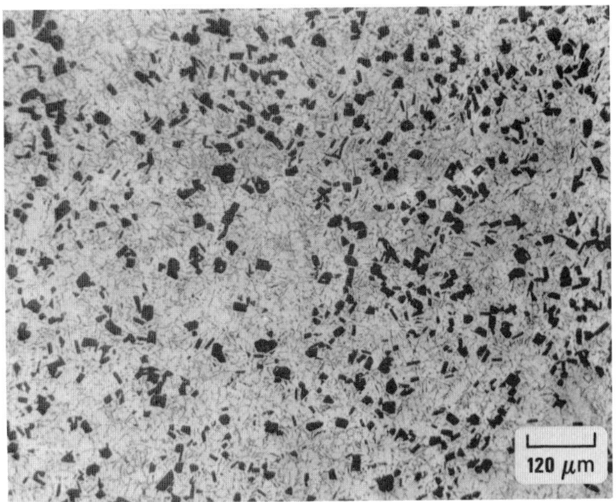

(a) no strontium

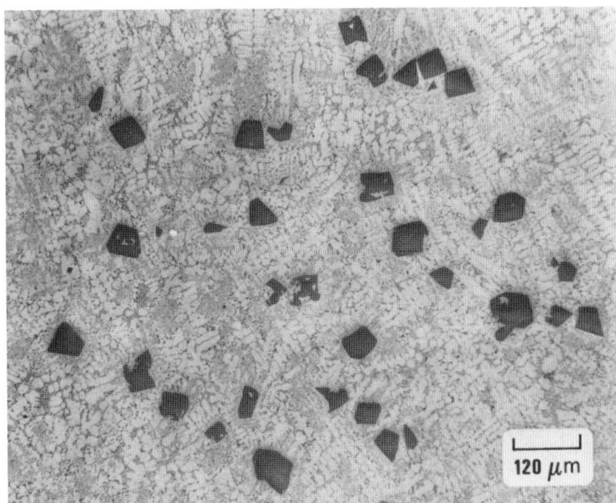

(b) 0.01%Sr

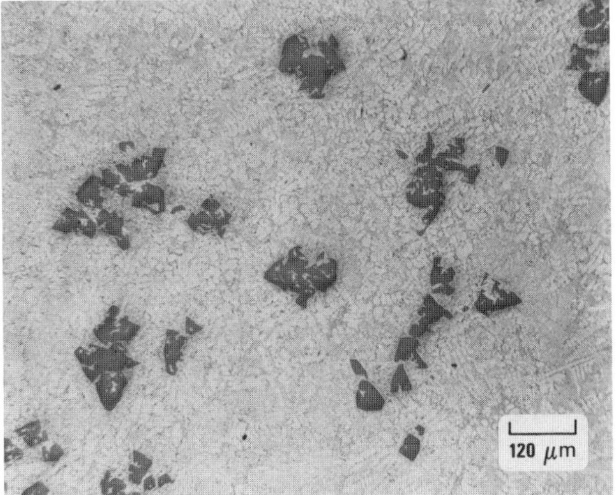

(c) 0.03%Sr

Figure 7.13. The effect of strontium additions to a phosphorus refined A390 alloy chill cast at 29°C sec^{-1} [8].

Table 7.3. Tensile Properties of Binary and Strontium-Treated Al-17%Si Synthetic Alloy

Strontium wt. pct.	UTS MPa	$YS_{0.2}$ MPa	Elongation in 50 mm (%)	E GPa
0	120	102	0.5	80
0.02	150	102	1.4	72
0.04	148	101	1.7	71
0.06	150	112	1.9	78

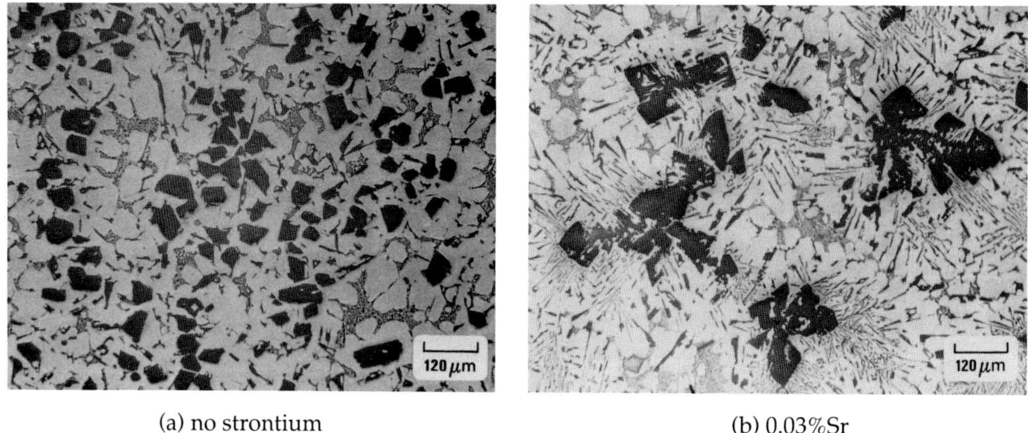

Figure 7.14. The effect of strontium additions to a phosphorus-refined A390 alloy sand cast at 1.4 °C sec⁻¹[8].

7.5   Sodium Treatment

Sodium, like strontium, can be added in sufficient quantity to a phosphorus treated alloy to neutralize the effect of phosphorus and to modify the eutectic. Very limited research has been done on sodium treated hypereutectic alloys, but two forms of primary silicon seem possible. Perhaps most interesting is the almost spheroidal shape (Figure 7.15a) first reported by Mascré[9]. This morphology is not perfectly spheroidal, and close examination will show that it is actually a multi-faceted form. With increasing sodium concentration, these primary silicon grains decrease in size and become more spheroidal. Very high concentrations of sodium are required to produce this structure and the level probably depends on the cooling rate. The structure shown in

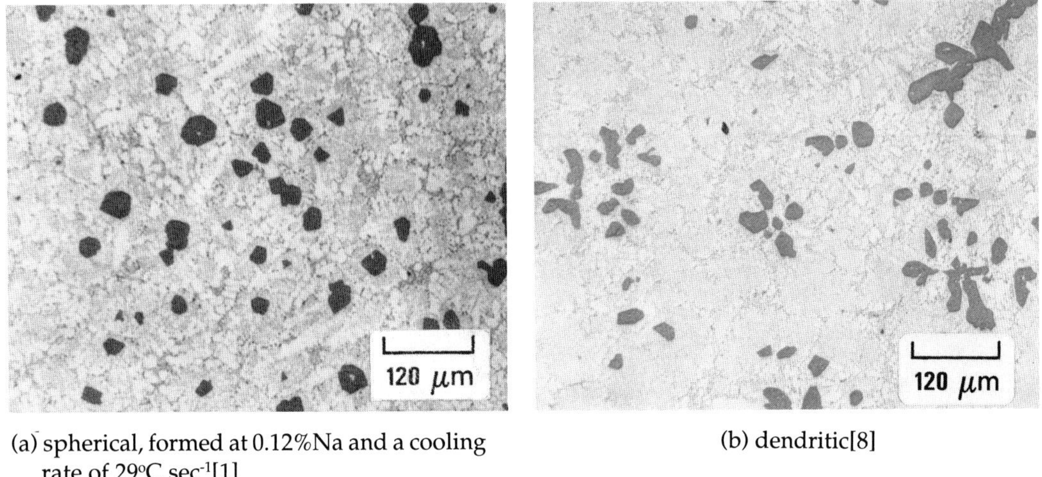

(a) spherical, formed at 0.12%Na and a cooling   (b) dendritic[8]
    rate of 29°C sec⁻¹[1]

Figure 7.15. Primary silicon shapes found in sodium treated A390 alloy.

Figure 7.15(a), required the use of over 0.1% sodium at a cooling rate through the solidification range of 29°C sec⁻¹. At this high concentration, the eutectic is always well modified.

Dendritic silicon (Figure 7.15b) is more readily produced with sodium treatment than is the spheroidal variety. The exact conditions required to form either type are not at all defined, but dendritic silicon is likely favored at lower freezing rates and lower sodium concentrations. Again, a well modified eutectic structure forms.

Microstructures of the type illustrated in Figure 7.15 should yield good tensile properties, probably better than those obtained with simple phosphorus treatment. Mascre[9], reported a deterioration of properties with the use of sodium, but his samples probably contained high levels of porosity. Some more recent work, under better controlled conditions, indicates that the dendritic morphology (Figure 7.15b) provides enhanced strength and ductility over phosphorus treated A390. Typical values are summarized in Table 7.4. If the usual rules of metallurgy are followed, the spheroidal type should yield even better properties.

In Chapter 6, we saw that sodium-strontium treatments could be of assistance in providing both short- and long-term modification. A combined addition to hypereutectic alloys will yield long-term eutectic modification, but the effect on the primary silicon is highly time dependent, due to the fact that strontium and sodium act in somewhat different ways on this phase.

In Figure 7.16, we show the structure of an A390 alloy treated with 0.045%Sr. An addition of 0.08%Na is then made and allowed to fade. The dendritic silicon present initially is typical of strontium treatment. This transforms to a multi-faceted, almost spheroidal shape with the initial sodium addition, but then degenerates to a dendritic morphology as the

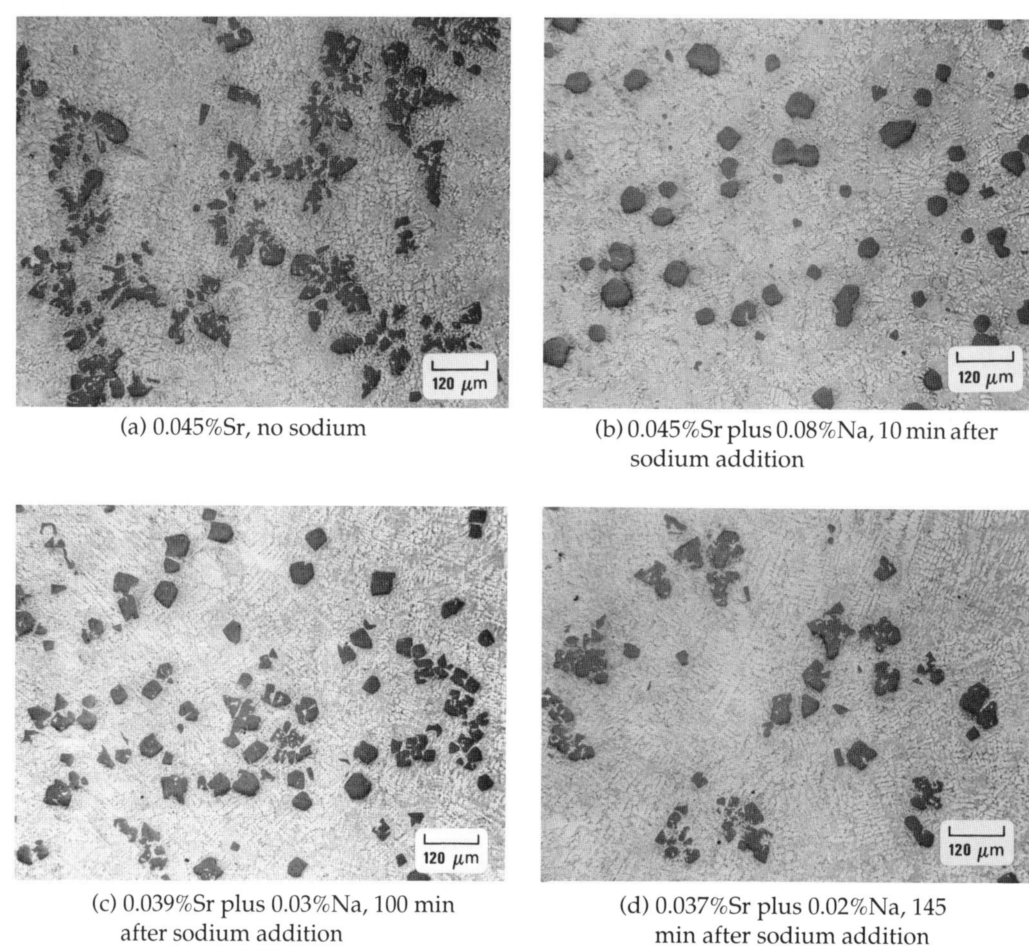

(a) 0.045%Sr, no sodium

(b) 0.045%Sr plus 0.08%Na, 10 min after sodium addition

(c) 0.039%Sr plus 0.03%Na, 100 min after sodium addition

(d) 0.037%Sr plus 0.02%Na, 145 min after sodium addition

Figure 7.16. Sodium-strontium treatment of an A390 alloy chill cast at 29°C sec⁻¹ [1].

sodium fades to 0.02%, 145 minutes after addition. It is evident that the effects are quite different from those which occur in hypoeutectic alloys. There is, however, some indication that combined sodium-strontium treatments can result in a greater number of fine spheroidal silicon particles than is obtained by a single sodium treatment[1].

In summary, current melt technology does not allow the coexistence of a modified eutectic with a primary silicon phase having the fineness and uniformity of distribution made possible by phosphorus treatment. The eutectic can, however, be modified by using larger quantities of classical modifiers than are needed in hypoeutectic alloys. This results in a primary silicon which is somewhat different in shape and coarser than that found in phosphorus refined alloys, but which is still finer than in untreated alloys. Because of the large amount of eutectic present, these treatments can have a beneficial effect on properties. The limited information available to date indicates that the tensile properties obtainable by modifier treatment are as good, and in some cases better, than those which result from phosphorus treatment alone.

Table 7.4. Tensile Properties of A390 Alloy—Phosphorus Treated and Sodium Treated[8]

Alloy Treatment	UTS (MPa)	$YS_{0.2}$ (MPa)	Elongation in 50 mm %
As received (0.003%P)	198	181	0.6
0.016%Na	237	169	0.8
0.14%Na	243	158	0.9
0.11%Na	232	160	0.9
0.09%Na	222	159	0.9
0.08%Na	220	151	0.8
0.06%Na	249	159	1.0

7.6   Refinement by Other Elements

Although phosphorus is the only element which is used commercially as a refiner, several others have been identified which act in a similar way. None of these has been shown to produce the high level of refinement possible with phosphorus, and at the same time to modify the eutectic. Therefore, none of these elements present any advantages over phosphorus, which is relatively easy to add to the melt, acts in low quantities, is long lasting and is inexpensive.

Sulphur is the additive which has received the most attention. Unfortunately, it is difficult to analyze for in aluminum. Elemental sulphur vaporizes at 445C (835F) and so will fume when added to foundry alloys. It is best added as a sulphide, and will produce excellent refinement once it is dissolved in the melt. Optimum levels have not been identified due to analytical problems, but if recoveries of 10%-20% are assumed, it appears that good refinement can be obtained in alloys containing 0.05% to 0.01%S. Like phosphorus, sulphur forms stable compounds with sodium and strontium, precluding refinement and modification at the same time. Another similarity with phosphorus is that higher melt temperatures result in better refinement.

Selenium is another element which has been shown to refine primary silicon. It too is very stable in the melt and yields finer primary silicon if the melt is heated at 900C (1650F) or even higher[10]. Given the noxious effects of selenium, it is difficult to see that this element would replace phosphorus.

Arsenic provides refinement comparable to phosphorus in sand cast alloys and even somewhat better in die casting. Recovery rates are unknown, but it appears that additions of up to 0.6%As are necessary[11]. Arsenic is probably very stable in aluminum melts. It is, however, toxic and its effect is cancelled by sodium treatment.

Several of the rare earth elements have also been said to refine primary silicon. Among these are cerium, lanthanum and neodynium. Not enough development work has been done on these to define the exact quantities needed and their effects on microstructure under production conditions.

Although there seems to be little advantage to replacing phosphorus by any other chemical additive, a technique which may prove of use in certain future applications is the intermittent stirring of the melt during freezing. Electromagnetic stirring every 30 seconds has been shown to be beneficial to refinement of both primary and eutectic silicon and to reduce gravity segregation tendency[12].

References

1. Tenekedjiev, N. "Strontium Treatment of Hypereutectic Al-17%Si Casting Alloys," M. Eng. Thesis, McGill University (1989).
2. Sulzer, J. "How to Grain Refine High Silicon Aluminum Alloys," *modern casting*, 39 (1961) pp. 38-43.
3. Bates, A.P. and D.S. Calvert. "Refinement and Foundry Characteristics of Hypereutectic Aluminum-Silicon Alloys," *The British Foundryman*, 59 (1966) pp. 119-33.
4. Adachi, M. "Modification of Hypereutectic Al-Si System Casting Alloys," Journal of Japan Institute of Light Metals, 34 (1984) pp. 430-36.
5. Tagami, M. and S. Yo. "Influence of Casting Temperature on Refinement of Primary Silicon Crystals in a Hypereutectic Al-20%Si Alloy by Phosphorus Addition," *Journal of Japan Institute of Light Metals*, 26 (1976) pp. 273-79.
6. Arnold, F.L. and J.S. Prestley. "Hypereutectic Aluminum-Silicon Casting Alloys Phosphorus Refinement," *AFS Transactions*, 69 (1961) pp. 61-69.
7. Schneider, K. "Aluminum Casting Alloys Properties Improvement by Grain Refining," *AFS Transactions*, 68 (1960) pp. 176-81.
8. Tenekedjiev, N. and J.E. Gruzleski. "Sodium, Strontium and Phosphorus Effects in Hypereutectic Al-Si Alloys," *AFS Transactions*, 97 (1989) pp. 127-36.
9. Mascré, C. "Modification of High-Silicon Aluminum Alloys and the Corresponding Structures," *Foundry Trade Journal*, 94 (1953) pp. 725-30.
10. Tagami, M. and Y. Serita. "Refinement of Primary Silicon Crystals in a Hypereutectic Al-20%Si Alloy by Addition of Metallic Selenium," *Journal of Japan Institute of Light Metals*, 22 (1972) pp. 489-93.
11. Clegg, A.J. and A.A. Das. "The Influence of Structural Modifiers on the Refinement of the Primary Silicon in a Hypereutectic Aluminum-Silicon Alloy," *The British Foundryman*, 70 (1977) pp. 56-63.
12. Momono, T. and K. Ikawa. "Structural Change of Hypoeutectic Al-Si Alloys by Convective Flow in Solidification Process," *Journal Japan Institute of Light Metals*, 29 (1979) pp. 240-45.

Chapter 8

Grain Refinement

8.1 The Definition of a Grain

In this chapter we treat the question of refining the size of the primary aluminum grains during solidification. Grain size in single phase metals has long been known to be of significance in determining, for example, the tensile properties. A general rule taught to all students of metallurgy is that finer grains are, for most applications, preferable to coarse grains. This rule is true only to the extent that the alloy is mainly single phase and applies to the 200 series of aluminum foundry alloys based on the Al-Cu system.

The properties of cast alloys which contain a large fraction of eutectic, such as the Al-Si alloys, are much less dependent on grain sizes, and grain refinement of these alloys is of lesser value. It is the brittle eutectic silicon phase which determines their properties, and hence processes of eutectic modification become much more important than grain refinement. In addition to the eutectic silicon size and shape, the dendrite arm spacing (DAS) determined by the cooling rate through the mushy zone is of prime importance.

It is important that the reader have a clear concept in his mind of the differences between these three features of the microstructure. They are all, more or less, independent of each other and for example, coarse or fine grained castings may occur with modified or unmodified silicon, and with either large or small dendrite arm spacings. The sketch in Figure 8.1 should help to make the distinction clear.

Each grain contains a family of aluminum dendrites which all originate from the same nucleus. The dendrite arm spacing (DAS) is determined by the cooling rate through the mushy zone, with slower cooling resulting in larger values of DAS. Between the dendrite arms is found the eutectic silicon phase which may or may not be modified. These three features of the microstructure are very different in size. Grains in aluminum foundry alloys are found in the range 1-10 mm; DAS values vary from 10 to 150 μm, and the eutectic silicon may be found in plates up to 2 mm in length or spheres of less than 1 μm. in diameter. The reader should keep in mind that a metallographic samples usually represents a random section through the microstructure and that proper statistical methods must be employed to quantify these features of the microstructure. A description of these methods is outside the scope of this book, but the basics can be found in reference 1.

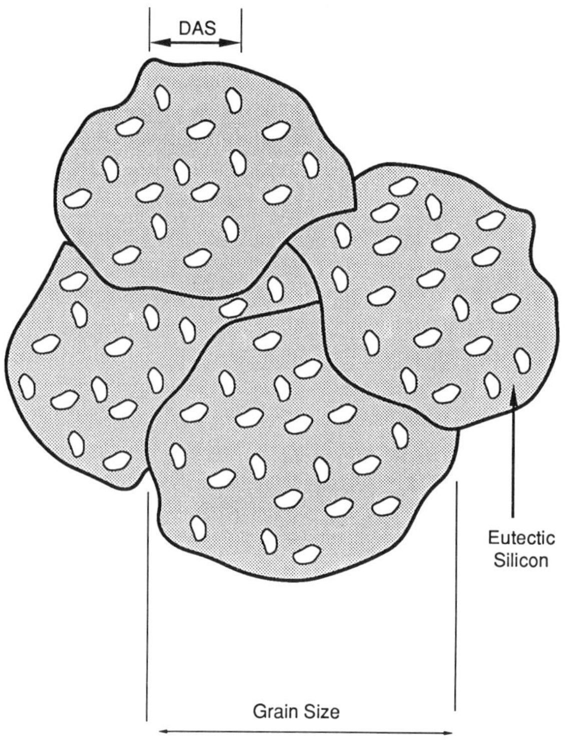

Figure 8.1. The basic elements of microstructure of an Al-Si type casting alloy.

8.2 Principles of Grain Refinement

The grain size of a cast material bears an inverse relationship to the number of nuclei present in the liquid which are able to act during the solidification process. If one accepts that each grain is nucleated by one foreign particle or nucleus—the process known as heterogeneous nucleation—then it is clear that a greater number of nuclei will allow more grains to form, resulting in a smaller grain size. Time also enters the process of grain nucleation as there is evidence to suggest that at least some nuclei require a certain time in the liquid phase before they become active. Thus the best combination of conditions to promote extensive nucleation, and hence a fine grain size, would seem to be the presence of a large number of nuclei coupled with a slow rate of freezing to provide the required time span for them to act.

It is well known that not all foreign particles present in a metallic melt are equally effective as nuclei. The question of what constitutes an effective nucleus and why, is still open to considerable debate despite decades of intensive research. An interesting review of the various facets of this question has been done by Mondolfo[2]. The simple and quite inadequate theory of heterogeneous nucleation suggests that the interfacial energy between the nucleant and the nucleus (the material which is solidifying), is of prime importance. Various possibilities are illustrated in Figure 8.2 and, of these, condition (c) is usually regarded as the optimum. Here, the interfacial energy between nucleus and nucleant is minimal and the nucleus is able to envelop the nucleant. In effect, it forms a film of high radius of curvature with little expenditure of energy. Mini-

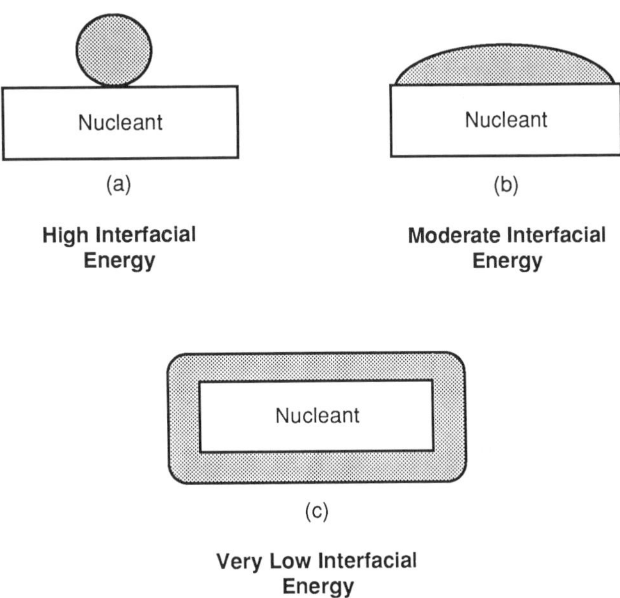

Figure 8.2. The effect of interfacial energy on the geometry of a nucleus (shaded) formed on a nucleant.

mal surface energy between nucleant and nucleus is usually achieved if a similarity of crystal structure exists between at least one atomic plane in the nucleus and one in the nucleant. Then very close atomic matching can occur across the interface separating the two, as seen in Figure 8.3. Since a large part of surface energy can be related to atomic mismatch across the surface, any minimizing of this mismatch will promote a low surface energy.

This very simple view of heterogeneous nucleation suggests that good nuclei should possess atomic planes in their crystal lattice similar to atomic planes in the lattice of the material to be nucleated, which in our case is aluminum. Thus a definite crystallographic relationship should exist between aluminum and its nucleant. This is very often found to be the case, but in the most studied example, that of $TiAl_3$, many different relationships are found depending on the investigator. Nevertheless, for the moment, a crystallographic compatibility would seem to be a necessary if not sufficient criterion for a good nucleant.

Liquid aluminum foundry alloys contain a multitude of foreign particles and substrates ranging from oxides and spinels to the wall of the mold itself. At a given temperature, or undercooling, below the melting point of the alloy, any foreign particle may or may not be effective as a nucleant. Particles which offer the best crystallographic similarity to aluminum will become effective nucleants at temperatures very close to the alloy liquidus temperature, i.e., at a small undercooling; particles which offer poorer crystallographic similarity will require higher undercoolings to become effective nuclei. There exists, therefore, a spectrum of possible nuclei in any liquid foundry alloy which can become effective nucleants over a range of

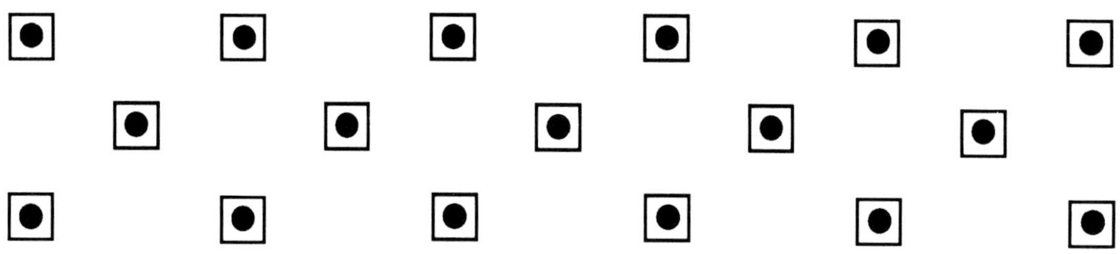

(a) one-to-one lattice matching; substrate is an excellent grain refiner for aluminum

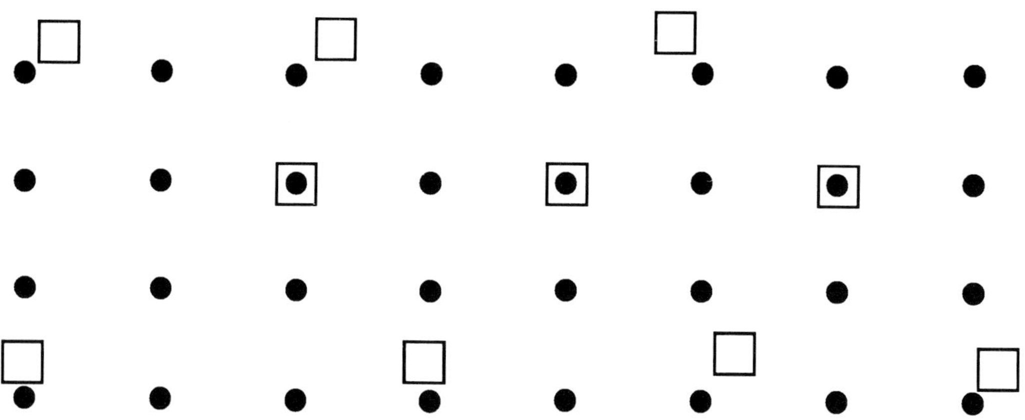

(b) very poor lattice matching; substrate is a poor grain refiner for aluminum

Figure 8.3. Two possible relationships between the atomic arrangement on a crystal plane of an aluminum nucleus ▢ and the atoms of a substrate ●. In case (a) there is excellent grain refinement while in case (b) there is very poor nucleation due to a high interfacial energy between aluminum and the substrate.

temperatures. If it is possible to undercool the liquid by a large amount, say several tens of degrees, then many different particles will act as nucleants, and a fine grain size casting should result. This grain refinement by chilling will be discussed in somewhat more detail in the next section of this Chapter; suffice to state here that most often it is simply not possible to chill the liquid to a sufficient degree that all nuclei can act, and hence very often, nucleating agents are added to the liquid alloy in a process called chemical grain refinement. The nuclei which are added, how they are added, and how they act will be dealt with in section 8.4.

8.3 Grain Refinement by Chilling and the Effect on Dendrite Arm Spacing

If a melt is rapidly cooled (chilled), the rate of heat extraction can greatly exceed the rate of heat generated by the freezing process (latent heat of solidification). As a result, the liquid undercools as its temperature falls below the liquidus temperature. If this undercooling is sufficient, the full range of heterogeneous nuclei present in the liquid can become active. This multiple nucleation results in a fine as cast grain size.

Few data exist in the literature on the effect of cooling rate on the grain size of aluminum foundry alloys, but chilling can result in grain sizes in the 0.1-1000 μm range. The finest sizes are, of course, achieved by the most rapid cooling which is possible in only very thin sections and using rapid solidification techniques. While the range of grain sizes achievable by chilling is greater than with any other grain refinement technique, the method itself is of only limited usefulness in practice, since the rate of cooling is determined by the molding method used. This in turn is usually determined by considerations other than grain size; considerations such as economics, casting size, number of castings to be produced, etc. In any event, the rapid chilling of large castings produced by any method is usually impractical due to the shear amount of latent heat which must be removed. At best, some local control of grain size can be achieved by the insertion of chills in critical locations.

The dendrite arm spacing (DAS), which is not related in any direct way to the grain size, can be varied considerably by cooling rate and by choice of casting method. Figure 8.4 illustrates the DAS values obtainable in a 356 alloy. As mentioned previously, the DAS is far more important in determining mechanical properties of multiphase alloys such as the family of Al-Si foundry alloys. Some typical properties expressed as a function of DAS are shown in Figure 8.5. It is probably this type of behavior which is responsible for much of the confusion associated with the effect of grain size on mechanical properties in these alloys. Smaller DAS values are caused by faster solidification rates which are usually associated with finer as cast grain sizes, but it is the DAS and not the grain size which is the determining

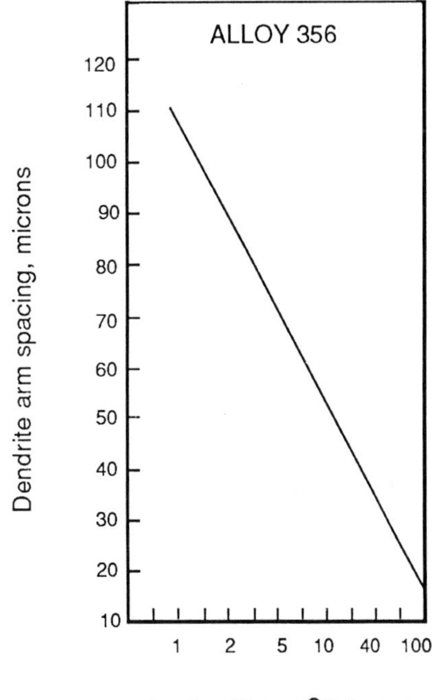

Figure 8.4. Variation of the DAS of a 356 alloy with cooling rate [7].

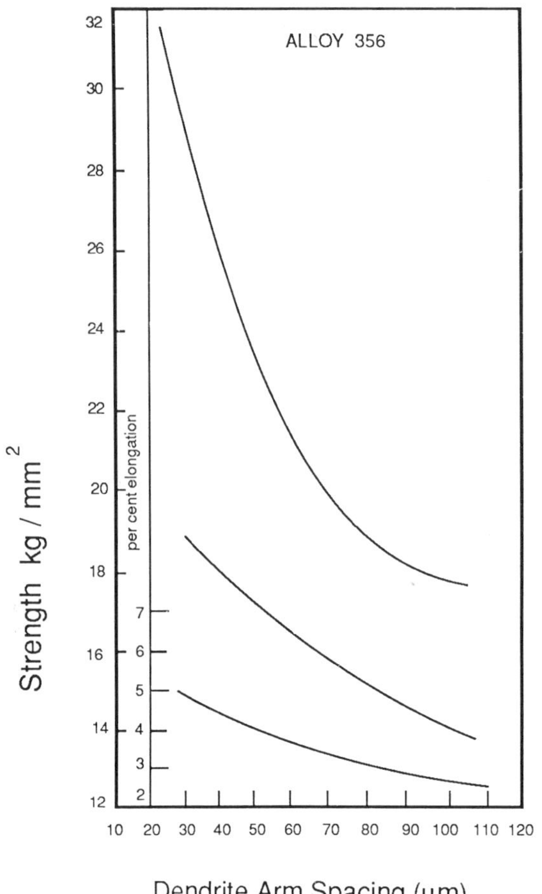

Figure 8.5. Variation of some tensile properties of an A356 alloy with DAS[7].

factor in the mechanical properties.

A very dramatic example of the effect of decreasing the DAS on the distribution of brittle intermetallic phases is given in Figure 8.6 taken from the work of Mondolfo[2]. In this aluminum bearing alloy, cast in sand and in a metal mold, the intermetallic network is broken up as the DAS decreases with more rapid cooling. The slowly cooled (sand cast) structure is poor due to the continuous network of brittle phases. It would be expected to be inherently brittle, while the more rapidly solidified structure will exhibit higher ductility and tensile strength due to the non-continuous nature of the brittle phases.

Finally, in the Al-Si alloys, it must be remembered that chilling can cause partial modification of the silicon phase (Chapter 3), and that this too can contribute to the mechanical properties. Again, the grain size itself is not a major factor.

8.4   Chemical Grain Refinement

This is the most widely practiced, and most foolproof method of grain refinement. In chemical grain refinement, additions of effective nuclei are made to the melt through either master alloys or fluxes. A fine grain size is promoted by the presence of an enhanced number of nuclei, and solidification proceeds at very small undercoolings. Since chemical grain refinement does not influence the freezing rate, it has no effect on the DAS which remains relatively large due to the slow solidification associated with the low undercooling. In this regard, chemical grain refinement exerts a less beneficial effect on the mechanical properties than does grain refinement by chilling discussed in the previous section. Nevertheless, chemical grain refinement is advantageous to mechanical properties, particularly those that are related to hot tearing and porosity. As discussed later, grain refinement reduces the hot tearing tendency, and leads to a finer distribution of porosity.

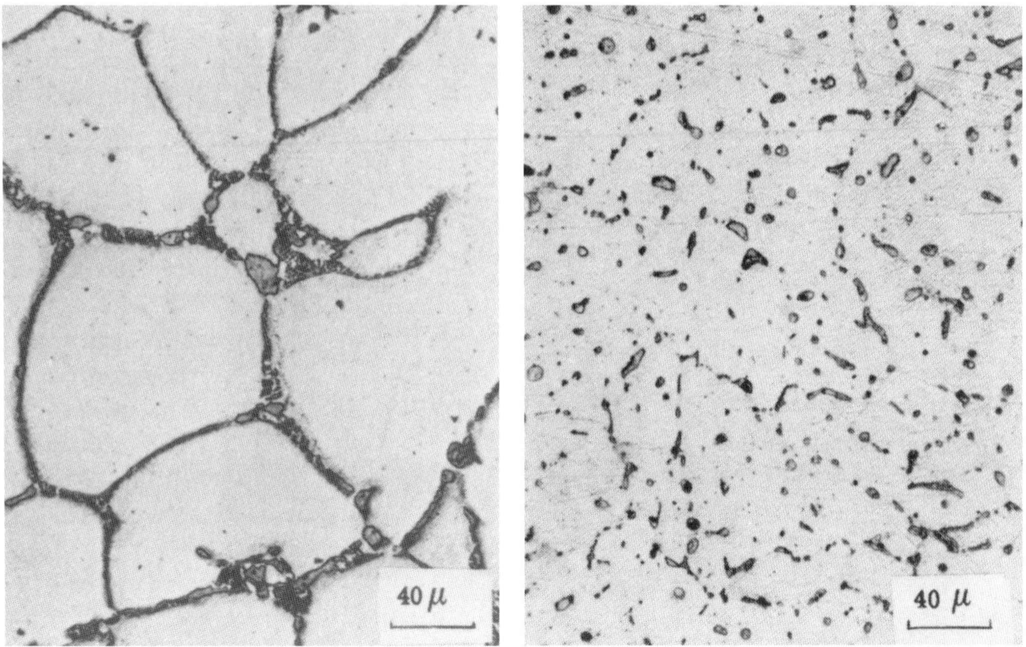

Figure 8.6. An aluminum bearing alloy cast in a sand (left) and a metal mold (right). The faster freezing rate caused by the metal mold results in a smaller DAS and a breakup of the brittle intermetallic phase [2]. (Reprinted with permission from *Grain Refinement in Castings and Welds*, edited by G. J. Abaschian and S.A. David, The Minerals, Metals & Materials Society, 420 Commonwealth Drive, Warrendale, PA 15086, 1983.)

The chemical grain refinement of aluminum and its alloys has been practiced for over 50 years, mostly by primary aluminum producers in ingot casting. The foundry industry has more or less borrowed the techniques developed by primary producers, but as we shall see, there is evidence to indicate that the best grain refiners for wrought alloys are not necessarily the best for casting alloys.

Titanium and Titanium-Boron Grain Refinement

Aluminum alloys are grain refined by the addition of typically from 0.02 to 0.15% Ti or Ti-B mixtures in the range of 0.01-0.03% Ti and 0.01% B. The titanium and boron are added via master alloys available in ingot or waffle form or as salt mixtures. Grain refinement by the addition of Ti alone is reasonably well understood; however, the role of boron which acts to make the titanium refinement more effective is still the subject of great controversy.

To understand how titanium works as a grain refiner, it is necessary to examine the aluminum-rich end of the Al-Ti phase diagram (Figure 8.7). Master alloys used for the treatment of liquid aluminum usually contain several percent (up to 10%) of titanium and hence will consist of the intermetallic phase $TiAl_3$ in a matrix of more or less pure aluminum. The microstructure of a commercial Al-Ti master alloy is shown in Figure 8.8. It is the $TiAl_3$ phase which is known to be the effective nucleant for aluminum, and several

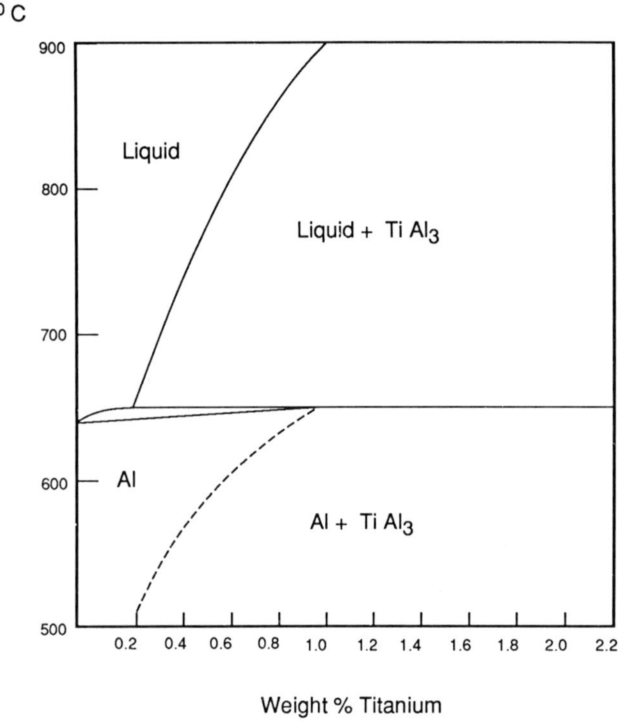

Figure 8.7. The aluminum-rich end of the Al-Ti phase diagram.

researchers have shown favored low energy relationships between TiAl$_3$ and aluminum nucleated on it. In addition to a favored crystallographic relationship, TiAl$_3$ particles will react with the liquid phase on cooling below 665C according to the peritectic reaction:

$$\text{Liquid} + \text{TiAl}_3 \longrightarrow \alpha(\text{solid})$$

The α(solid) phase which forms is almost pure aluminum. It envelops the TiAl$_3$ particle as illustrated in Figure 8.9 and acts as the site for further growth of the aluminum grain. As a result of this mechanism the number of nuclei and hence the final as-cast grain size is related to the microstructure of the original master alloy. An Al-Ti alloy which contains many small TiAl$_3$ phase particles will be a better grain refiner than one which contains fewer but larger TiAl$_3$ particles. This, in fact, is one of the difficulties associated with the use of master alloys; the effectiveness depends on the microstructure of the alloy and can vary from batch to batch and from supplier to supplier.

Since only 0.01-0.03% Ti is added, it is obvious from the Al-Ti phase diagram that the TiAl$_3$ particles from the grain refiner will dissolve, and the grain refining ability will decrease with time. This process, known as fading, is illustrated in Figure 8.10. Curve (a), which shows the

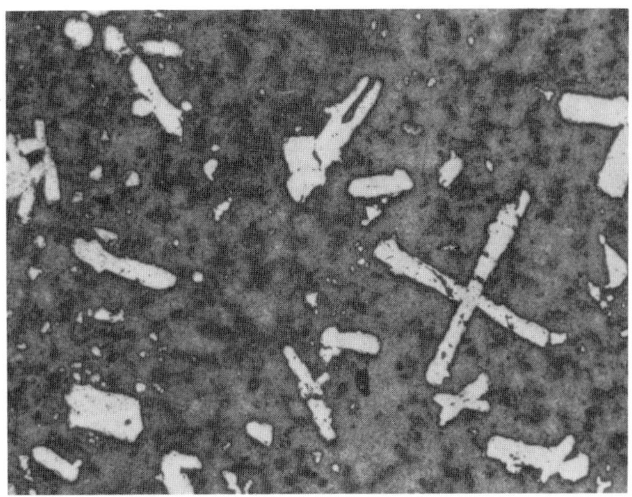

Figure 8.8. Microstructure of a commercial Al-Ti master alloy. (x125).

behavior of a binary Al-Ti master alloy addition, indicates that after about 40 minutes of holding time the grain size reverts to that of an untreated alloy. This figure also shows that, if Ti and B are used in concert, not only is the grain size finer, but the fading time is very significantly prolonged. This is particularly important in large electric furnaces which

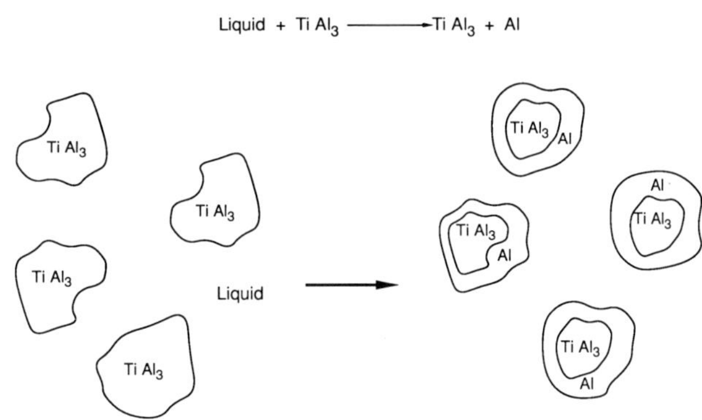

Figure 8.9. Nucleation by the peritectic reaction in the Al-Ti system.

have relatively low rates of throughput. Melt pouring rates of 800-1000 lb/hr are common with furnace capacities of up to 10,000 lb. Most commercial grain refiners contain a mixture of both titanium and boron.

While the effect of boron is well known, the reasons for this effect are the subject of great controversy in the literature, and the reader is referred to reference 3 for a brief but thoughtful discussion of the various hypotheses.

It is well known that the processing conditions of Al-Ti-B master alloys affect the final performance and that not all master alloys are created equal. Grain refining ability seems to depend on the morphology of the intermetallic phases which are present in the alloy. Figure 8.11 taken from the work of Guzowski *et al* [3], illustrates some typical morphologies found in good and bad Al-Ti-B grain refiners. Poor grain refiners contain blocky type TiAl₃ crystals, (Figure 8.11a), which are similar to those found in binary Al-Ti grain refiners. Here the boron has apparently exerted no influence on the intermetallic phase during manufacture of the grain refiner. Good grain refiners contain duplex intermetallic phases, (Figure 8.11b), which consist of TiAl₃ particles whose surface is covered with small boride particles, perhaps TiB_2 or $(Ti,Al)B_2$. The conditions favoring the formation of the duplex particles in the master alloy are quite spe-

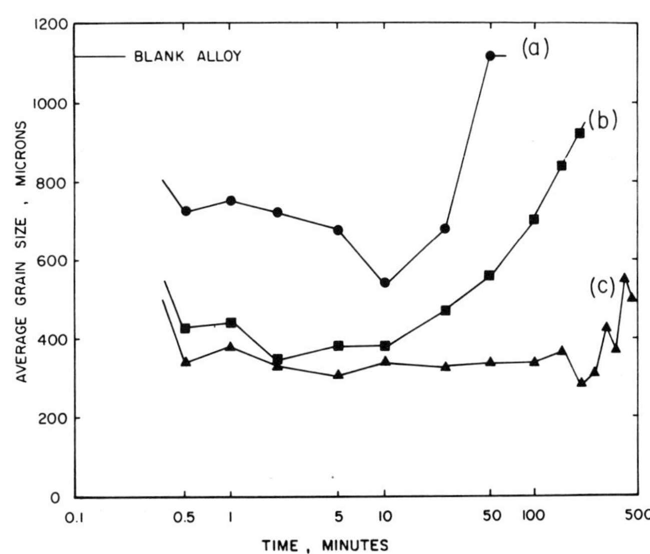

Figure 8.10. Grain refining ability in 99.7% Al with time of [3]:
(a) 0.01% Ti added as a 5.35% Ti master alloy;
(b) 0.01% Ti added as a 5.4% Ti-0.034% B master alloy;
(c) 0.01% Ti added as a 5.0% Ti-0.2% B master alloy.

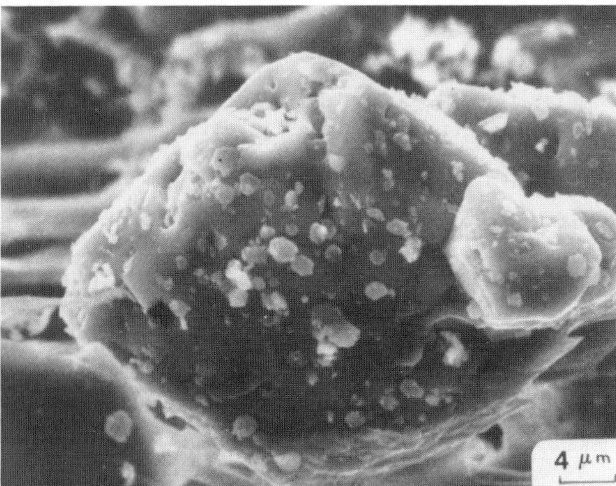

Figure 8.11. Morphologies of some intermetallic phases found in Al-Ti-B grain refiners [3]: blocky type TiAl₃ (top); duplex intermetallics consisting of TiAl₃ studded with borides (bottom).

cific and are related to the holding time in the liquid state after the master alloy has been manufactured, but before it is cast. Given the complexity of Al-Ti-B grain refiners, it is no wonder that the process is so poorly understood, and that significant variations in product perfomance exist.

It has already been mentioned that the foundry industry has more or less borrowed its grain refining technology from from the primary producers. Primary alloys contain little silicon (less than 0.05%), and in these alloys, titanium or titanium-boron mixtures provide powerful grain refinement. Boron, by itself, will not grain refine at all. In foundry alloys, the picture is quite different, since silicon, copper and zinc actually hinder titanium grain refinement. Surprisingly, boron acting on its own is a much better grain refiner. These effects are illustrated in Figure 8.12 where the benefits of an increase in the boron content of the grain refiner is obvious. The discovery of

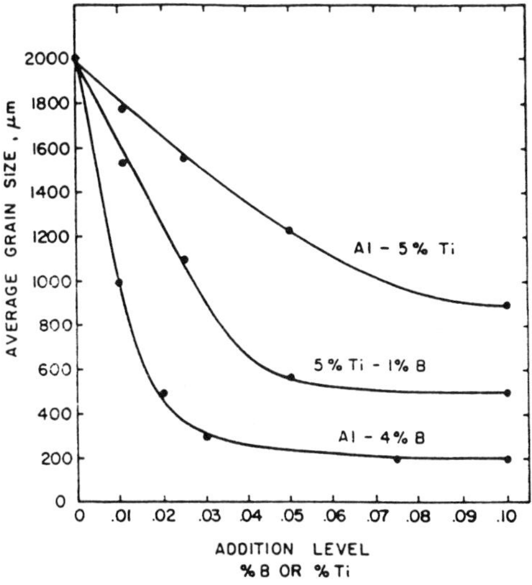

Figure 8.12. The beneficial effect of boron on the grain refinement of 356 alloy [4].

this important difference in grain refining behavior between wrought and cast alloys underscores once again the importance of melt chemistry and the need for research and development on casting alloys. Recently, an improved grain refiner for cast alloys has been put on the market. It contains 2.5% Ti and 2.5% B and results in improved refinement as shown in Figure 8.13 for 356 and 319 alloys. Although high-boron grain refiners appear to offer many advantages to the aluminum foundry industry, one of their negative aspects appears to be a tendency to contribute to hard spots in castings. This effect is not well documented at present.

8.5   The Effect of Grain Refinement on Properties

Grain refinement exerts a positive influence on several properties of cast alloys; notably, hot tearing tendency and porosity

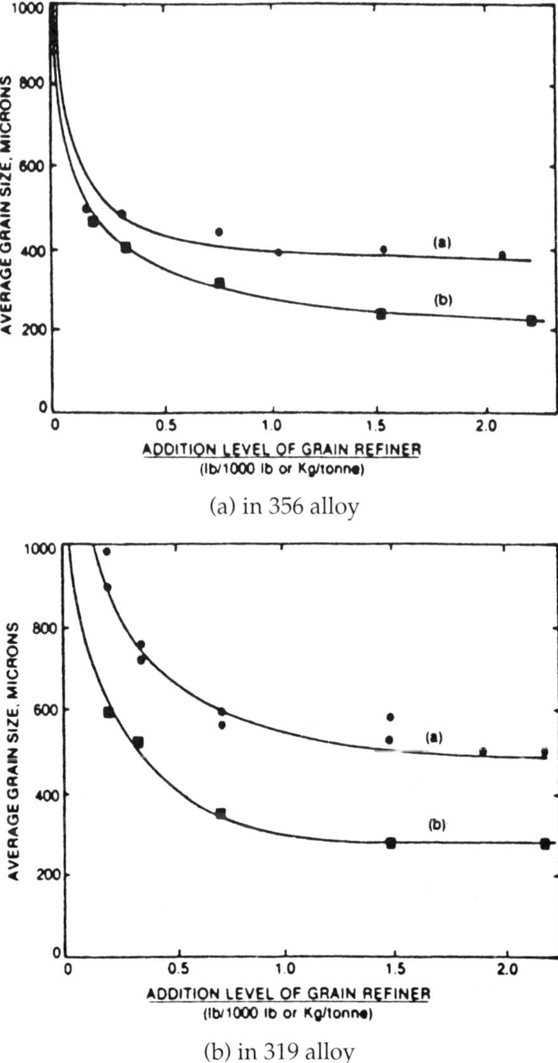

(a) in 356 alloy

(b) in 319 alloy

Figure 8.13. Comparison of grain refinement with Al-5% Ti-1% B master alloy (curve a) and Al-2.5% Ti-2.5% B (curve b) alloy. (The casting temperature is 725C [1350F] and the contact time is 5 minutes.) (Courtesy KB Alloys, Inc.)

and shrinkage distribution. Only very limited data is available for alloys of the Al-Si type. The more popular research alloy seems to have been Al-4% Cu, and data obtained on this system should be applicable to 200 series alloys. The Al-4%Cu alloys are largely single phase materials containing a minimum of eutectic. They are therefore quite different from Al-Si casting alloys which are used because of their high eutectic content. Much of what is presented in the remainder of this chapter is gleaned from literature on Al-Cu alloys, and the reader should bear in mind that it is not necessarily applicable to Al-Si alloys.

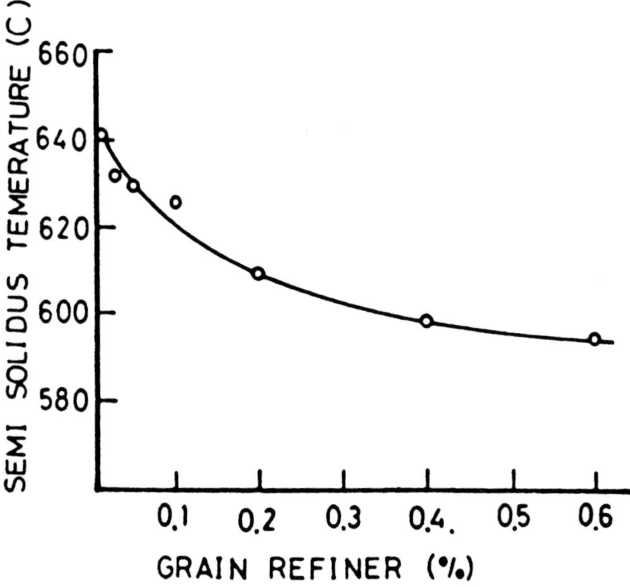

Figure 8.14. *The lowering of the semi-solid temperature of Al-4% Cu alloy with addition of grain refiner* [5].

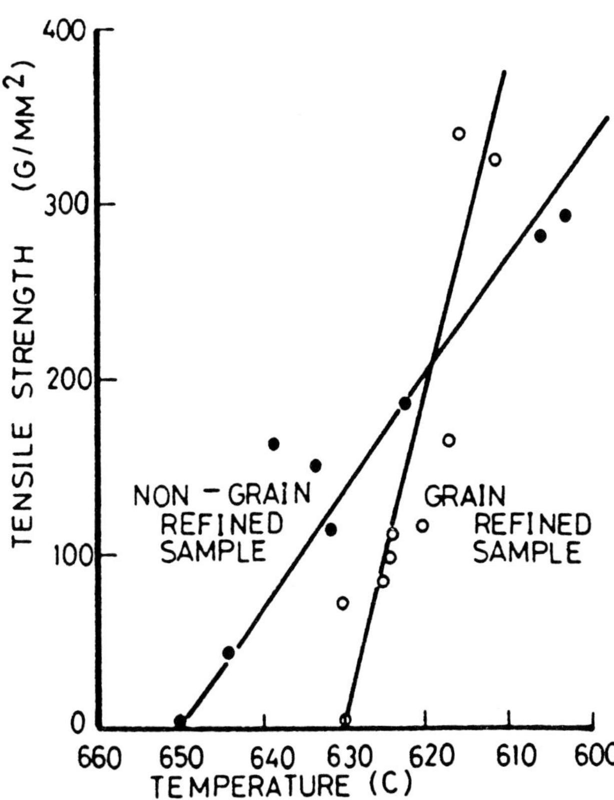

Figure 8.15. *Hot tensile strength of an Al-0.5%Mg-0.4%Si alloy, grain refined and not grain refined*[5].

Hot Tearing Tendency

There is a considerable evidence to indicate that grain refined castings are less prone to hot tearing. Grain refinement results in a more general nucleation, but a less contiguous solid as films of liquid remain to surround each individual grain. This has the effect of lowering the temperature at which contraction of the whole solid begins, and consequently the temperature range over which hot tearing can occur is narrowed.

Work in Japan in the 1970's[5], defined the temperature at which general contraction of the solid begins as the semi-solidus temperature. Figure 8.14 shows the effect of grain refinement on this temperature for an Al-4% Cu alloy. The lowering of the semisolidus temperature reduces the hot tearing tendency in two ways, which can be seen from the results of hot tensile tests on two partially solidified alloys (Figure 8.15). Firstly, the temperature at which the semi-solid is able to sustain a measurable load is reduced (in this example from 650C to 630C), and hence the time interval over which cracking can occur is shortened. Perhaps even more important is the fact that grain refined castings gain mechanical strength at a faster rate than non-grain refined ones. This can be seen by comparing the slopes of the lines in Figure 8.15; the increase in tensile strength is 3.5 gm/mm^2°C^{-1} if no grain refiner is used, but rises to 9.1 gm/mm^2 °C^{-1} with a grain refiner. Grain refined castings, therefore, have a shorter temperature range over which they can hot tear, and they develop strength faster in this range.

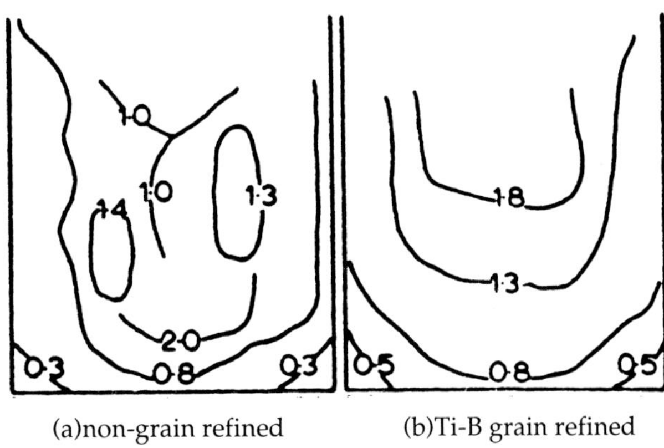

Figure 8.16. Isopore distribution in an Al-4.5% Cu ingot casting. Isopores are lines of equal volume percent porosity[6].

Porosity Distribution

The effect of grain refinement is to smooth out porosity into a more even distribution of small pores. Non grain-refined alloys tend to solidify with pipes or regions of highly concentrated internal porosity. The difference in porosity distribution between a grain-refined and non grain-refined Al-4.5% Cu alloy is apparent in Figure 8.16. The isopore distribution when the material is not grain refined indicates highly concentrated porosity; with grain refinement the structure is much more uniform and the isopores reflect the settling out of the fine equiaxed grains during solidification. This more even distribution of porosity can result in grain refined castings being more pressure tight.

Intermetallic Distribution

In alloys which contain a high volume fraction of eutectic, such as the Al-Si alloys, it is not expected that grain refinement would influence the distribution of intermetallics. These are found in the interdendritic spaces and hence the DAS rather than the grain size is of more importance.

On the other hand, the Al-Cu type alloys do contain some brittle intermetallic which can be distributed in a more favorable manner by grain refinement. For these alloys, grain refinement provides definite advantages.

Surface Appearance

Castings which are to be anodized should be grain refined, since large grains are evident on the surface, appearing as areas of different reflectivity.

Fluidity

As mentioned in Chapter 5, the measurement of fluidity is difficult, and most testing techniques are capable of revealing only large differences. Data on grain refined alloys is sparse, but that which is available shows a 10-15% decrease in fluidity of an Al-4.5% Cu alloy on grain refinement. Although differences of this magnitude are difficult to determine with certainty, it is reasonable to expect that grain refinement would lower fluidity somewhat. The early nucleation caused by grain refinement results in slurry flow (solid plus liquid) from virtually the moment of pouring, and since slurries flow with more difficulty than simple liquids, fluidity should be reduced.

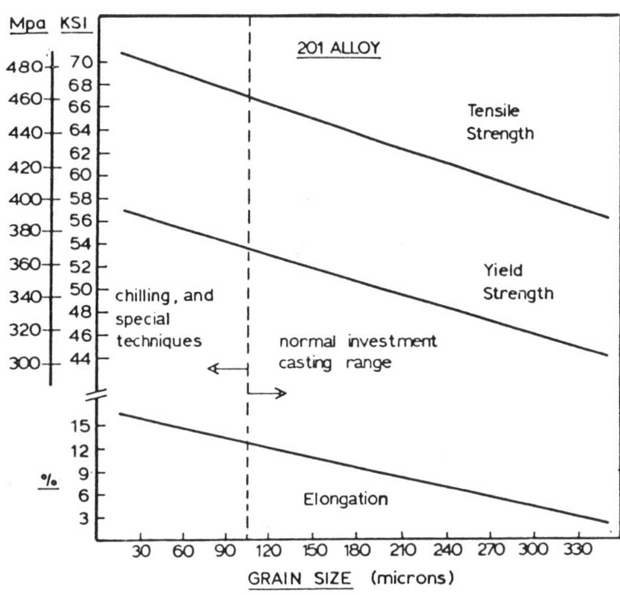

Figure 8.17. The variation of tensile properties of a 201 alloy with grain size[8].
(The original version of this material was first published by the Advisory Group for Aerospace Research and Development, North Atlantic Treaty Organisation [AGARD/NATO] in Conference Proceedings CP 325 "Advanced Casting Technology," August 1982.)

Mechanical Properties

Properties such as tensile strength and elongation, are usually improved by grain refinement. In Al-Si alloys, this is mainly due to improvements in porosity distribution rather than to any decrease in grain size. As pointed out earlier in this chapter, the properties of these alloys are controlled by the eutectic and any influence of grain size, *per se*, is secondary. In Al-Cu alloys, however, grain size is particularly important in determining the distribution of intermetallics and in reducing hot tearing tendency. It is these factors, coupled with improved porosity distribution, which lead to enhanced mechanical properties, such as those shown in Figure 8.17 for a 201 alloy.

8.6 The Interrelationship between Chemical Grain Refinement, Cooling Rate and Modification

In this chapter we have discussed how chemical grain refinement and cooling rate (chilling) can be used to change the metallurgical structure of a casting. Chapters 3-6 have explored the process of modification which can influence the structure of the eutectic silicon in casting alloys which contain silicon. The final as-cast structure is determined by a complex interplay between the cooling rate of the casting and the effect of additives such as grain refiners and/or modifying elements. Often, the same structure can be achieved by different processing routes. It is appropriate at this time to summarize these routes as we have done in Figure 8.18, which hopefully will clarify the effects of certain casting variables on grain size, DAS and eutectic silicon structure.

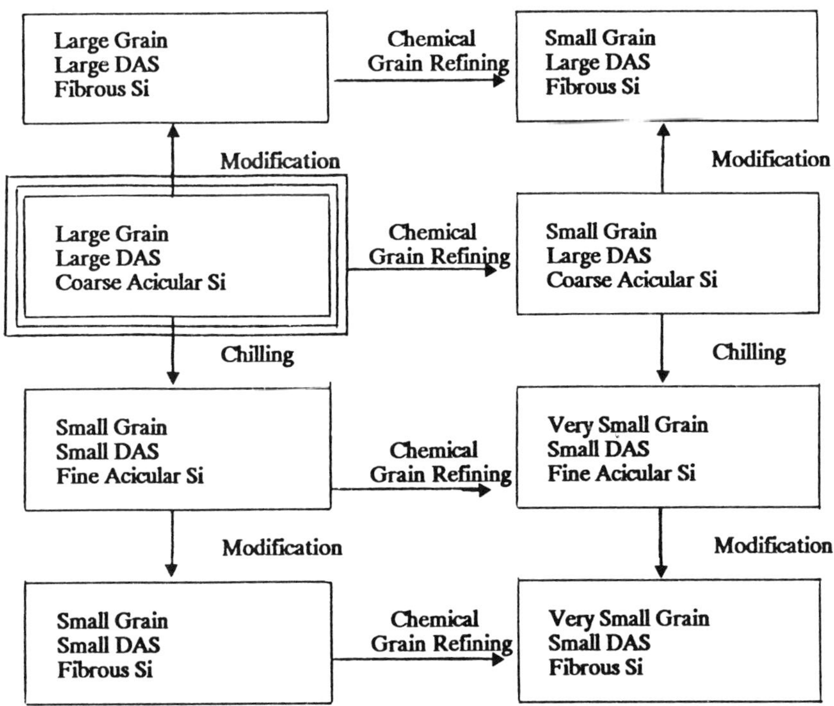

Figure 8.18. Relationship between chemical grain refinement, chilling and modification on structure.

References

1. Modin, H. and S. Modin. *Metallurgical Microscopy,* John Wiley and Sons (1973).
2. Mondolfo, L.F. "Grain Refinement in the Casting of Non-Ferrous Alloys," *Grain Refinement in Castings and Welds,* Metallurgical Society of AIME (1983) pp. 3-50.
3. Guzowski, M.M., G.K. Sigworth, D.A. Senter. "The Role of Boron in the Grain Refinement of Aluminum with Titanium," *Metal. Trans.,* 18A (1987) pp. 603-19.
4. Sigworth, G.K. and M.M. Guzowski. "Grain Refining of Hypoeutectic Al-Si Alloys," *AFS Transactions,* 93 (1985) pp. 907-12.
5. Kubota, M. and S. Kitaoka. "Solidification Behavior and Hot Tearing Tendency of Aluminum Casting Alloys," *AFS Transactions,* 81 (1973) pp. 424-27.
6. Entwistle, R.A., J.E. Gruzleski and P.M. Thomas. "Development of Porosity in Aluminum-base Alloys," *Solidification and Casting of Metals,* The Metals Society, publication 192 (1979) pp. 345-49.
7. Radhakrishna, K., S. Seshan and M.R. Seshadri. "Dendrite Arm Spacing in Aluminum Alloy Casting," *AFS Transactions,* 88 (1980) pp. 695-702.
8. Kennerknecht, S. "Metallurgical Aspects of Quality Control in the Production of Premium Quality Aluminum Investment Castings for the Aerospace Industry ," *Advanced Casting Technology,* AGARD Conference Proceedings No. 325.

Chapter 9

Gassing of Aluminum Foundry Melts

9.1 Hydrogen Solubility

Solubility in Pure Aluminum

Hydrogen is the only gas with any significant solubility in molten aluminum. Unfortunately, it is easily dissolved and is abundant, with the result that virtually all molten aluminum contains some level of dissolved hydrogen. This gas plays a major role in the development of unsoundness due to porosity in castings, and considerable effort is expended in controlling hydrogen dissolution and in removing it from the liquid alloy.

The solubility of hydrogen in pure aluminum at one atmosphere pressure is shown in Figure 9.1. In the liquid phase, the solubility, as determined by Opie and Grant[1], is given by the equation:

$$\log_{10} S = \frac{-2550}{T} + 2.62$$

Here the solubility, S, is in milliliters of hydrogen at standard temperature and pressure per 100 grams of aluminum, and the temperature, T, is expressed in degrees Kelvin.

The nature of the hydrogen problem in aluminum is very evident from Figure 9.1 which shows that the hydrogen solubility has three distinct characteristics:

1) a strong temperature dependence in the liquid state, implying that increases in superheat result in increases in hydrogen. In fact, the hydrogen solubility doubles for each 110°C (200°F) increase in superheat;
2) a low solubility in solid aluminum;
3) a large change in solubility at the melting point.

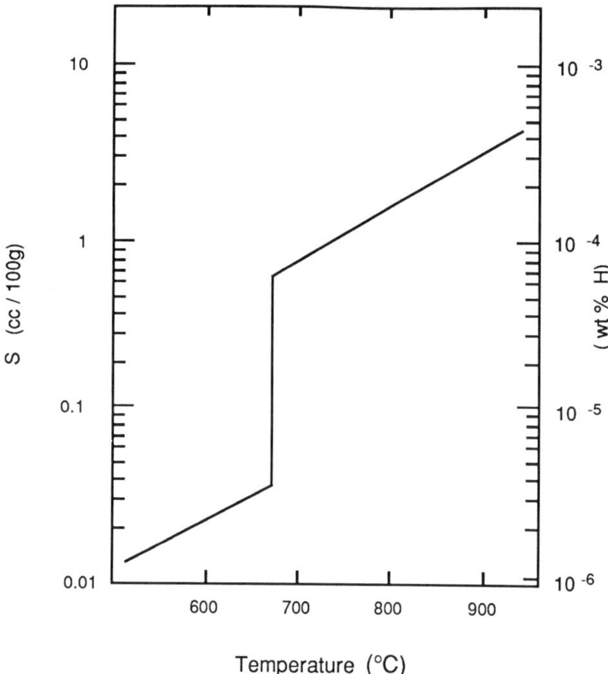

Figure 9.1. The solubility of hydrogen at one atmosphere pressure in pure aluminum.

These last two factors mean that virtually all of the relatively large amount of hydrogen which is present in the liquid is unable to dissolve in the solid on freezing. Hydrogen, therefore, accumulates in the liquid at the solidifying interface and may eventually assist in porosity formation as described in section 9.3. A simple calculation based on Figure 9.1 serves to illustrate the problem. At the melting point, the solubility in the liquid is approximately 0.7 ml/100 gm Al. Thus, only between 4% and 5% of the hydrogen dissolved in the liquid can remain in the solid after freezing.

Solubility in Alloys

The addition of alloying elements to aluminum changes the hydrogen solubility, although the solubility curves retain the same basic shape as that shown in Figure 9.1. Some elements such as silicon, zinc, copper and manganese decrease the solubility, while others such as magnesium, titanium, nickel and lithium increase it. The most important alloying elements in foundry alloys are, of course, copper and silicon, and their effects on hydrogen solubility have been fairly well documented. The solubility relationships of some liquid Al-Cu and Al-Si alloys are given in Table 9.1.

No precise measurements of hydrogen solubility have been carried out on the complex chemistries of commercial foundry alloys. However, it is possible to estimate the solubility in these alloys from data such as that given in Table 9.1.

In Figure 9.2, we present estimated solubilities for a 356 and 319 alloy. In drawing these curves, it is assumed that the alloying elements do not change the solubility relationships in the solid, nor do they alter the temperature dependence in the liquid. It is seen that liquid alloys will dissolve somewhat less hydrogen than pure aluminum, but due to the depression of the freezing point by the alloying elements, approximately the same relative decrease in solubility occurs on freezing. Therefore, while pure aluminum can retain 4%-5% of its dissolved hydrogen on solidification, this value increases to only 6%-7% for the alloys. Thus, hydrogen rejection into the liquid is almost as severe a problem for foundry alloys as it is for pure aluminum.

Aluminum-lithium alloys, which are only now beginning to be used in foundry applications, provide an interesting example of alloys which have a high hydrogen solubility, and yet

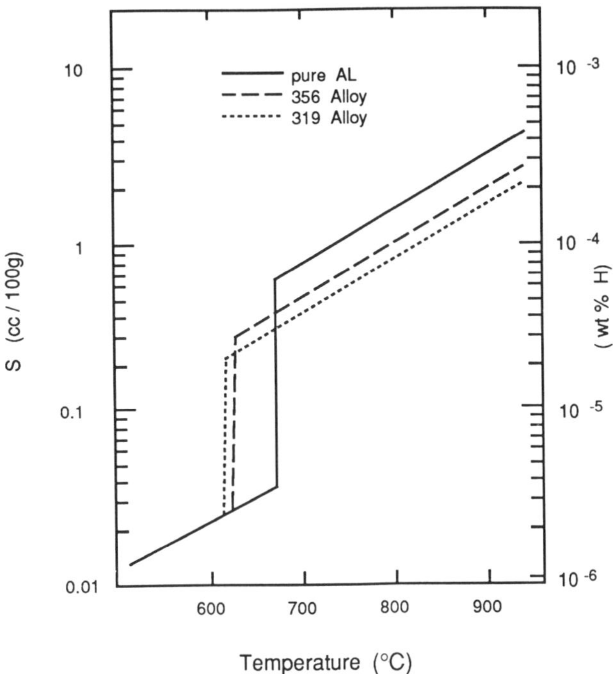

Figure 9.2. The estimated solubility of hydrogen in 356 and 319 alloys compared to the solubility in pure aluminum.

are considerably more tolerant of hydrogen than the classical foundry alloys. Lithium increases the hydrogen solubility in both the liquid and the solid alloys, but the increase in the solid is far greater. Therefore, on freezing, an Al-Li alloy can retain between 30% and 50% of its dissolved hydrogen in the solid, compared to only 4% to 7% for the classical Al-Si and Al-Cu alloys. Al-Li alloys, therefore, exhibit a lowered tendency to porosity.

The equilibrium hydrogen solubilities in foundry alloys range from about 0.6 to 1 ml/100 gm of Al depending on the alloy and the temperature. It is, however, extremely rare to find melts, which have been reasonably well handled containing these levels of hydrogen. For example, remelted primary or secondary ingot will usually contain between 0.2 and 0.3 ml/100 gm Al. There is a definite barrier to excessive hydrogen dissolution in aluminum alloys, and it is probably the oxide film which forms so readily on the melt surface. This film serves to isolate the melt from the atmosphere above it, and as we shall see in the next section, it is this atmosphere which often is the source of hydrogen. Any melt handling procedure designed to minimize hydrogen pickup should aim to maintain this oxide film as intact as possible. Unfortunately, even the normal levels of hydrogen encountered are too high in most cases, and degassing procedures must be employed. (Degassing is the subject of Chapter 10.)

Table 9.1. Hydrogen Solubility in Some Binary Alloys[1]

Alloy	Solubility*
Al-2%Cu	$\log_{10} S = \dfrac{-2950}{T} + 2.90$
Al-4%Cu	$\log_{10} S = \dfrac{-3050}{T} + 2.94$
Al-8%Cu	$\log_{10} S = \dfrac{-3150}{T} + 2.94$
Al-16%Cu	$\log_{10} S = \dfrac{-3150}{T} + 2.83$
Al-32%Cu	$\log_{10} S = \dfrac{-2950}{T} + 2.57$
Al-2%Si	$\log_{10} S = \dfrac{-2800}{T} + 2.79$
Al-4%Si	$\log_{10} S = \dfrac{-2950}{T} + 2.91$
Al-8%Si	$\log_{10} S = \dfrac{-3050}{T} + 2.95$
Al-16%Si	$\log_{10} S = \dfrac{-3150}{T} + 3.00$

* in ml (STP) per 100 gm Al temperature, T (in °K)

9.2　Sources of Hydrogen

Most hydrogen which finds its way into molten aluminum comes from the dissociation of water vapor at the surface of the liquid aluminum according to the reaction:

$$2Al_{(l)} + 3H_2O_{(g)} \longrightarrow Al_2O_{3(s)} + \underline{6H}$$

This reaction is highly favored at foundry operating temperatures, and it is safe to assume that all water vapor coming into contact with molten aluminum will dissociate in this way. The contact must be with metallic aluminum, not the oxide. The reaction cannot proceed if the molten bath is protected by the oxide (dross) film on its surface. It is, however, important to note that above about 930C (1700F), the oxide film loses its protective nature through a change in its crystal structure, and water vapor can readily penetrate the film. While this temperature is well above normal foundry melt temperatures, care should be taken to see that the local temperature of the melt surface does not reach this value (for example, during melting). If it does, rapid increases in dissolved hydrogen can be expected. Several common sources of water vapor will now be discussed

The Atmosphere

The atmosphere is one of the most important sources of water vapor. For example, air at 26C (79F) and 65% relative humidity contains almost 16 gm/m^3 of water. These atmospheric conditions are found in North America and Europe throughout much of the year. Experienced founders know of the difficulties with gassing in the hot, humid conditions of summer, and usually find many fewer such problems during a cold, dry winter.

Fluxes

Foundry fluxes are, in general, hygroscopic salts which will naturally pick up water from the atmosphere. If added to a melt, a pronounced increase in hydrogen level can be experienced. Fluxes should be carefully packaged and stored, and can be heated to above 100C (212F) to drive off absorbed water. They will, however, quickly reabsorb water, and so should be used immediately after heating. Some fluxes also contain water of crystallization which is bonded into their crystal structure. This is more difficult to remove and may involve heating to temperatures of 200C (390F) or even higher.

Crucibles

New crucibles always contain some moisture in their pores, and so a melt produced in a new crucible will be gassier than one made in an old crucible. The water vapor can, at least in part, be driven off by preheating the crucible. Even a used crucible will pick up some moisture if it is allowed to sit for several days without being heated, and a more gassy melt may be produced when the crucible is eventually put into service.

Combustion Gases

The products of combustion of most fuels contain 10% to 20% water vapor. Natural gas will produce up to $2m^3$ of water vapor for each cubic meter of gas burned. In order to minimize hydrogen pickup from this very important source, it is recommended that a slightly oxidizing flame be used. The advantages of electric melting to minimize hydrogen are obvious.

Refractories

Like crucibles, new refractories contain some moisture which will add hydrogen to the melt until it is completely driven off. This may require several melts and several days of operation.

Foundry Tools

Tools, such as plungers and ladles, which have been insufficiently preheated can be important sources of hydrogen.

Charge Materials

Charge materials may add hydrogen through moisture contained on the surface of the charge or in pores and crevices. They may also provide hydrogen through sources other than water vapor. For example, scrap may be contaminated with grease or cutting oil. Such contaminants can be partly burned off, but complete removal is very difficult. Partly corroded aluminum can add hydrogen by decomposition of the corrosion product, aluminum hydroxide, through the reaction:

$$Al(OH)_{3(s)} + Al_{(l)} \longrightarrow Al_2O_{3(s)} + \underline{3H}$$

Avoiding Hydrogen Pickup

The avoidance of hydrogen dissolution requires constant vigilance during melt handling. Sources of hydrogen should be eliminated as much as possible through care in drying of fluxes, tools, crucibles and charge materials. Melt dross is an effective barrier to hydrogen pickup and effort should be expended to see that breaking of the dross occurs as infrequently as possible. Thus, stirring actions or melt additions should be made as gently as possible. The oxide layer is usually broken during any pouring operation, and if the melt has been previously degassed, a regassing will occur. Pouring should always be conducted in such a way as to minimize the height of fall and turbulence. The use of pumps to pump liquid metal through conduits so that all contact with air is minimized can serve to combat hydrogen pickup during molten metal transfer. Finally, overheating the metal should always be avoided, as the hydrogen solubility increases rapidly with temperature (Figures 9.1 and 9.2).

9.3   Gas and Shrinkage Porosity

Porosity is the most common defect found in cast metals, and is the major cause of rejection of castings. In this section, we describe its relationship to the hydrogen concentration and its effect on the properties of cast alloys.

Metal casters often attempt to classify porosity as due either to shrinkage or gas. While this is sometimes valid it is most often incorrect, since virtually all porosity found in reasonably good castings is due to a combination of gas and shrinkage. There are, however, a few cases where pure shrinkage or pure gas porosity can be identified.

Pure Shrinkage Porosity

Macroscopic voids found in a casting are almost always due to shrinkage caused by poor feeding. Large pores encompassing many dendrites or even grains such as the one shown in Figure 9.3a, can be ascribed to shrinkage on the basis of their size. The largest shrinkage pore of all in any casting is usually the pipe which ideally occurs in the riser. Shrinkage porosity can also likely occur on the surface of a casting, and it is probable that much of the pinhole surface porosity which is often ascribed to gas is, in fact, due to shrinkage. During the solidification of long freezing range alloys, channels of interdendritic liquid may extend to the casting surface. As these channels freeze, they contract and the liquid shrinks away from the surface to leave a pinhole pore in the manner sketched schematically in Figure 9.4.

Pure Gas Porosity

Pores which are due solely to gas are caused by the entrapment of a gas bubble in the solidifying solid. The nucleation of hydrogen bubbles in liquid aluminum is an extremely difficult process, necessitating very high levels of dissolved hydrogen. Any casting which has been prepared using even average methods of melt handling and degassing is unlikely to contain free hydrogen bubbles, and so pure hydrogen porosity occurs much less frequently than is normally supposed. An example of a free gas pore found in all Al-8%Si alloy which contained 0.8 ml H_2/100 gm Al is shown in Figure 9.3b.

Pores Due to Gas and Shrinkage

The vast majority of microporosity encountered in castings is due to a

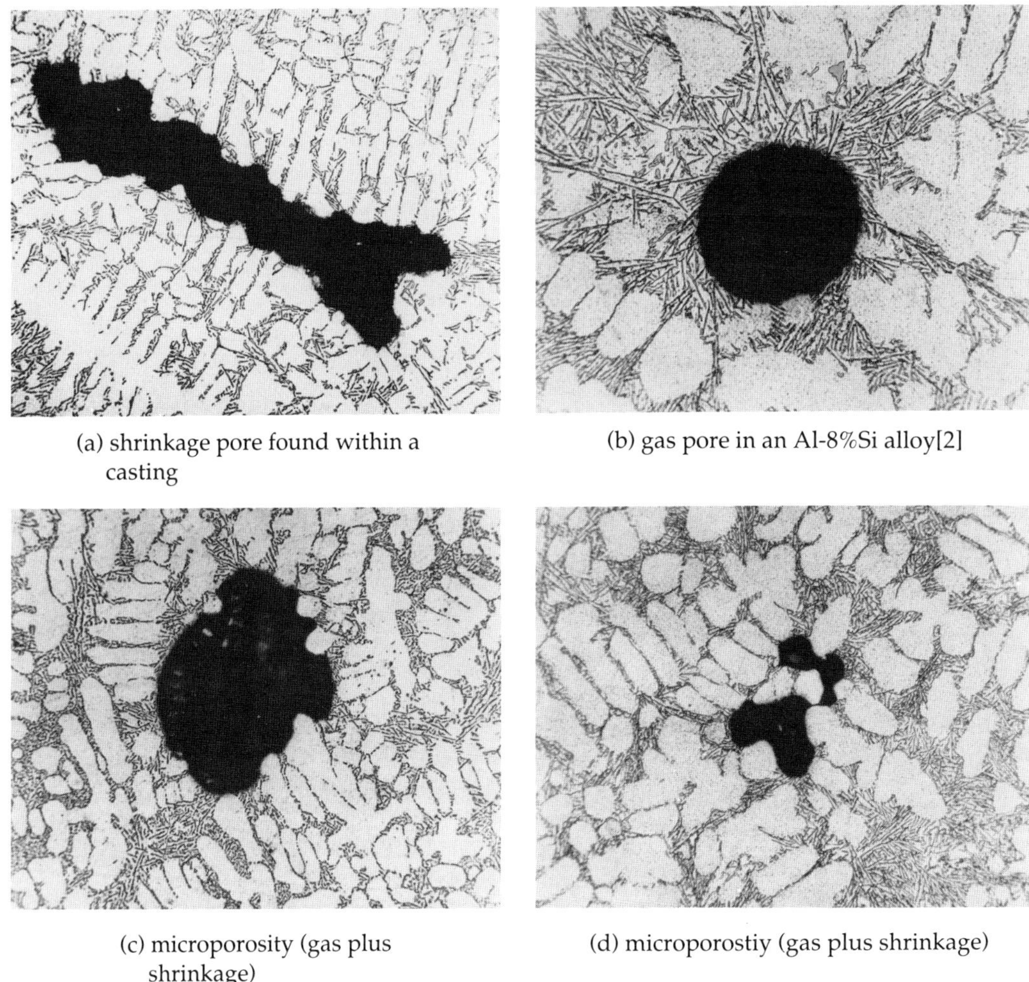

(a) shrinkage pore found within a casting

(b) gas pore in an Al-8%Si alloy[2]

(c) microporosity (gas plus shrinkage)

(d) microporostiy (gas plus shrinkage)

Figure 9.3. Typical porosity in cast aluminum alloys.

combination of gas and shrinkage. These pores typically occur in the interdendritic regions which are the last parts of the structure to freeze. They usually impinge on only one or two dendrite arms and can vary in size from a few up to several hundred microns. Some examples are shown in Figure 9.3c and d. Such porosity occurs in the very last liquid regions to freeze, and both shrinkage and dissolved hydrogen are necessary for its formation because of the extreme difficulty of nucleating a pore in a solidifying metal system.

The reasons why most porosity forms only very late in the solidification process and why it must involve both gas and shrinkage can best be seen by considering a simple equation to describe the conditions under which a pore is stable. In order for a pore to exist, its internal pressure, (P_i), must be sufficient to overcome all of the external forces which can act to make it collapse. These are:

- the atmospheric pressure acting on the melt surface (P_{atm});
- the metallostatic head pressure (P_H);
- the forces due to the surface tension of the solid or liquid phases which contact the pore. These equal $\gamma\,(1/r_1 + 1/r_2)$ where γ is the surface tension and r_1 and r_2 are the principal radii of curvature of the pore.

Thus:

$$P_i \geq P_{atm} + P_H + \gamma\,(1/r_1 + 1/r_2). \tag{1}$$

While P_{atm} and P_H are not excessively large in most casting operations, the surface tensions of metals are very high, and of course when a pore begins to form, the radii of curvature, r_1 and r_2, are of necessity extremely small. Thus, the term $\gamma\,(1/r_1 + 1/r_2)$ is very large, and high values of the internal pressure are necessary to form a pore. This internal pressure is made up of two components:

1) P_G, the internal gas (hydrogen) pressure. This is significant only in the very last stages of freezing when large amounts of hydrogen have accumulated in the pockets of interdendritic liquid due to the decrease in solubility in the solid as discussed in 9.1.

2) P_S, the shrinkage pressure, best viewed as a tensile stress in the pockets of interdendritic liquid due to the solidification shrinkage of this liquid, and the inability of this shrinkage to be fed through the almost completely frozen dendrite mesh.

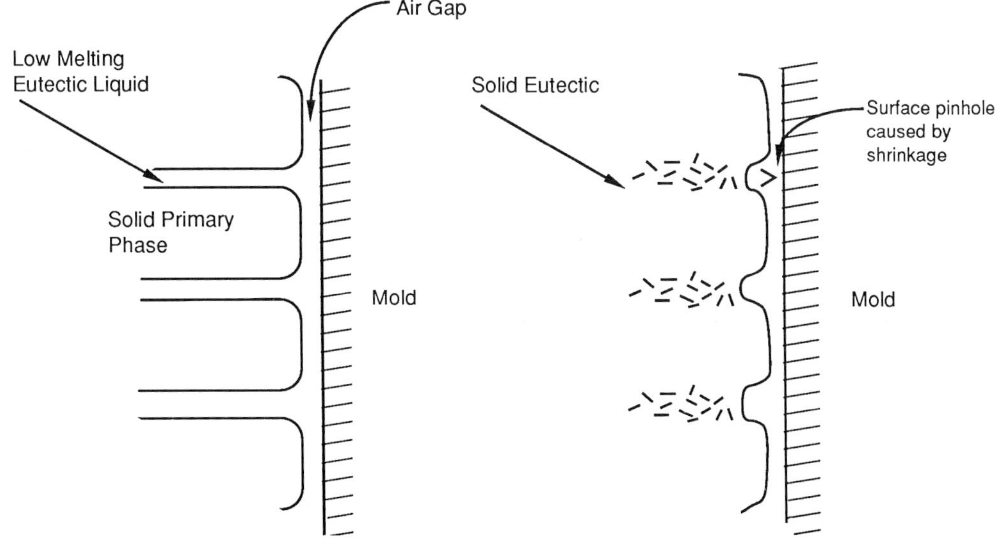

Figure 9.4. A possible shrinkage mechanism for the formation of pinhole surface porosity.

In order for a pore to be stable and to grow, the following condition must prevail:

$$P_G + P_S \geq P_{atm} + P_H + \gamma(1/r_1 + 1/r_2), \tag{2}$$

and as this can occur only in the final stages of the freezing process, we find most porosity to be of the fine interdendritic type shown in Figure 9.3c and d. Even this relationship is a bit of an over-simplication, since usually the combined sum of the gas pressure (P_G) and the shrinkage pressure (P_S) is insufficient to cause pore nucleation, and some foreign particle is necessary to assist in heterogeneous nucleation of the pore. Therefore, inclusions such as those described in Chapter 11 can assist in porosity formation. Dirtier melts have a greater tendency to result in more porous castings than do cleaner melts processed by filtration.

Casting Variables and Porosity Formation

In Figures 9.5-9.7 we present some experimental results obtained by Fang and Granger[3] on microporosity formation in A356 alloys. These results, which are among the most comprehensive available, illustrate graphically the effects of casting variables on porosity. We will consider each melt processing or casting variables separately.

Cooling Rate

An increase in cooling rate has the effect of decreasing both the total amount of porosity (Fig. 9.5) and the average pore size (Fig. 9.7). With increased solidification rates, less time is available for hydrogen to diffuse into the interdendritic spaces and so P_G is reduced. In addition, once a pore is nucleated, the time factor will inhibit hydrogen from diffusing into it and smaller sized pores will result. Fast cooling rates also have an effect on the shrinkage

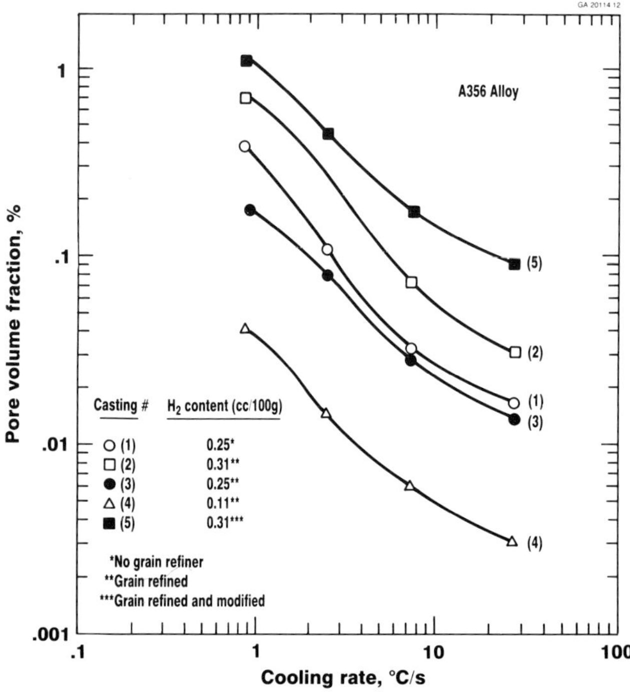

Figure 9.5. Pore volume fraction as a function of cooling rate for A356 alloy[3].

pressure, P_s. The higher temperature gradients, present with more rapid solidification, tend to limit the length of the mushy zone and to make feeding easier. The value of P_s is reduced and porosity formation made more difficult.

Melt Hydrogen Concentration

Figure 9.6 illustrates that, for a given cooling rate, an increase in hydrogen concentration increases the amount of porosity. This effect becomes greater as the cooling rate decreases due to the effects described above. It is significant that at very rapid cooling rates (28°C/sec, in Fig. 9.6), there is little effect of hydrogen concentration on pore volume. A critical hydrogen level below which no porosity will form has often been mentioned. A more realistic approach is to define such a threshold value for a given fraction of porosity. This is done in Figure 9.8 for cooling rates up to 30°C/sec.

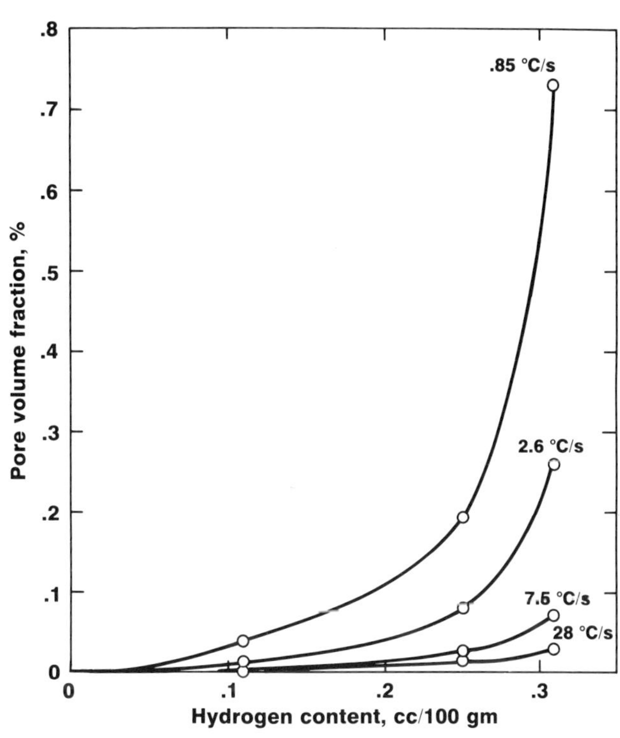

Figure 9.6. Pore volume fraction in grain-refined A356 alloy as a function of melt hydrogen content at different cooling rates[3].

If, for example, a porosity level of 0.005% is acceptable in a given application, then a hydrogen level in excess of 0.1 ml/100 gm Al is tolerable, provided all parts of the casting solidify at rates faster than 30°C/sec. Threshold values fall sharply at cooling rates below about 5°C/sec, illustrating the difficulty of producing pore-free sand castings. On the other hand, permanent mold castings require less stringent hydrogen control.

Grain Refinement

While grain refinement may result in a slightly lower total amount of porosity, its main benefit is to produce a more uniform distribution of finer pores. This may be due to a somewhat improved feeding which accompanies grain refinement, but is more likely related to enhanced nucleation of the pores caused by the intermetallic phases introduced via the grain refiner. For a detailed discussion of grain refiner structures, the reader is referred to Chapter 8.

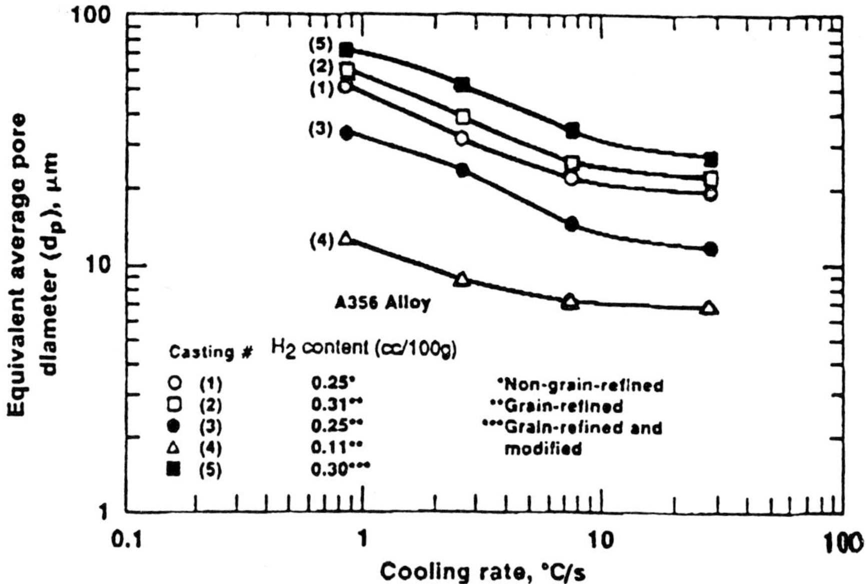

Figure 9.7. Equivalent average pore diameter as a function of cooling rate in A356 alloy [3].

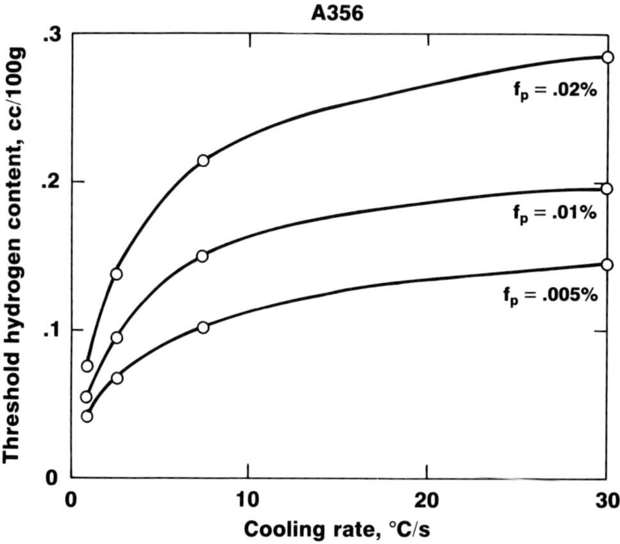

Figure 9.8. Measured melt hydrogen content to yield pore fractions, f_p, of 0.005%, 0.01% or 0.02% in A356 alloy cooled at rates up to 30C/sec [3].

Modifier Additions

As discussed in Chapter 4 there is a redistribution of porosity from primary piping into microporosity. Modified castings tend to contain a higher level of dispersed microporosity than do their non-modified counterparts. The curves of Figure 9.5 (curves 2 and 5) indicate that this is true for a range of cooling rates at a constant dissolved hydrogen level. The hydrogen level used here is quite high (0.31 ml H_2/100 gm Al), and the effect at much lower levels, eg., less than 0.1 ml H_2/100 gm Al, has not been studied. Also evident in Figure 9.5 is the important role of hydrogen concentration in determining the amount of porosity. Degassing from 0.31 to 0.11 ml H_2/100 gm Al has a much greater effect on the amount of porosity than does the addition or removal of a modifier. In Chapter 4, it was postulated that the effects of modifiers on porosity distribution are due to a reduction in the liquid-gas surface energy. Such a reduction would serve to reduce the right hand side of equation 2, and to make pore growth more favorable for the same hydrogen concentration and shrinkage pressure.

The Effects of Porosity on Properties

In most cases, microporosity is an undesirable feature of the cast structure, because it degrades certain important properties. Sometimes, however, it can be beneficial, for example in reducing hot tearing or in decreasing the amount of risering required.

Tensile Properties

Much data exists to demonstrate the negative effect of hydrogen and the resultant porosity on tensile properties. In Figure 9.9, we show the effect of hydrogen on the tensile properties of a sand cast 356 alloy. Both the tensile strength and the elongation are seriously decreased as the hydrogen concentration increases. The yield strength is only very slightly affected as would be expected, since this property is more related to the metallurgical state of the aluminum matrix than to defects in the structure. The elongation is particularly affected at high freezing rates, where all of the benefits usually resulting from finer microstructures are lost due to porosity.

Surface Appearance

Porosity becomes particularly evident on a machined surface, and can be the cause of rejection of castings in applications where surface appearance is critical. Polished and anodized surfaces are particularly susceptible to defects caused by porosity.

Fatigue and Impact Properties

Although not well documented in aluminum castings, such porosity is known to be the origin of fatigue cracks under appropriate loading conditions. It is reasonable to expect, therefore, that porosity will reduce the fatigue resistance of castings. Similarly, surface or internal pores can act as stress raisers during impact loading and will serve to decrease the impact strength of the alloy.

Pressure Tightness

Interconnected porosity can often run through the complete section of a casting from wall to wall. This can result in a loss of pressure tightness in castings which are required to contain liquids or gases under pressure, and is usually associated with a combination of film oxides and gas.

Hot Tearing Tendency

Resistance to hot tearing is improved by the presence of hydrogen. The gen-

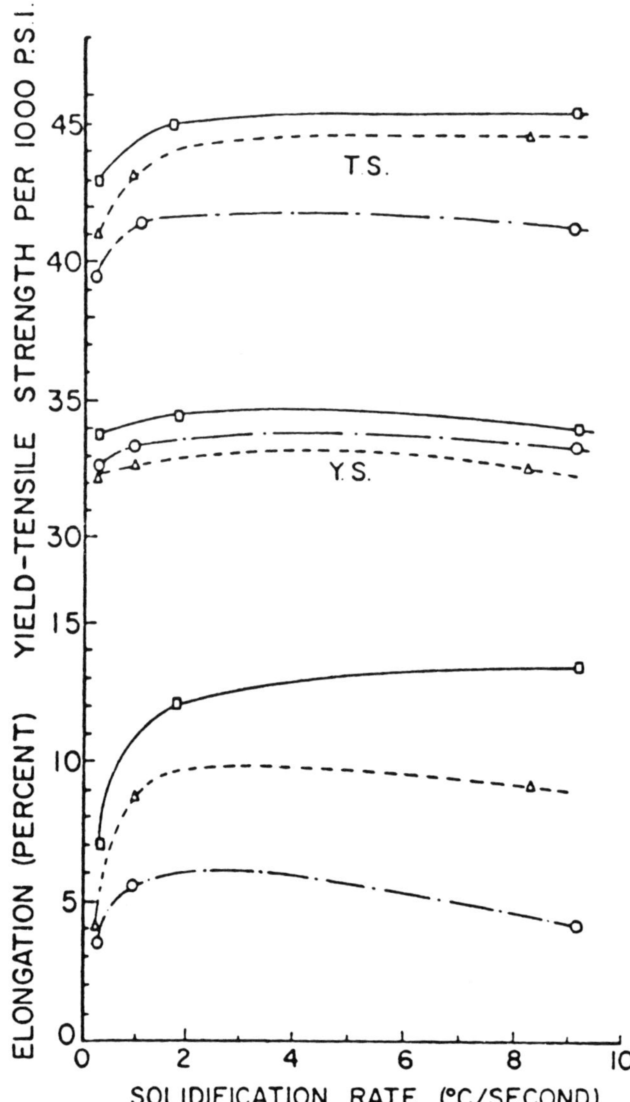

Figure 9.9. Tensile properties of 356 alloy sand cast with various hydrogen levels [4]:

   ○   *0.37 ml H_2/100 gm Al*
   △   *0.26 ml H_2/100 gm Al*
   □   *0.16 ml H_2/100 gm Al.*

eration of porosity creates local stresses within the casting which counterbalance the tensile stresses that normally cause hot tearing.

Blistering

Castings which contain very large quantities of porosity may blister when heat treated due to expansion of the gas trapped within the pore. While not a problem in properly treated sand or permanent mold castings, this is a significant difficulty in die castings where the porosity is due not so much to dissolved hydrogen, but to air which is trapped in the die during injection.

Macroshrinkage

The evolution of large quantities of hydrogen during solidification can compensate for some, or even all, of the normal solidification shrinkage of the alloy. Risering requirements can, therefore, be minimized and casting yield increased. Hydrogen is sometimes added to, or at least not removed from, melts which are used to produce castings for which mechanical or surface properties are not critical, eg., ornamental castings. Because of this potentially beneficial effect of hydrogen, it is important when degassing melts to know exactly which hydrogen level is best for the particular application.

9.4　Measuring the Hydrogen Concentration

Proper control of the melt hydrogen concentration requires some method of measurement. Several techniques have been developed which range from the very simple to the quite complex. While not all of these are of use in the foundry, they will be described here, since from time to time the foundryman may find it useful to use one of the more complex methods for control or calibration purposes.

Of the available methods, three (recirculating gas, sub-fusion analysis and vacuum fusion) provide a direct measurement of the hydrogen concentration. The others are indirect techniques in that they involve the measurement of some physical property usually related to density or bubble formation, and from this the hydrogen concentration can be inferred. We shall begin our description with the indirect methods, most of which are only semi-quantitative in nature.

Simple Density Techniques

These are the simplest and least expensive ways to indirectly measure the hydrogen level. A standard sample is cast and its density is measured by the Archimedes Principle—assuming all of the porosity in the sample is due to hydrogen evolution. The volume of hydrogen is then given by:

$$\text{ml } H_2/100 \text{ gm Al} = 100 \, (1/D_s - 1/D_t)$$

$$\text{where} \quad D_s = \text{sample density}$$
$$D_t = \text{theoretical density of the alloy.}$$

The theoretical density can be determined by a simple calculation based on the alloy chemistry, or it can be estimated by measuring the density of a chill-cast sample poured from well degassed metal.

At best, simple density techniques are only semi-quantitative ways of measuring the melt hydrogen. One of the biggest assumptions is that all internal porosity is due to hydrogen. As we have seen, most porosity is due to a combination of gas and shrinkage, therefore density measurements of this type will tend to overestimate the hydrogen concentration of the melt. Another problem, common to all density techniques, is that pores are often heterogeneously nucleated on solid inclusions, and samples cast from filtered metal can be up to 25% more dense than those poured from unfiltered metal having the same hydrogen concentration. Finally, the density differences found from different hydrogen levels are relatively small, and so the precision of this technique is not great.

The Reduced Pressure Test (Straube-Pfeiffer Test)

This is the most widely used method of hydrogen control in the foundry industry. The test itself consists of solidifying a sample under a partial vacuum in order to encourage pore formation through a reduction of the atmospheric pressure over the sample. When pores are nucleated, they will expand due to the lowered pressure, and a much more porous sample is produced than if freezing is allowed to proceed at normal atmospheric pressure. This has the effect of magnifying gas effects and increasing the sensitivity of density measurements to the quantity of hydrogen actually dissolved in the melt. This magnification is evident in the photograph shown in Figure 9.10 of a sample containing 0.22 ml H_2/100 gm Al solidified at 76 mm Hg pressure and at atmospheric pressure.

The equipment required for this test is very simple, and can be purchased commercially or constructed in-house. A schematic of a typical reduced pressure test apparatus is shown in Figure 9.11. It consists of a simple mechanical vacuum pump connected to a sturdy bell jar having a removable lid fitted with a vacuum seal and a window. Samples are poured in thin-walled steel cups or ceramic crucibles, usually of about 200 gm capacity. Pressures of from 5 to 100 mm Hg have been used, with 50 to 100 mm being the most common. The effects which are observed in the solidified sample are very sensitive to the pressure, and so it is essential that results which are to be compared

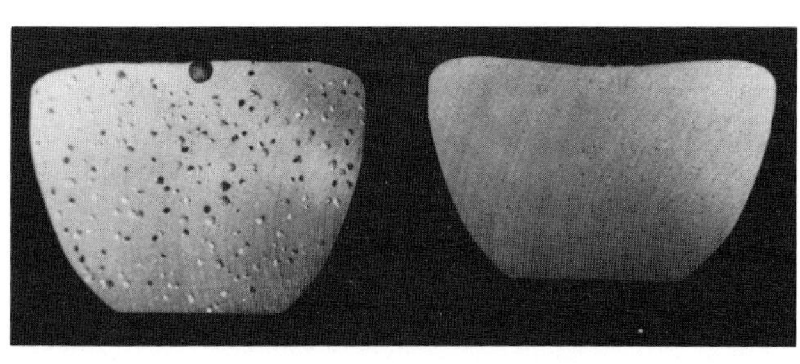

(a) sample solidified under 76mm Hg   (b) sample solidified at atmospheric pressure (760mm Hg)

Figure 9.10. The effect of pressure on porosity in A356 alloy containing 0.22 ml H_2/100 gm Al [5].

be obtained from samples produced at identical pressures. Pressure control can be achieved either through use of a surge tank or through incorporation of a bleed valve into the vacuum line. Solidification of the sample takes only about five minutes, and an indication of the degree of gassing of the melt can be obtained within this time.

Depending on the information desired, the sample may be evaluated in any one of several ways, such as:

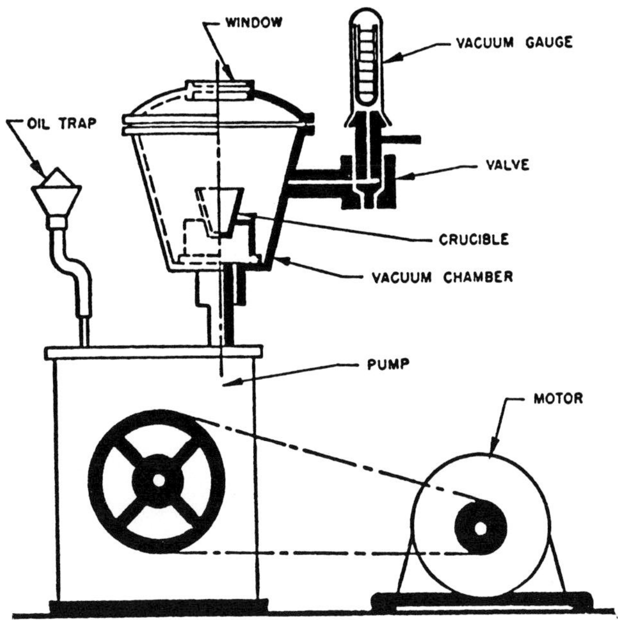

Figure 9.11. A typical reduced pressure test apparatus.

1) observation of bubble evolution during solidification. This is indicative of a higher gas content coupled with abundant heterogeneous nucleation of gas bubbles on dirt particles.
2) observation of the surface of the solidified sample. A puffed-up or convex surface indicates a high degree of internal porosity caused by gas nucleation. On the other hand, a concave surface due to piping indicates a low gas concentration.
3) sectioning the sample and rough polishing of one cut face will allow for a visual assessment of the degree of internal porosity. A very porous sample is due to a high gas content coupled with pore nucleation. Several sectioned samples are shown in Figure 9.12, which demonstrates that melts containing higher hydrogen levels lead to reduced pressure samples which have both greater amounts of internal porosity and exhibit a convex top surface.
4) measuring the sample density and using this, after suitable correction, to calculate the melt hydrogen concentration. This aspect of the test is dealt with in more detail later.

In its simplest form, this is a comparative, non-quantitative test. In-house standards can be developed to compare, for example, the degree of convexity of the top surface or the amount of porosity seen on a cut face. For a given application, the operator can, with experience, develop the expertise necessary to assess the gas concentration of the melt from the appearance of the reduced pressure sample. Since this test is used primarily in a compara-

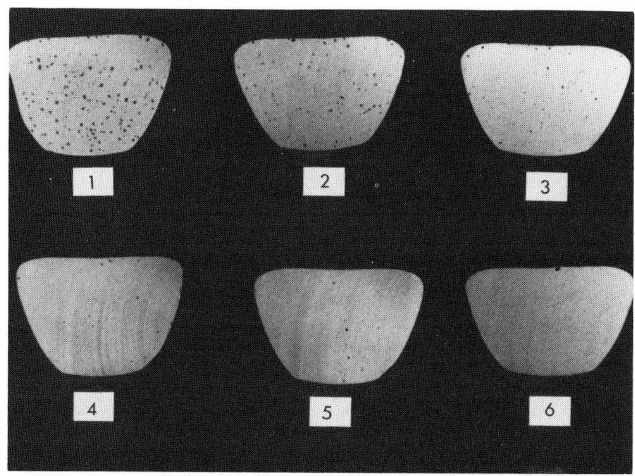

Figure 9.12. Sections of reduced pressure test samples, A356 alloy solidified at 76mm Hg pressure[5].

Sample No.	Hydrogen concentration by Telegas (ml/100 gm Al)	Density (gm/cm³)
1	0.23	2.486
2	0.16	2.550
3	0.13	2.584
4	0.10	2.609
5	0.09	2.619
6	0.07	2.636

tive way on a "go-no-go" basis, it is essential that it always be conducted under identical conditions.

Attempts have been made to make the reduced pressure test quantitative by calculating the pore volume and relating this to the hydrogen concentration [6-8]. This variation of the reduced pressure test is not used extensively in commercial practice, but it does, in principle, offer an inexpensive and rapid method for the shop floor determination of dissolved hydrogen. The method is most successful if a constant volume reduced pressure sample is used which incorporates a riser. Shrinkage is then concentrated in the riser which can be removed to leave a sample which contains mainly gas-related porosity. If the theoretical and actual densities (D_t and D_s) are measured as before, the hydrogen gas concentration can be determined from the formula:

$$\text{ml } H_2/100 \text{ gm Al} = P_2/P_1 \times T_2/T_1 \times 100 \times (1/D_s - 1/D_t)$$

where P_2 = the solidification pressure, eg., 75 mm Hg
P_1 = atmospheric pressure, 760 mm Hg
T_2 = the alloy solidus temperature in °K
T_1 = 273°K.

Use of this formula assumes that all of the hydrogen dissolved in the melt is present as porosity in the reduced pressure sample. Two problems are apparent in such an assumption. First, the amount of porosity is known to be determined by the melt cleanliness. A dirtier melt will solidify with more porosity and give a higher apparent hydrogen concentration. A recent, more sophisticated test is the so-called "vibrated vacuum gas test," in which the solidifying sample is vibrated at about 6000 cps in order to assist gas bubble nucleation, and to reduce the effects of variations in melt cleanliness. A second factor is the removal of some hydrogen by the vacuum pump during solidification. This will tend to give hydrogen values which are too low.

This latter factor is very important as seen from the data presented in Figure 9.13. Here the density of reduced pressure samples (triangular points) is plotted against the hydrogen concentration of an A356 alloy melt measured by the direct recirculating gas technique described later in this section. Calculated hydrogen levels (round points), determined from the preceding equation, are considerably lower than the real measured values, due to hydrogen loss during the reduced pressure test. However, these values can be corrected by multiplication with a correction factor which is determined by freezing a sample at atmospheric pressure, measuring its density and determining its hydrogen concentration (H_a). A sample from the same melt, which has presumably the same hydrogen concentration, is then frozen under vacuum, and its hydrogen level (H_v) calculated from its density. The correction factor is then defined as:

correction factor = H_a / H_v

As seen in Figure 9.13, multiplication of the reduced pressure values by the correction factor yields good agreement (square points) with the recirculating gas technique values. The reduced pressure test can, therefore, be reasonably quantitative, provided a constant correction factor can be determined. This method is made more reliable if an independent means of measuring hydrogen (such as the recirculating gas technique) is available to allow an occasional check on the accuracy of the determinations. It is to be expected that the correction factor will vary from alloy to alloy. More hydrogen should be pumped out of a short freezing range alloy, such as 413, because there is less of a dendrite mesh to contain it. The correction factor, which can only be determined experimentally, will be larger than that for a longer freezing range alloy, such as 356.

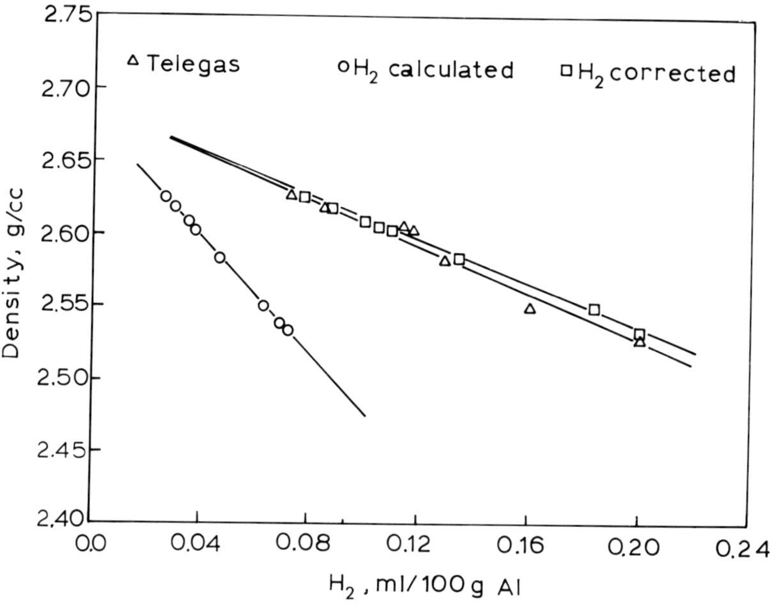

Figure 9.13. A comparison between the density of reduced pressure samples, the calculated hydrogen concentration of these samples and the measured (Telegas) hydrogen concentration of the liquid A356 alloy [5].

Some care should be used in the interpretation of reduced pressure tests on modified alloys. Tests done on strontium and sodium modified A356 show that modification does not change the density-hydrogen relationship for this alloy (Figure 9.14 a, b). The visual interpretation of samples cast with higher gas levels is, however, misleading: while modified alloys may contain fewer pores they are certainly larger (Chapter 4). These samples *appear*, therefore, to have more porosity, although the density, hydrogen concentration and total amount of porosity are unchanged by modification. The sections illustrated in Figure 9.15 show this effect. This visual impression is important only at high hydrogen concentrations, eg., greater than 0.2 ml H_2/100 gm Al. At concentrations lower than 0.1 ml H_2/100 gm Al no pores are obvious in reduced pressure samples of either unmodified or modified alloys. This is reflected also in the density of these samples and illustrates the importance of proper degassing for pore-free castings. If the reduced pressure test is to be used as a qualitative evaluation of modified alloys, a set of standards separate from those used for unmodified alloys should be developed and used.

Quantitative Reduced Pressure Test (Severn Science)

This is a variation of the Straube-Pfeiffer test in which a constant amount of metal (about 100 gm) is allowed to solidify in an isolated vacuum chamber. As solidification proceeds, hydrogen gas is evolved and raises the pressure in the chamber. The amount of hydrogen given off can be readily calculated

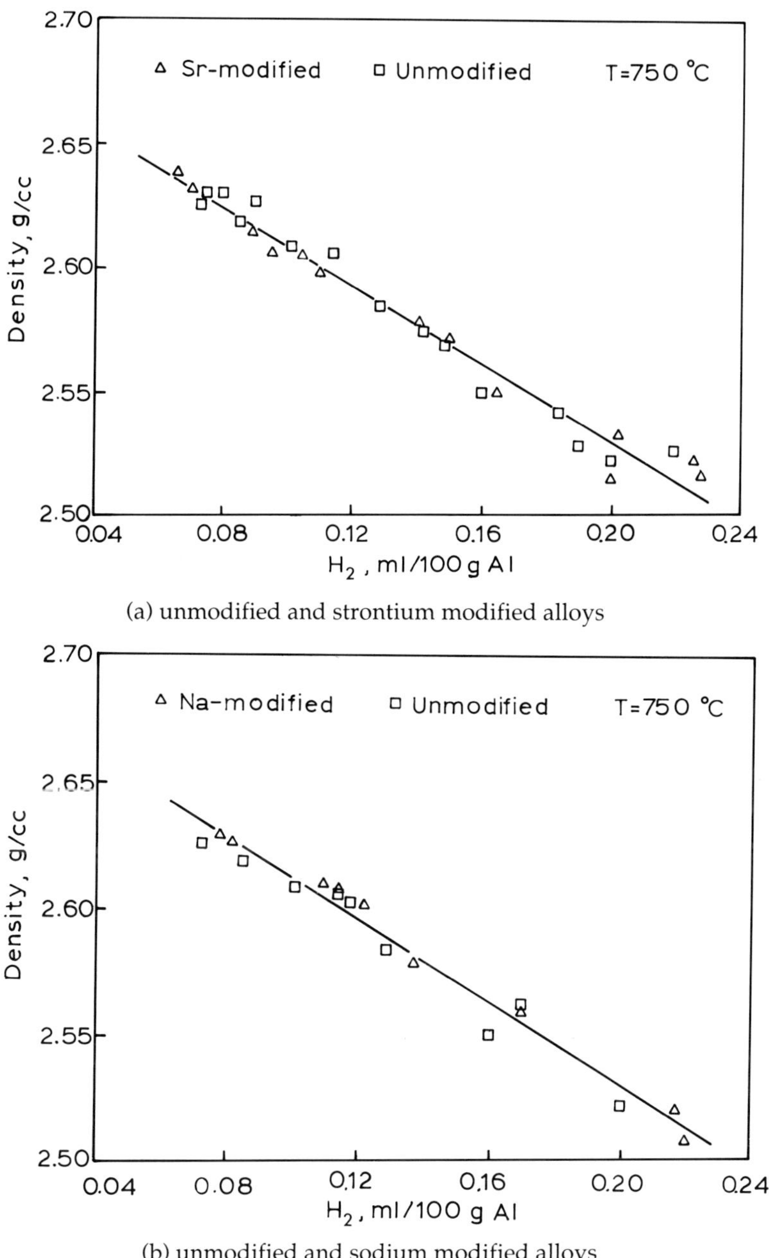

(a) unmodified and strontium modified alloys

(b) unmodified and sodium modified alloys

Figure 9.14. A comparison of reduced pressure densities and actual Telegas measured hydrogen concentrations in an A356 alloy melt [5].

from the pressure rise and displayed on a digital readout. While this method is rapid, it does assume that all hydrogen present in the melt is released from the sample on solidification. Sensitivity is good at higher hydrogen levels, but difficulties are experienced in detecting hydrogen concentrations less than 0.1 ml/100 gm Al. This is a relatively new technique which is currently under evaluation by the industry.

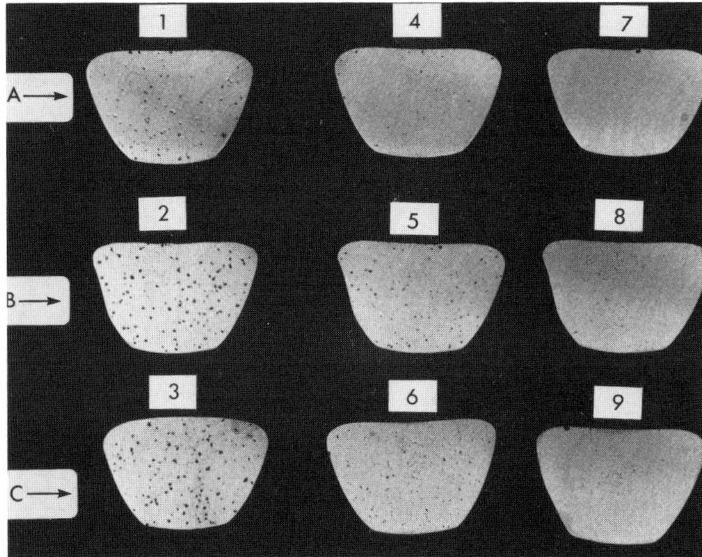

Figure 9.15. The visual appearance of sectioned reduced pressure samples of A356 alloy [5]:

A: *unmodified*
B: *strontium modified*
C: *sodium modified.*

Sample No.	Hydrogen concentration by Telegas (ml/100 gm Al)	Density (gm/cm³)
1	0.21	2.504
2	0.22	2.484
3	0.22	2.497
4	0.12	2.585
5	0.13	2.578
6	0.13	2.580
7	0.07	2.636
8	0.07	2.633
9	0.08	2.637

First Bubble Test

In the first, or initial, bubble technique, a small quantity of alloy is held molten within a vacuum chamber. The pressure within the chamber is gradually reduced while the surface of the bath is observed visually. At some point, a bubble of hydrogen will break the surface, and the temperature and pressure at which this occurs are noted. It is assumed that the chamber pressure at which the first bubble occurs, equals the partial pressure of hydrogen in the melt. If this is true, the hydrogen concentration can be calculated from Sievert's Law,

$$S = \sqrt{P}\ e^{\frac{-k}{T}}$$

where S = hydrogen concentration in ml/100 gm Al
      P = chamber pressure
      k = constant
      T = bath temperature

Values of the constant, k, have been measured experimentally and so it should be a simple matter to calculate the hydrogen concentration. Although this method offers rapid analysis, and several commercial versions of it are on the market, it is plagued by lack of reproducibility. The first bubble is frequently difficult to detect, and a given melt sample will often exhibit large scatter in attempts to repeat the conditions for first bubble formation. Poor correlation has been found with the methods of direct measurement of hydrogen such as recirculating gas and sub-fusion analysis.

The main reason for the difficulties with the first bubble test lie in the assumption that bubble nucleation will occur as soon as the solubility limit for hydrogen has been exceeded. This is simply not true, for as we have seen earlier in this chapter, surface energies present a large barrier to bubble formation in liquid metals. Heterogeneous nucleation becomes necessary, and the whole test becomes more dependent on melt cleanliness than on pressure. Melts with different dirt levels, but having the same hydrogen concentration, will exhibit bubble nucleation under different conditions of chamber pressure and melt temperature. Even attempts to repeat nucleation on one sample will lead to different results, as any two bubbles will form on two different nuclei at different pressures and temperatures. This is a theoretically attractive test, but it appears that on practical grounds the first bubble test is defeated by the problem of pore nucleation in liquid metal alloys.

Recirculating Gas

This method was developed over 20 years ago. It forms the basis for two types of commercial equipment: Telegas and Alscan (Fig. 9.16), and remains the only method available for direct measurement of hydrogen in liquid aluminum and its alloys. The apparatus recirculates a small volume of nitrogen through the molten alloy bath. As the nitrogen contacts the bath, it picks up hydrogen, until eventually the hydrogen content of the nitrogen stream comes into equilibrium with the hydrogen concentration of the bath. The partial pressure of hydrogen in the nitrogen, which then equals the partial pressure of hydrogen in the molten alloy, is determined by measuring the thermal conductivity of the nitrogen-hydrogen mixture. Calibration is done beforehand using bottled premixed hydrogen-nitrogen gases. Once the

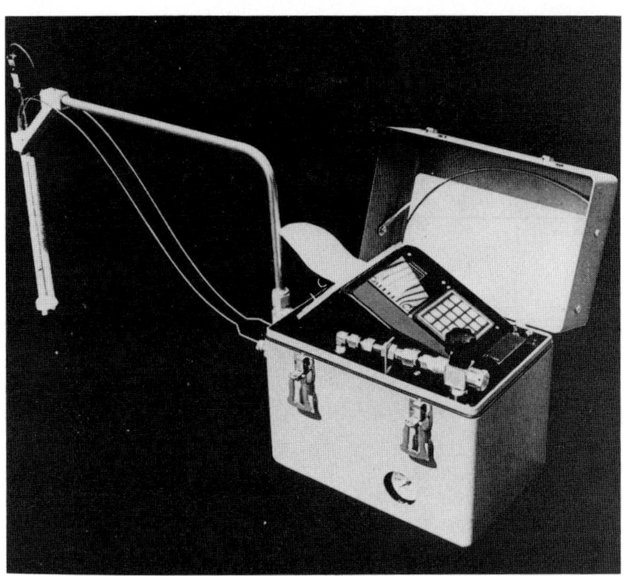

Figure 9.16. A hydrogen measurement instrument based on the recirculating gas technique. (Photo courtesy of Alcoa.)

partial pressure is known, application of Sievert's Law allows determination of the dissolved hydrogen concentration.

The method is used in several primary aluminum plants and in laboratories, but foundry use has been limited because certain parts of the instrument are delicate. This is true, particularly of the ceramic probe which is immersed in the melt and through which the nitrogen gas circulates. In addition to being fragile, it is costly and prone to plugging by oxides and other particles in the melt. Use with strontium or sodium treated alloys is particularly difficult. Recent improvements in this probe may make recirculating gas methods more attractive for foundry use.

Another difficulty is that, while the hydrogen solubility in pure aluminum is well known, it is less well known in foundry alloys. Hydrogen concentrations obtained from these methods are based on pure aluminum, and correction factors are then applied to give the values in the various foundry alloys. For some alloys, such as 356, these factors are well known, but for some of the more complex alloys, the solubility relationships are not so well understood and the values obtained are less certain. Nevertheless, recirculating gas techniques remain the most reliable, although certainly neither the cheapest nor the most practical, means of measuring actual hydrogen concentrations in the liquid. They are free of all the nucleation and solidification effects which plague the other methods.

Sub-Fusion and Vacuum Fusion

These are mainly laboratory techniques capable of providing very accurate hydrogen analysis. They are used to calibrate the other methods and to occasionally check their performance. A commercial version of the vacuum fusion method (the Leco technique) is available and has found use in several primary aluminum plants. So far, it has not been used on the foundry floor due to the capital cost, the length of time required for analysis, and the sample preparation necessary to obtain good results.

In both sub-fusion and vacuum fusion, the hydrogen is extracted under vacuum from a highly chilled sample. A special copper or graphite mold is used to produce this sample, which must be solidified fast enough to retain all hydrogen in the solid without porosity formation. With sub-fusion analysis, the sample is heated to about 50°C below its melting point; in vacuum fusion it is melted. Several hours are required to extract all of the hydrogen from the solid sample, while 10-15 minutes are needed in the case of vacuum fusion. The extracted hydrogen is separated from background gases by allowing it to diffuse through a heated palladium tube into an isolated chamber in which the hydrogen concentration can be determined from a pressure rise or some physical measurement.

Sample preparation is important when using either of these methods as hydrogen contamination of the surface can occur readily. While vacuum fusion is much faster than sub-fusion analysis, the latter is preferred for alloys, such as foundry alloys, which contain volatile elements (magnesium, zinc, sodium) which can vaporize and condense in the apparatus.

Evaluation of Hydrogen Measurement Techniques

The reader can see that several methods are available to determine the gas level of a foundry melt. These range form the highly sophisticated and quantitative to the simple and qualitative. Foundry melts are usually subjected to a relatively large number of melt treatments (modification, grain refining, fluxing, degassing) and proper hydrogen control necessitates some knowledge of the gas level throughout the processing sequence. With these various treatments, gas levels can change quickly, and any useful hydrogen test should therefore be both simple and rapid, requiring no more than a few minutes to obtain an indication of the hydrogen level. Both the reduced pressure test and the recirculating gas methods are capable of providing this response time. Of the two, the recirculating gas technique gives more useful information, since it yields the actual hydrogen concentration of the liquid. Unfortunately, in its present form, it may not be a practical instrument for use in the foundry, both from the point of view of ease of use and operating cost. The reduced pressure test is left as the alternative. At present, it is only qualitative in nature, but with a renewed interest in this technique, we hope to see more sophisticated versions on the market. Ideally these could incorporate both vibration and some method of calibration so as to make the test more quantitative and more reproducible.

References

1. Opie, W.R. and N.J. Grant. "Hydrogen Solubility in Aluminum and Some Aluminum Alloys," *Trans. AIME*, 188, (1950) pp. 1237-41.
2. Entwistle, R.A., J.E. Gruzleski, P.M. Thomas. "Development of Porosity in Aluminum-Base Alloys," *Solidification and Casting of Metals*, The Metals Society, publication no. 192 (1979) pp. 345-49.
3. Fang, Q.T. and D.A. Granger. "Porosity Formation in Modified and Unmodified A356 Alloy Castings," *AFS Transactions*, 97 (1989) pp. 989-1000.
4. Chamberland, B. and J. Sulzer. "Gas Content and Solidification Rate Effect on Tensile Properties and Soundness of Aluminum Casting Alloys," *AFS Transactions*, 72 (1964) pp. 600-607.
5. Mulazimogolu, M.H., N. Handiak, J.E. Gruzleski. "Some Observtions on the Reduced Pressure Test and the Hydrogen Concentration of Modified A356 Alloy," *AFS Transactions*, 97 (1989), pp. 225-32
6. Rosenthal, H. and S. Lipson. "Measurement of Gas in Molten Aluminum," *AFS Transactions*, 64 (1955) pp. 301-305.
7. Sulinski, H.V. and S. Lipson. "Sample for Rapid Measurement of Gas in Aluminum,"*AFS Transactions*, 67 (1959) pp. 56-64.
8. Church, J.C. and K.L. Herrick. "Quantitative Gas Testing for Production Control of Aluminum Casting Soundness," *AFS Transactions*, 78 (1970) pp. 277-80.

Chapter 10

Degassing Aluminum Foundry Melts

10.0   Introduction

In this chapter we discuss the methods for removing hydrogen from liquid aluminum alloys. A basic step in hydrogen control is prevention of hydrogen dissolution in the first place. It is expensive to remove hydrogen from liquid alloys, and so no effort should be spared to minimize its pickup by the melt. A careful consideration of the sources of hydrogen, outlined in the previous chapter, along with a concentrated effort to minimize these sources whenever possible, can go a long way to controlling gas levels.

The use of degassing and choice of a degassing technology will depend on the level of hydrogen desired. In some applications, such as high quality aerospace castings, specifications on casting soundness are particularly stringent and mechanical properties must be reproducible and guaranteed. These castings are heavily risered, casting yield is low, and the cost of the metal is only a small fraction of the total value of the casting. Such considerations dictate as low a hydrogen level as possible (less than 0.1 ml/ 100 gm Al) and the use of some of the more sophisticated degassing techniques.

With other castings, such as many commercial castings, shape rather than structural requirements is of prime importance. Casting yield is of concern in these relatively low cost products, and a certain level of gas related porosity is desirable in order to reduce riser size. Minimum gas levels may not be required and some intermediate concentration must be achieved in order to obtain the desired widely dispersed micro-porosity. If structural considerations are of no importance, quite high gas levels can be used to minimize or even eliminate risering through the production of large amounts of internal porosity.

The required gas level depends very much on the application and on the details of the specific casting, since solidification conditions will vary greatly from one casting design to another. It is important, therefore, that some idea be formed of the desired hydrogen level. Degassing should always be used in conjunction with one of the methods of measuring hydrogen described in the last chapter. If this is not done, the foundryman has no reliable and reproducible method of knowing when the required hydrogen level has been reached.

10.1   Degassing Methods

Although many processes exist to degas liquid aluminum and its alloys, they all fall into one of three general types:
- natural degassing
- gas purging
- vacuum degassing

Only gas purging is used extensively in the foundry industry, and so we will concentrate on this family of processes. A brief description of the other two is given for completion.

Natural Degassing

A liquid aluminum bath will lose hydrogen to the atmosphere if the melt is supersaturated. This natural degassing occurs very slowly, usually requiring a few hours to achieve useful degrees of degassing. It is therefore too slow to be of much practical use. Natural degassing does not occur in all cases, since the melt must be supersaturated. Liquid aluminum alloys do not often reach their hydrogen saturation limits under common melt processing conditions, and so hydrogen supersaturation and subsequent natural degassing can only occur in special situations, such as if the melt temperature is lowered to a low enough value that it becomes supersaturated with hydrogen; or if an extreme amount of hydrogen has been added, for example by use of a wet flux.

Natural degassing is favoured by lower temperatures and by dry atmospheres over the melt, i.e., those free of hydrogen. One method of encouraging natural degassing is to lower the melt temperature for a period of several hours.

Gas Purging

This is the treatment with an inert gas, a reactive gas or a combination of these. Virtually all foundry degassing uses a process which falls into this category. As with many developments in aluminum liquid metal technology, degassing processes were first perfected for use in primary aluminum smelters. In many cases these processes could be applied directly to foundry alloys. In some cases, some modification, usually down-scaling, has been necessary.

The principle of gas treatment is shown in Figure 10.1. Hydrogen atoms in a liquid metal will diffuse to a region where no hydrogen is present (this is the principle of natural degassing mentioned above). Hydrogen exists in liquid aluminum in atomic form, not as molecular hydrogen (H_2). Effective removal requires that the hydrogen atoms combine to form hydrogen gas molecules. However, as we have seen in Chapter 9, formation of a gas bubble in liquid metals is not an easy matter. To solve this problem we introduce hydrogen-free bubbles of some gas into the melt. Hydrogen atoms can diffuse into these bubbles, and since the gas phase is already present, there is no need to nucleate separate bubbles of hydrogen. The reaction:

$$\underline{H} + \underline{H} \longrightarrow H_2 \text{ (gas)}$$

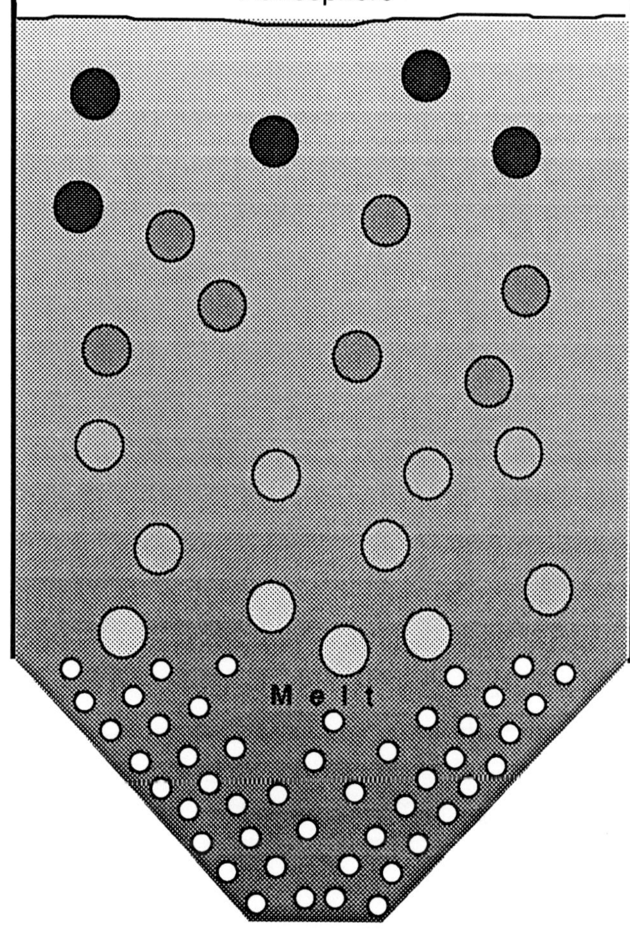

Figure 10.1. The principle of gas purging.

can proceed easily with the gas bubble and the hydrogen lifted out of the melt as the bubble rises to the surface.

Differences between the various degassing techniques lie in the type of gas which is introduced and in how the bubbles are formed. As it turns out, the efficiency of degassing is strongly dependent on bubble size, which in turn depends on how the bubbles are produced in the melt. Efficiency can also depend on the type of gas used. We will deal with these two questions in the following sections, but for the moment let us just consider some of the practical details of the various gas bubbling techniques.

Several gases and gas mixtures can be employed. Of course, the gases must be completely free of water vapor or they will introduce hydrogen into the melt. Among the more common are pure, dry nitrogen or argon. Pure chlorine was used extensively in the past, but has now been largely discontinued due to its toxicity and corrosive effects on equipment and ventilation ducting. Various mixtures of an inert gas and chlorine are in use. These include:

$$99\% Ar\text{-}1\% Cl_2$$
$$90\% Ar\text{-}10\% Cl_2$$
$$80\% N_2\text{-}10\% Cl_2\text{-}10\% CO$$
$$90\% N_2\text{-}10\% Cl_2$$
$$70\% N_2\text{-}30\% Cl_2.$$

Mixtures of nitrogen and freon-12 (CCl_2F_2) have come into common usage with the freon concentration varying from 5% to 30%. Sulphur hexafluoride, SF_6, has also been shown to be effective.

The treatment gas can be introduced into the melt in several ways. The simplest, but least effective, is to use a straight graphite lance with an outside diameter of from 2.5-5 cm (1-2 in.), and an inside diameter of 0.3 cm (0.5 in). Such a lance produces large bubbles with diameters of the order of 2-3 cm (1 in.) which rise close to the lance surface, and so contact a minimum of liquid metal (Fig. 10.2). In order to effectively degas a large melt, several such lances must be used, or if only one is employed it must be constantly moved about.

A much finer dispersion of gas bubbles can be achieved with use of a porous graphite head. These heads are usually about 15 cm (6 in.) in diameter, and they screw onto a graphite lance. The bubbles tend to follow the lance up to the surface, and so either a gentle movement of the plug should be used in the metal, or several such assemblies should be introduced in order to contact the most metal in the minimum time.

The primary aluminum companies and commercial gas producers have developed a number of systems (MINT, SNIF, ALPUR, DMC, etc.) to degas and in some cases to filter the large quantities of aluminum with which they work. The most sophisticated and most efficient degasser available for foundry use is an adaptation of some of this technology. The rotary impeller degasser introduces the treatment gas into the melt through a special impeller head which rotates rapidly, chops the gas stream into very fine bubbles (3-6 mm, 1/8-1/4 in.), and then disperses them throughout the body of the liquid. The rotation speed and flow rate can be varied according to the size of the melt. Figure 10.3 shows the dispersion achieved with such a device used in a water-model study. A schematic illustrating the essential parts of a rotary impeller degasser is presented in Figure 10.4. Units have been designed to sit on the top of crucible furnaces or ladles for batch degassing (Fig. 10.5a) or for continuous operation in the dipout well of a reverberatory furnace (Fig.10.5b). Final hydrogen levels achieved depend on initial hydrogen content, purge gas flows, impeller rotation speed, size of vessel, and treatment time employed. Hydrogen levels down to 0.05 ml/100 g Al can often be achieved by this process.

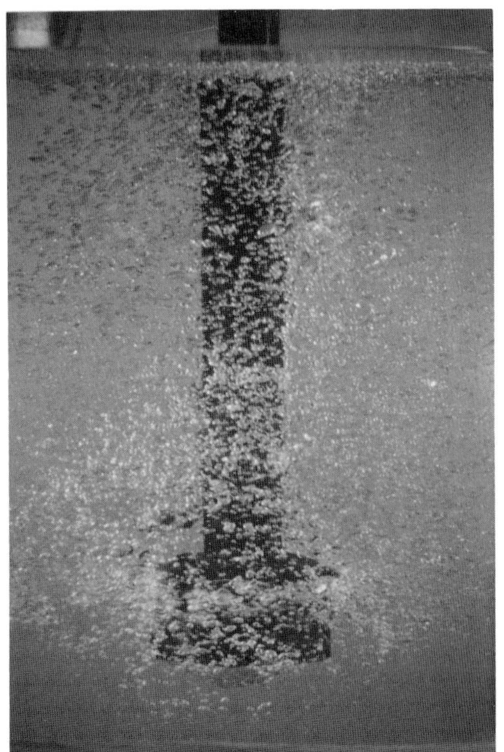

Figure 10.2. Large bubbles produced by use of a graphite lance in a water model study of degassing[1].

Figure 10.3. Dispersion of fine bubbles obtained in a water model study using a rotary impeller[1].

In tablet (or flux) degassing, the gas is introduced into the melt in solid pill form. The pills, added by a perforated bell, are composed of chemicals which thermally decompose to release gas bubbles. Most chemical degassers depend on chlorine generating compounds. The most popular solid degasser is hexachloroethane, C_2Cl_6, which decomposes at temperatures above 700C (1290F). Like all solid fluxes, these materials are hygroscopic and must be stored in a dry atmosphere, otherwise their use will add hydrogen to the melt rather than remove it. This is the least controllable method of degassing since decomposition is rapid. As a result, several treatments may be necessary in order to provide a long enough reaction time. Dross formation is promoted by this method and obnoxious fumes are generated. Despite these disadvantages, tablet degassing is often quite suitable to the treatment of small melts and finds application where precise control of degassing is not required. Fluxes for degassing can also be blown into the melt using a carrier gas. This flux injection technique is described in detail in Chapter 12.

Vacuum Degassing

Vacuum degassing is used in some large European foundries. The elimination of atmospheric pressure over the melt encourages formation of hydro-

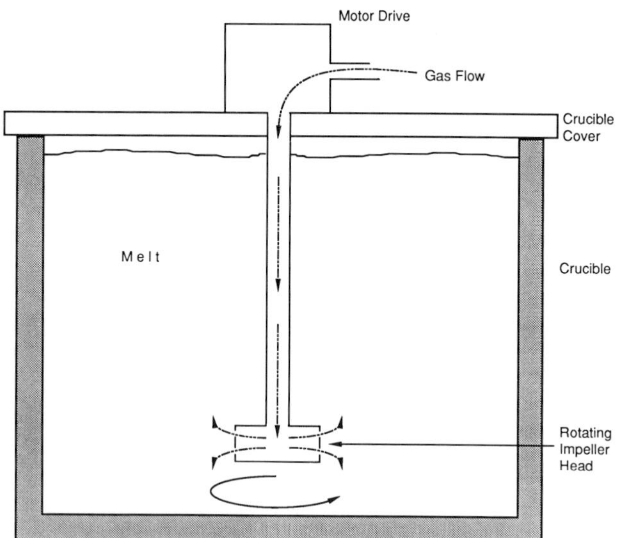

Figure 10.4. Schematic of a rotary impeller degasser, used on a crucible furnace.

(a) for batch treatment in crucible furnaces

(b) for continuous operation in the dipout well of a reverberatory furnace

Figure 10.5. Rotary impeller degassing units[1].

gen bubbles which are removed through the pumping system. Degassing is also promoted through provision of a completely hydrogen-free atmosphere. In order to degas in a reasonable period of time, a slow flow of pure, dry nitrogen gas can be introduced by a lance into the bottom of the melt. This stirs the melt and helps the rising bubbles to overcome the metallostatic head. A schematic of a vacuum treatment apparatus capable of treating 1200 kg of aluminum appears in Figure 10.6. While extensive capital investment is required, this is a non-polluting degassing method which can provide hydrogen concentrations down to about 0.08 ml H_2/ 100g Al.

Methods of Adding Hydrogen

In some cases it may be desired to add hydrogen to the melt, for example, if some dispersed microporosity is

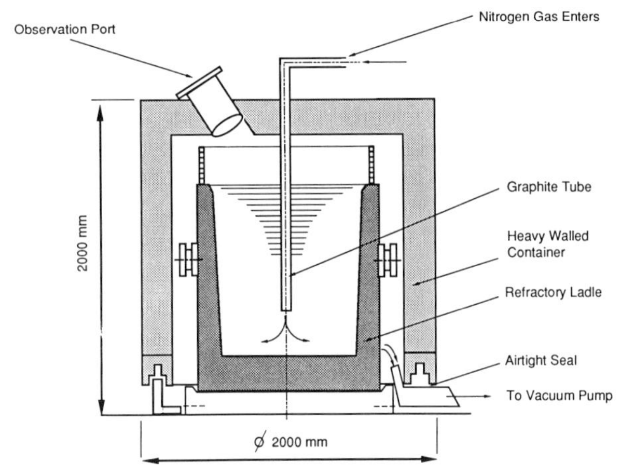

Figure 10.6. A foundry vacuum degassing unit.

wanted to combat macroshrinkage, and degassing has lowered the hydrogen level to below the target value. The common method to do this is to add organic agents such as green wood or potatoes. Fluxes are also available which add hydrogen through decomposition of their water of crystallization.

10.2   Degassing Efficiency

Since gas purging is the only widespread method used in foundry applications, our discussion will be confined to this family of techniques. Some excellent detailed mathematical analyses of hydrogen removal can be found in the literature (see references 2-4). This section presents a summary of the major factors determining degassing efficiency, so that those in the foundry can easily understand what determines how quickly, or how slowly hydrogen can be removed from a melt.

In order to appreciate what determines the efficiency of a degassing process, it is necessary to consider the process illustrated in Figure 10.1 in a bit more detail. Hydrogen removal can be thought of as a process requiring several successive steps:

1) hydrogen transport in the molten alloy to the surface of a purge gas

bubble, usually by a combination of diffusion and convective motion;

2) transport of hydrogen atoms by diffusion through a thin stagnant boundary layer of liquid at the bubble surface;

3) chemical absorption of hydrogen atoms onto the bubble surface;

4) reaction of hydrogen atoms to form hydrogen gas molecules ($\underline{H}$ + $\underline{H}$ —> H_2) and desorption of these molecules from the bubble surface;

5) diffusion of gaseous hydrogen into the main volume of the bubble;

6) removal of the hydrogen containing bubbles from the melt surface.

Depending on the degassing arrangement, almost all of these steps can control the efficiency of the process. The exceptions are steps 5 and 6. Diffusion of gaseous species is rapid at liquid aluminum degassing temperatures, and the physical set-up of a degassing system is usually such that removal of the hydrogen at the melt surface is rapid.

As it turns out, the single most important factor in determining degassing efficiency is the bubble size. Calculated degassing efficiencies for two melt hydrogen concentrations as a function of bubble diameter are presented in Figure 10.7, taken from reference 4. Here, a system operating at 100% efficiency is defined as one in which the partial pressure of hydrogen in the purge gas bubble equals the partial pressure of hydrogen dissolved in the melt. In other words, the rising gas bubble is saturated with respect to hydrogen; it cannot absorb anymore. Similarly, a system at 50% efficiency is one in which the rising gas bubbles contain only one-half of the hydrogen thermodynamically possible. If the efficiency could be doubled, twice as much hydrogen could be removed using the same amount of purge gas.

It is seen that very high efficiencies, even 100%, can be achieved at small bubble sizes. Bubbles of 5 mm diameter or less are necessary. Large bubble sizes result in very poor efficiencies (20% to 30%). These trends can be easily interpreted in terms of the steps in degassing listed previously. For a given flow rate,

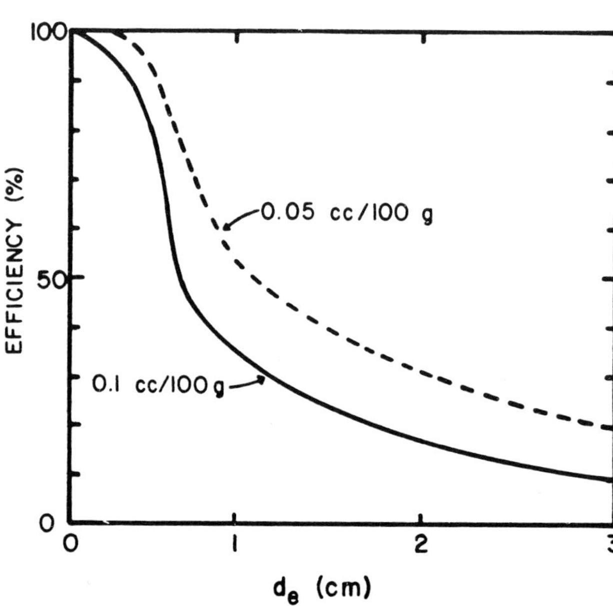

Figure 10.7. Degassing efficiency as a function of purge gas bubble size for two melt hydrogen concentrations[4].

large bubbles are far apart, and hence the hydrogen atoms have far to diffuse to reach the bubble surface. The system is now controlled by transport processes in the liquid aluminum (step 1). Because the bubbles are large and the hydrogen atoms have difficulty reaching them, they contain only a small fraction of the hydrogen which is thermodynamically possible, and so the degassing efficiency is low.

A large number of small bubbles will be close together and hydrogen atoms will have only a short distance to travel in the melt to arrive at a bubble surface. Liquid phase transport is no problem, but the small volume bubbles quickly become saturated with hydrogen (efficiency is 100%), and step 4 ceases. At this point, an increase in the degassing rate can be achieved only by increasing the purge gas flow rate. Happily, the degassing efficiency for a given bubble size improves as the melt hydrogen concentration is lowered. This is a reflection of the fact that it takes less hydrogen to equilibrate a purge gas bubble as the melt concentration decreases. In addition to ease of saturation and facility of hydrogen transport, large and small bubbles have other important differences which relate to the degassing efficiency. Most important is their rate of rise in the melt. Small bubbles float more slowly and have a longer contact time during which they can absorb hydrogen. Large bubbles, being more buoyant, rise rapidly and so have a much shorter contact time with the melt.

What does all of this mean to the founder trying to degas an aluminum alloy melt? Obviously, considerable effort should be expended in trying to produce the finest possible bubble dispersion in the melt. Not only will this yield low hydrogen concentrations in a minimum of time, but it will result in savings on the cost of purge gas. In a system operating at 50% efficiency, half of the gas used absorbs no hydrogen and is simply dispersed into the atmosphere.

Of the four commonly used methods for introducing degassing agents into aluminum melts, lance degassing and tablets are the most inefficient. Lance degassing produces large bubbles which rise only in the vicinity of the lance and tend to collide and coalesce, thus becoming even larger. Decomposition of tablet degassers is uncontrolled, bubble size is variable, and the bubbles are confined to a volume of metal around the plunger which is used to introduce the degasser. A considerable refinement of the bubble size occurs with use of a porous plug. However, bubbles again tend to rise only in a plume above the plug. The volume of melt contacted is limited, and bubble coalescence occurs reducing efficiency.

By far the most uniform pattern of fine bubbles is obtained with a rotary impeller degasser. Not only is bubble size smaller than with other techniques, but a properly sized rotor disperses the bubbles throughout the melt to achieve maximum contact. Close to 100% efficiency can be reached with this method. The degassing performance of a lance, porous plug and

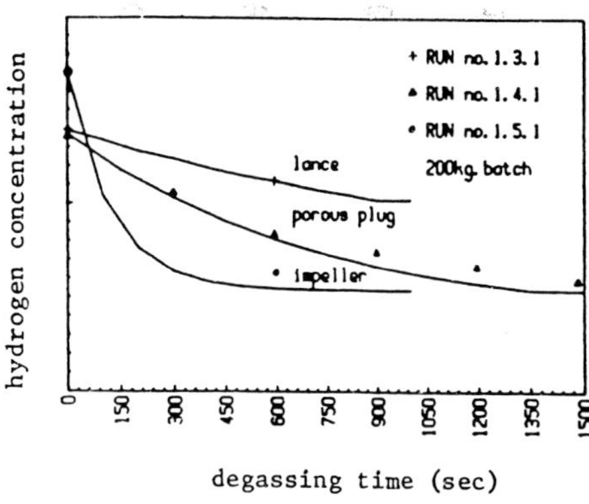

Figure 10.8. The degassing performance of a lance, porous plug and an impeller on a batch of 357 alloys[3]. (Reprinted with permission from *Light Metals 1984*, edited by J.P. McGeer (1984), The Metallurgical Society, 420 Commonwealth Drive, Warrendale, PA.)

impeller is compared in Figure 10.8 on a 200 kg (500 lb) bath of 357 alloy. The advantages of impeller degassing are evident compared to the other two methods. Lance degassing never yields the same final hydrogen level as does rotary impeller, and porous plus degassing requires three times as long, and of course three times as much purge gas to reach the same hydrogen concentration.

Temperature also plays a role in degassing efficiency. Hydrogen solubility and the partial pressure of hydrogen in the melt decrease rapidly as the melt temperature is decreased. It is not surprising, therefore, that degassing efficiency is improved at lower melt temperatures. In view of this, degassing should always be carried out at the lowest temperature which is practical for a particular operation. Calculated degassing curves for rotary

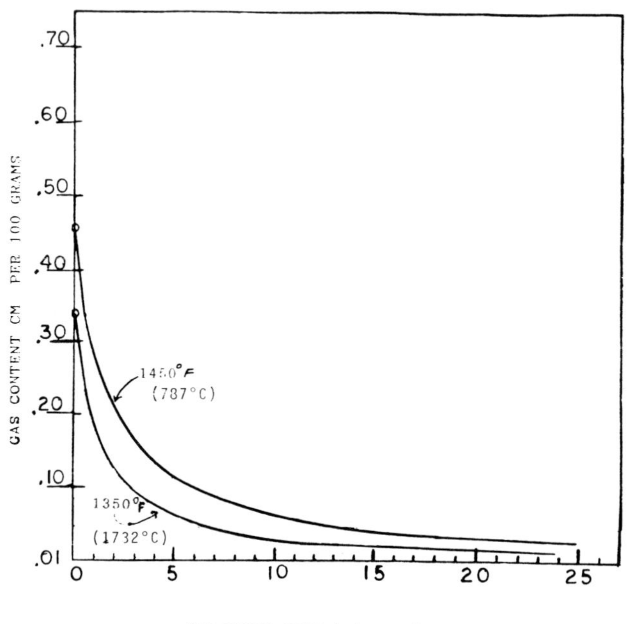

Figure 10.9. The effect of melt temperature on rotary impeller degassing[5].

impeller degassing at two different temperatures illustrate this effect in Figure 10.9.

10.3   Inert or Reactive Gas Treatment

The gases employed for degassing treatments fall into two types: inert gases such as argon or nitrogen which have no or very little reactivity with liquid aluminum, and reactive gases such as chlorine, freon, fluorine or mixtures of these with nitrogen or argon. The reactive gases undergo a chemical reaction with liquid aluminum and several of the alloying elements or additives used in melt treatment. In the next section we will deal with the compositional changes brought about by reactive gas treatment. Here we confine ourselves to the differences between reactive and inert gas treatment as related to hydrogen removal.

With reactive gas treatment, the most important chemical reaction involves chlorine acting to form aluminum chloride, which is a gas above 183C (361F) according to the reaction:

$$2Al_{(l)} + 3Cl_{2(g)} \longrightarrow 2AlCl_{3(g)}$$

As aluminum chloride gas bubbles rise through the melt they pick up hydrogen by the mechanism outlined previously. One of the great disadvantages of chlorine use is the pollution which it causes and the corrosion of metal structures. This is caused by the aluminum chloride coming into contact with damp air above the melt and hydrolizing by the reaction:

$$2AlCl_{3(g)} + 3H_2O_{(g)} \longrightarrow 6HCl_{(g)} + Al_2O_{3(s)}$$

The effluent is a fine white fume—a mixture of hydrochloric acid and aluminum oxide. The toxicity and destructive effects of chlorine treatment can be reduced by diluting it with an inert gas, usually nitrogen. Therefore, mixtures of nitrogen with 10%-30% of chlorine are often used.

Chlorine need not be introduced directly into the melt. It can be formed, *in situ*, by chemical decomposition of hexachloroethane tablets or of freon-12 gas. Hexachloroethane also has the disadvantage of generating obnoxious fumes. Care must be taken when using freon-12, since it decomposes to form carbon, chlorine and fluorine. The volume of carbon formed is such that it can quickly plug any porous diffuser system and render degassing ineffective.

One of the great benefits of using a reactive gas or a mixture of reactive and inert gases, is that they exert a cleaning effect on the melt. Chlorides formed at the gas bubble-melt interface change the surface tension of the aluminum so that oxide inclusions stick to the bubbles and are literally floated out of the melt. Inert gas also has a cleaning effect, but it is much enhanced by the use of some reactive gas.

  The advantages of using reactive gas for hydrogen removal are far less clear. Sometimes, use of a reactive gas produces better degassing while under

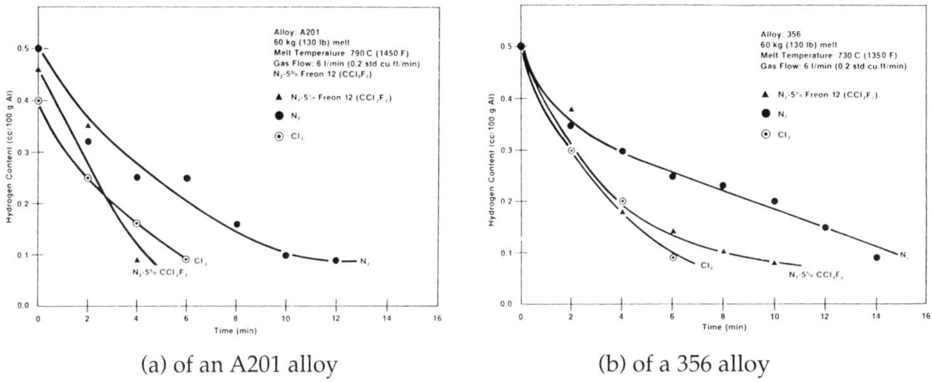

(a) of an A201 alloy   (b) of a 356 alloy

Figure 10.10. The influence of purge gas composition on degassing[6].

other conditions it appears to impair degassing. An improvement in hydrogen removal occurs if degassing is controlled by mass transfer of the hydrogen into the bubbles, i.e., if bubbles are larger and degassing efficiency is low, as for example, with lance degassing. Chlorine in this case assists the transfer of hydrogen into the purge gas bubble. The advantages of chlorine or nitrogen-5% freon over pure nitrogen are shown in the lance degassing of A201 and 356 alloys in Figure 10.10.

  When bubbles are small and gas removal is highly efficient, there is evidence that a reactive gas can actually slow down hydrogen removal. This has been observed in the Alcoa 622 Process which is essentially a rotary impeller process. It is proposed that magnesium chloride forms as a layer on the bubble surface and impedes the ability of hydrogen atoms to cross the surface and to enter the bubble[7].

  Similarly, degassing with a reactive gas is not possible if the melt contains a strontium modifier. It has been shown[8] that if nitrogen-5% freon is used, no degassing takes place until all of the strontium has been removed by the degasser. Once again, the mechanism is thought to involve the formation of a solid layer of strontium chloride on the bubble surface which prevents transfer of hydrogen into the bubble. Since freon also contains fluorine, it is quite possible that strontium fluoride plays a role in preventing degassing. Accordingly it is necessary to degas prior to modification if a reactive gas is to be used.

  The choice between an inert or a reactive gas is not a simple one. If melt cleaning during degassing is a priority, then a reactive gas should be used, but it should be kept in mind that excellent filtration systems are also available to clean the liquid (see Chapter 11). Should melt cleaning not be important, then the choice seems to be based on the efficiency of the degassing system used. A highly inefficient system can benefit from the use of a reactive gas, while a very efficient system will not benefit, and may actually become less efficient. In the case of an inefficient system, the benefits of changing to a rotary impeller degasser and using an inert gas should be considered. Finally, no matter what the efficiency of the degassing, only an inert gas should be employed in the presence of strontium.

10.4 Chemical Changes During Degassing

It is possible that other elements in addition to hydrogen will be removed during degassing. Among the alloying elements which may be affected, magnesium is the most important. Others which are commonly influenced by degassing are the modifiers sodium and strontium.

Although the greatest potential for chemical change occurs during reactive gas treatment, even inert gas treatment can lead to changes, particularly in modifiers. The increased melt agitation brought on by degassing accelerates the fading rate of both sodium and strontium. Vaporization of sodium is facilitated and some sodium is probably lost through the gas bubbles themselves. The effect is so severe that it is not recommended that degassing take place after a sodium addition. Rather, sodium-modified melts should be degassed prior to the modification treatment. Inert gas treatment also increases the rate of strontium fading through a greater opportunity for oxidation. Increases in the fade rate of up to three times have been measured. However, the fade rate of strontium is so low that an increase of even this magnitude is of little practical importance.

The largest changes in melt chemistry can occur when reactive gases are used to remove hydrogen. Primary aluminum producers frequently use a reactive gas treatment to refine molten aluminum, for example to remove alkali metal impurities (lithium, sodium and calcium). In such cases, these chemical changes are sought. However, in the foundry industry these changes are usually undesirable since alloy compositions are fixed with specifications. Elements which are lost during degassing may often have to be replaced in order to maintain the alloy specification, or the desired effects of a melt treatment such as modification may be lost.

As reactive gases contain some proportion of chlorine or fluorine, chemical changes brought on by degassing involve the formation of chlorides or fluorides. These are either solid or liquid at the degassing temperature and will float or be carried into the dross. Whether or not a given element in the alloy will react with the degassing agent depends on its free energy of formation with respect to aluminum chloride or aluminum fluoride. The free energies of formation of several metal chlorides and fluorides are shown in Figures 10.11 and 10.12 as a function of temperature. It can be seen that magnesium, calcium, sodium and potassium are all likely candidates for removal.

Magnesium removal has been most studied in primary and secondary aluminum production, and is removed through reaction with aluminum chloride gas bubbles as follows:

$$2Al_{(l)} + 3Cl_{2(g)} \longrightarrow 2Al_{3(g)}$$
$$2AlCl_{3(g)} + 3\underline{Mg} \longrightarrow 3MgCl_{2(l)} + 2Al_{(l)}.$$

In addition to thermodynamics, kinetics is also important in determining how much magnesium (or any other element) will be removed. The same kinetic factors are important in removing metallic elements from the melt as in removing hydrogen. Any highly efficient degassing process will, unfortunately, also be highly efficient in changing the alloy chemistry with respect to certain elements other than just hydrogen. A list of important elements which can easily be removed is given in Table 10.1 along with the chemical form in which they are removed. There has been virtually no systematic study of such chemical changes during the degassing of foundry alloys, and so it is not possible to state how fast the melt chemistry will be changed or what will be the final composition.

Magnesium removal by chlorine is very efficient above 710C (1310F) and over 90% of the magnesium can be removed. Below this temperature the reaction mechanism changes and the removal falls to 20%[9]. It is worth noting that the thermodynamic minimum for each of the elements listed in Table 10.1 is in the parts per million range. Very low concentrations can in theory be reached. The fastest rate of composition change

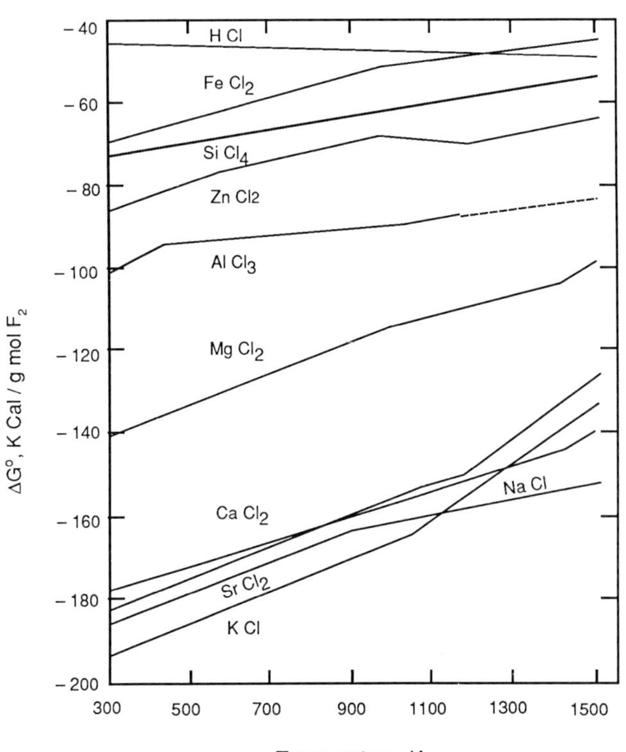

Figure 10.11. The free energies of formation of chlorides.

will occur if the pure reactive gas is used for purging. Dilution with an inert gas will slow the reaction considerably.

Experience with vacuum degassing has shown that alloy compositional changes are in general less severe than those occurring with reactive gas treatment, but more severe than with inert gas purging. Sodium loss is still significant and sodium treatment after degassing is again recommended. Strontium vaporizes, but relatively slowly—a loss of 0.005% from an initial addition of 0.05% has been reported. One of the main advantages of vacuum treatment is the quite small loss of magnesium (0.02% to 0.04%) at operating pres-

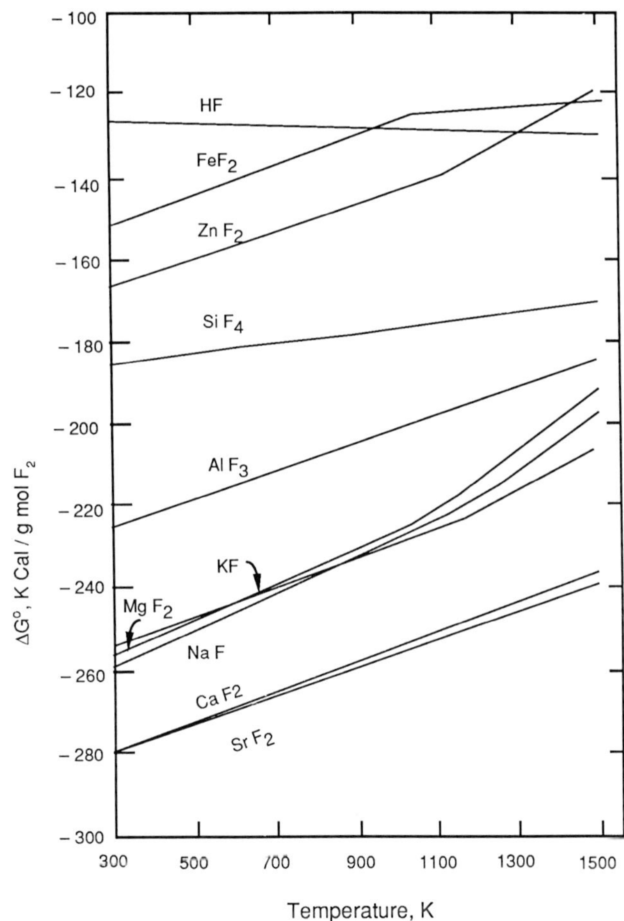

Figure 10.12. The free energies of formation of fluorides

Table 10.1. Element Removal with Reactive Gas Treatment

Reactive Gas	Element Removed	Reaction Product	Equilibrium State at 730C (1345F)
Chlorine	Be	$BeCl_2$	vapor
	Ca	$CaCl_2$	liquid
	Li	$LiCl$	liquid
	Mg	$MgCl_2$	liquid
	Na	$NaCl$	solid
	Sr	$SrCl_2$	solid
Fluorine	Be	BeF_2	liquid
	Ca	CaF_2	solid
	Li	Li_3AlF_6	solid
	Mg	MgF_2	solid
	Na	Na_3AlF_6	solid
	Sr	SrF_2	solid

sures above 1 mm Hg. On the other hand, zinc vaporization is very pronounced, so much so that the process is not suitable for zinc containing alloys.

References

1. Sigworth, G.K. "Practical Degassing of Aluminum," *modern casting* (March 1988) pp. 42-44.
2. Sigworth, G.K. and T.A. Engh. "Chemical and Kinetic Factors Related to Hydrogen Removal from Aluminum," *Met. Trans.*, 138 (1982) pp. 447-60.
3. Engh, T.A. and T. Pedersen. "Removal of Hydrogen from Molten Aluminum by Gas Purging," *Light Metals 1984* (1984) pp. 1329-43.
4. Sigworth, G.K. "A Scientific Basis for Degassing Aluminum," *AFS Transactions,* 95 (1987) pp. 73-78.
5. Anderson, A.R. "Rotary Impeller Degassing: Practical Observations," *AFS Transactions,* 95 (1987) pp. 533-36.
6. Mollard, F.R. and N. Davidson. "A Non-Polluting Degassing Technique for Aluminum Alloy Melts," *AFS Transactions,* 86 (1978) pp. 501-504.
7. Stevens, J.G. and H. Yu. "Mechanisms of Sodium, Calcium and Hydrogen Removal from an Aluminum Melt in a Stirred Tank Reactor—The Alcoa 622 Process," *Light Metals 1988* (1988) pp. 437-43.
8. Dimayuga, F.C., N. Handiak, J.E. Gruzleski. "The Degassing and Regassing Behaviour of Strontium Modified A356 Melts," *AFS Transactions* 96 (1988) pp. 83-88.
9. Whiting, L.V. and J.O. Edwards. "Removal of Magnesium from Aluminum by Fume less Chlorine Treatment," *AFS Transactions,* 82 (1974) pp. 189-92.

Chapter 11

Filtration

11.0   Introduction

Inclusions, or dirt, have long been recognized as a problem in the manufacture of castings. They are very often cause for rejection of a part by the customer, since their presence can lead to reduced mechanical properties, poor surface finish, increased porosity, or a tendency to increased corrosion. Early methods of inclusion control often focused on the molding process—the incorporation of pouring basins and runner extensions whose function was to trap dirt particles before they entered the mold cavity. Strainer cores which are refractory discs with discreet holes were also introduced into the runner system with the aim of trapping foreign particles.

None of these early attempts to remove inclusions was particularly effective, and at best only the largest particles were removed. With the increased use of aluminum alloy castings for structural applications in the aerospace industry, and for mass produced automotive parts, came the realization that improved inclusion control techniques were necessary, and hence effective methods of filtering liquid aluminum have been developed.

Filtration was used in primary aluminum production long before its application in the foundry industry. Early attempts at filtration of foundry alloys used techniques borrowed from the primary producers. More recently, manufacturers have developed, in parallel, filters for both primary and foundry aluminum casting.

In this chapter we will first examine the origin of inclusions in foundry melts and the types which may be expected. The effects of inclusion removal by filtration on mechanical and some physical properties of aluminum alloys will be discussed and we will then examine the types of filters available, how they work, and the factors determining their efficiency. Finally, a brief discussion of the ways to assess the cleanliness of a foundry melt will be given.

11.1 Inclusion Types and Origins

Inclusions in aluminum foundry alloys may be of two general types: exogenous and/or indigenous. Exogenous inclusions originate from outside the melt itself and include refractory particles such as alumina, silica, and silicon carbide, which result from the wear and erosion of crucible materials. These are found as discrete particles and may range in size from about 1 μm up to several millimeters. Indigenous inclusions arise from either chemical reactions within the melt itself, or else remain from some deliberate melt treatments such as fluxing or grain refinement. These inclusions may be either solid or liquid. Some important characteristics of indigenous inclusions are summarized in Table 11.1. Most of the data in this table is taken from reference[1].

It can be seen from Table 11.1 that inclusions may be present in several physical forms as discrete particles, clusters of particles, or as films. The oxides, in particular, may be found in both film form and as discrete particles. The photomicrographs of Figure 11.1 present the appearance of some typical inclusions.

11.2 Why Remove Inclusions?

As mentioned in the introduction to this chapter, inclusion removal has until recently been of greater concern to the primary aluminum producer than to the founder. Consequently, a much greater body of knowledge exists on the benefits of filtration of wrought alloys compared to cast alloys. However, the basic problems are often quite different. For example, a major concern with clean metal for wrought applications arises from the need to produce very thin strip for beverage can manufacture. As the inclusion size approaches the strip thickness, the probability of producing cans which leak is increased, and hence the drive to manufacture cleaner and cleaner metal. In foundry applications clean metal is desired for a variety of reasons related to mechanical properties, porosity and metal fluidity.

Tensile Properties

Filtered metal in general produces improved tensile properties with slightly increased values of UTS, and with significant improvements in ductility. Increases in percent elongation of from 25-100% can be obtained by filtration. It is, of course, always necessary to remember that aluminum casting alloys are never very ductile materials, and so an improvement of say 50% in the tensile ductility may not be large in absolute terms. In Figure 11.2 we see the

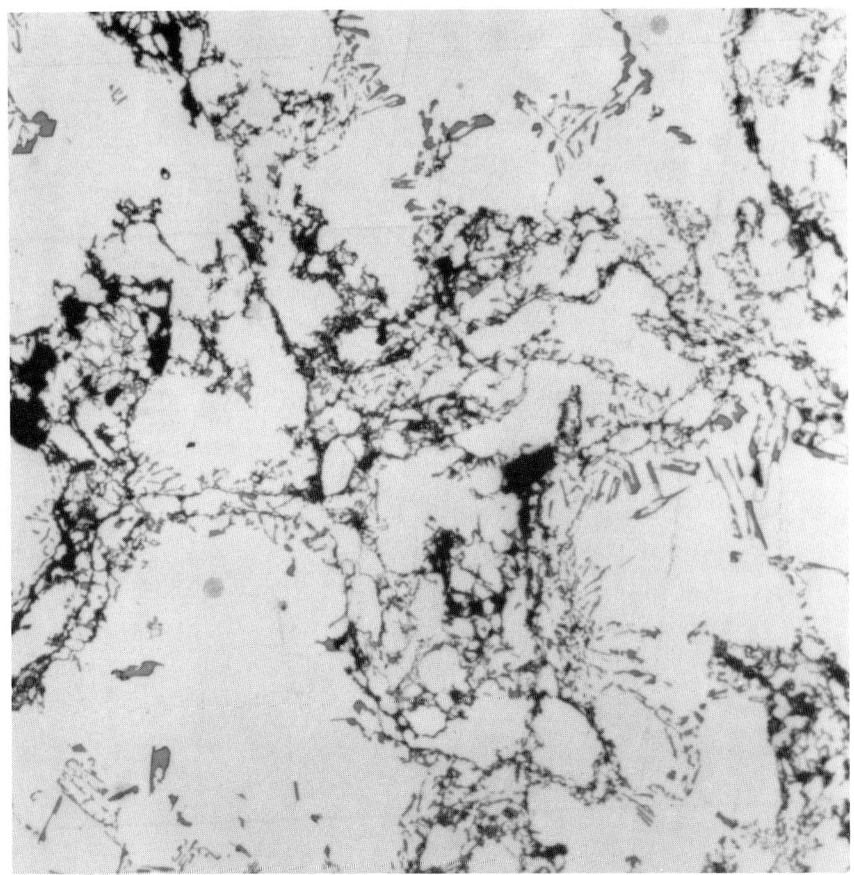

(a) Subsurface oxide skin inclusion found near the top surface of a runner in A356 alloy, X130. (Photo courtesy of Mr. Leonard Aubrey, Selee Corp.)

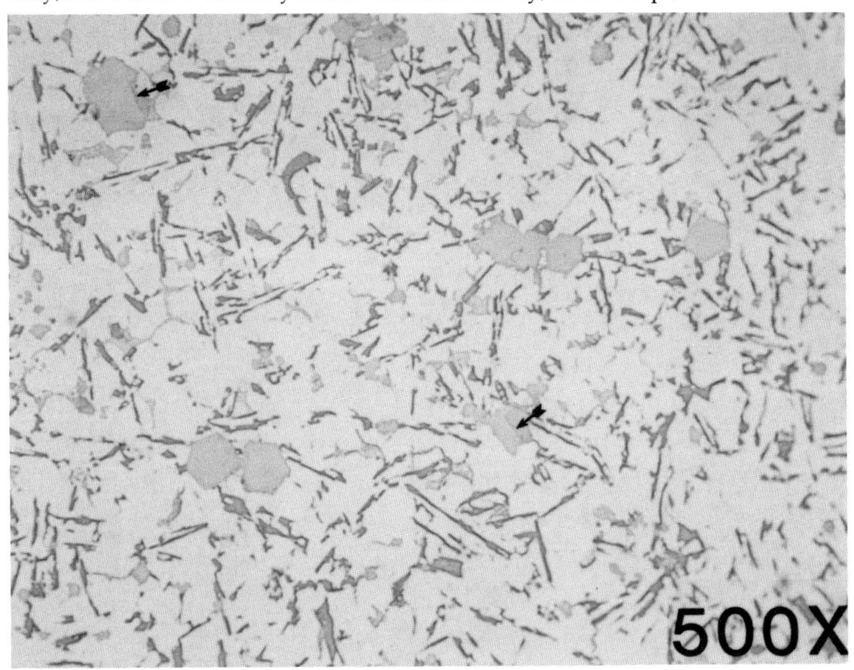

(b) sludge (AlSiFeMn(Cr)) in a die casting alloy[3].

Figure 11.1. Typical inclusions in aluminum foundry alloys.

Table 11.1. Indigenous Inclusions in Aluminum Foundry Alloys

Type	Origin	Shape	Size (μm)
Al_2O_3 (alumina)	dross	particles	0.2-30
		films	10 - 5000
MgO	dross	particles	0.1 - 5
		films	10 - 5000
$MgAl_2O_4$ (spinel)	dross crucible scale	particles	0.1 - 5
		films	10 - 5000
Chlorides Fluorides	fluxes	particles	0.1 - 5
TiC	grain refiner additions	particles	0.1 - 5
TiB_2 AlB_2	grain refiner additions	clusters	1 - 30
		particles	0.1 - 3
Fe-Cr-Mn (sludge)	melt reaction at low temperature in die casting alloys	particles	1 - 50

results of filtering an Al-4.5%Cu-1.5%Mg alloy through a bed filter. Filtration improves with bed depth as do the tensile properties.

The scatter obtained in tensile properties of cast aluminum alloys is well known to anyone who has done such testing. Inclusions which help to initiate fracture are a major contributor to this scatter, and their removal results in more consistent mechanical properties. Thus cleaner metal yields material with more reproducible and well defined properties.

Fatigue Properties

The role of inclusions in initiating fatigue cracks under dynamic loading conditions is well known. It has been demonstrated, for example, that TiAl$_3$ particles nucleate cracks in A201 alloys[5], and it is reasonable to expect that other types of inclusions act similarly, particularly if they are of large size. Filtered metal should, therefore, produce castings having an increased fatigue resistance. The compressor impeller illustrated in Figure 11.3 is an example of a part in which

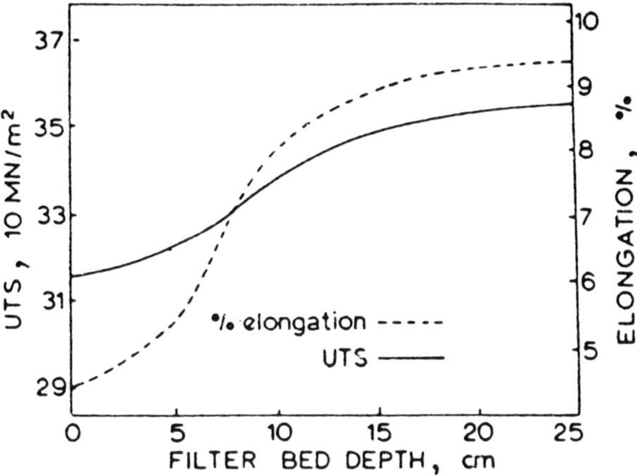

Figure 11.2. Variation in tensile properties with filtration for an Al-4.5% Cu-1.5% Mg Alloy[4].

fatigue properties are important. The impeller on the left was poured with unfiltered metal and failed dye penetrant inspection due to inclusions in the blade area. Filtration reduces the incidence of this type of defect and produces acceptable castings such as the one on the right.

Machinability

Inclusions are often ceramic particles. As such, the metal-nonmetal compounds formed are significantly harder than the aluminum alloy matrix and adversely affect machinability. Although little quantitative data exists to illustrate the deleterious effect of dirty metal on tool wear and breakage, anecdotal information is common, and clean metal pro-

Figure 11.3. Compressor impeller for heavy diesel equipment. The A201 part is manufactured by plaster investment casting. The dye penetrant indications on the unfiltered casting (left) are indicative of inclusions. The casting produced from filtered metal (right) exhibits no inclusions. (Photo courtesy of Selee Corporation.)

duced by filtration can be expected to improve machinability quite significantly.

Porosity

During the discussion in Chapter 9 of the formation of porosity in cast aluminum, it was mentioned that dirt particles can aid in pore nucleation. All other factors being equal, dirty metal may result in more porosity defects than clean metal. It is often supposed that dirty metal absorbs more hydrogen. Evidence for this is scanty, but aluminum melts may contain some hydrides which decompose on cooling to yield dissolved hydrogen. Whether due to an absence of nuclei, or to a lower total hydrogen content, it is certainly recognized that cleaner melts are more resistant to porosity formation in the final casting.

Fluidity

Oxide films exist as solids within the liquid phase and as such impair metal fluidity, particularly if present in large quantities. Removal of these films through filtration can significantly improve fluidity as illustrated in Figure 11.4.

It is very difficult to obtain specific data on the effects of filtration on a wide spectrum of cast alloys. Much remains to be done in this field, for example on the influence of clean metal on impact and fracture properties.

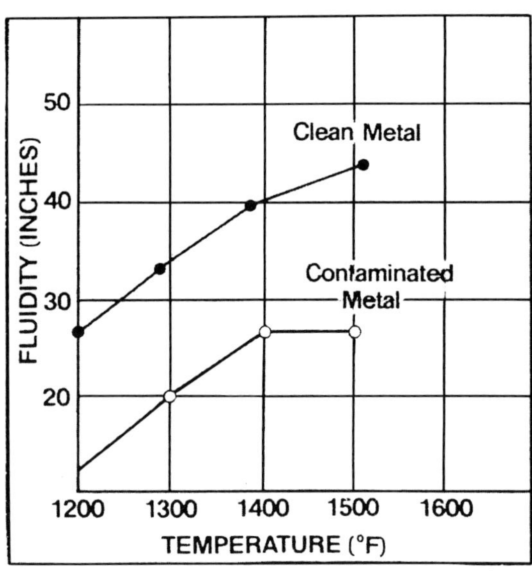

Figure 11.4. The effect of filtration on metal fluidity[6].

11.3   Filter Types and Placement

Filter Types

The earliest filters used for aluminum were woven fiberglass cloth and screens made from steel wire. These were placed in the gating system and acted to sieve out inclusions. As will be shown later, a sieving action is not a very efficient mechanism for inclusion removal, and these early filter devices removed only the largest inclusions from the melt.

Bed filters have been used in primary production plants for several years. In these, the molten aluminum is poured through a bed of tabular alumina balls and inclusions are trapped as they travel through the interstices between the balls.

Inclusions can also be removed by fluxing with a sparging gas as described in Chapter 10. In this method, the reactive gas wets the surface of the inclusion, allowing it to be attached to the gas bubble. Bubbles and inclusions rise to the melt surface, i.e., the inclusions are floated and they can be removed in the dross.

Today, the foundry industry standard is the ceramic foam filter, which for aluminum is manufactured of alumina (Al_2O_3) and contains 85-90% open porosity. These filters are available in a variety of shapes and sizes for many applications. Figure 11.5 shows a scanning electron micrograph of the filter surface showing the tortuous nature of the pore structure.

Ceramic foam filters are produced by impregnating reticulated polyurethane foam with a slurry of alumina mixed with a binder. The organic material is subsequently burned out and the alumina fired to develop strength. The result is a replication, in alumina, of the original foam. Their high porosity gives these filters a very light weight, but for certain applications, they can lack strength, and their high degree of porosity makes filtration of fine particles ($<20\mu m$) difficult. The porosity of a ceramic foam filter is designated by the number of pores per linear inch (ppi) measured on the filter surface. Common sizes are in the range 15-30 ppi.

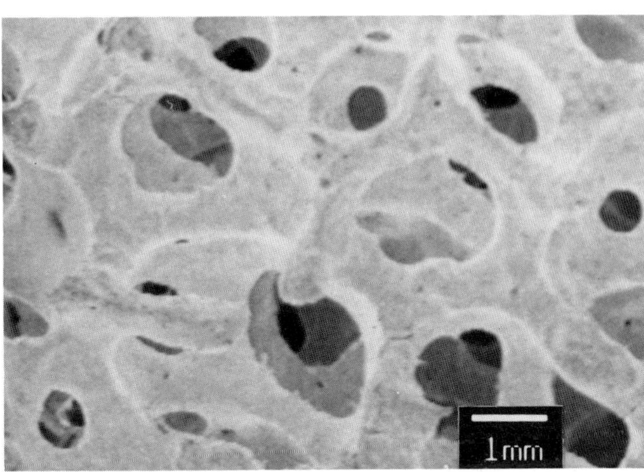

Figure 11.5. Scanning electron micrograph of the surface of a ceramic foam filter.

An alternative to the ceramic foam filter in some applications is the bonded particle filter (Fig.11.6). These filters are manufactured by mixing refractory grains of materials such as alumina or silicon carbide with a binder and firing to develop strength. The amount of porosity is varied by changes in the grain size of the ceramic particles. In general, these filters are denser than the ceramic foam filter and hence possess less porosity. Size designation is given as a grit size with approximate equivalences to ceramic foam filters, as in Table 11.2.

Figure 11.6. Bonded particle filters.(Photo reprinted with permission of the Metaullics Systems Div., Carborundum Co., Niagara Falls, NY.)

Filter Placement

Filters are placed either in the furnace or melting crucible or in the mold itself. Choice of location is dictated by the casting operation and by the cost per unit weight of metal filtered. For example in die casting, the metal is filtered as it leaves the melting furnace or as it is held in a ladle or crucible. The nature of sand or permanent mold casting permits inclusion of filters in the running system of the casting itself. Since metal flow into a mold cavity requires high flow rates, it is necessary to use quite open filters, such as the ceramic foam type, for this application. Bonded particle filters with their greater filtration efficiency, more robust nature, and slower associated flow rates, find application in furnaces and crucibles. Some typical applications of filters in furnace and melting crucible situations are shown in Figure 11.7.

Table 11.2. Grit Size Equivalent to ppi

Grit Size No.	Pores per inch (ppi)
6	30
8	40
10	50

The in-mold use of filters requires that they be placed as close to the mold cavity as possible—where they will do the most good. One major advantage of the use of a filter in the running system is that it virtually eliminates turbulent flow of the metal. Thus, the incidence of oxide formation or mold erosion after the filter is greatly reduced. Since mold cavities should fill from the bottom up, a horizontal filter position is usually recommended. Several filter locations, both horizontal and vertical, are illustrated in Figure 11.8.

11.4 How Filters Work

Two fundamental types of filtration have been identified: deep bed filtration and cake filtration. Modern metal filters operate by each of these mechanisms, although in a given filter type one may be more important than the other.

Deep Bed Filtration

In deep bed filtration, foreign particles are trapped within the body of the filter itself as the liquid metal flows through the filter. Usually, these particles are smaller than the filter pore size, and entrapment depends on their contact-

Figure 11.7. Some typical applications of filters in furnace and melting crucible situations. ([a] & [c] reprinted with permission of The Carborundum Co., Niagara Falls, NY.)

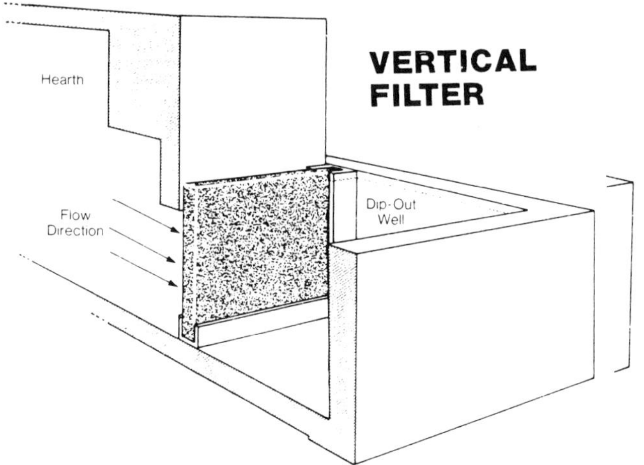

(a) the vertical gate filter

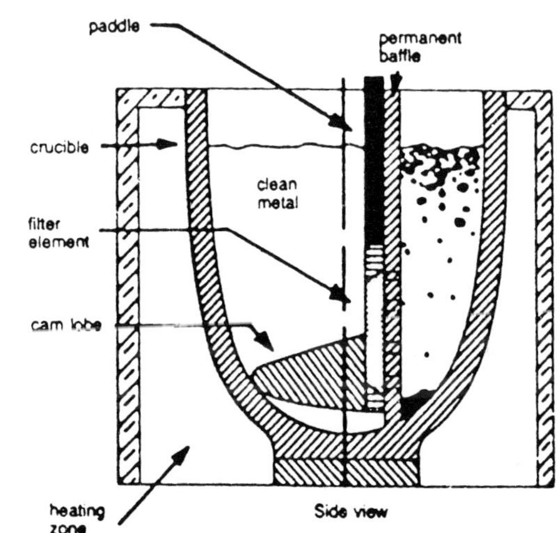

(b) melting crucible with filter baffle

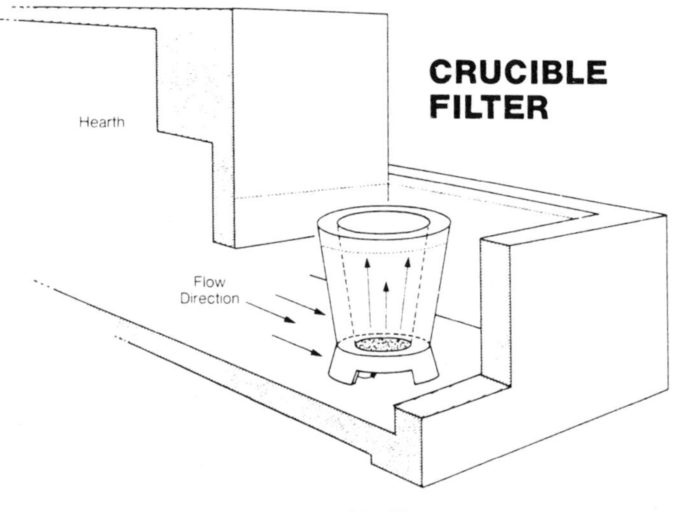

(c) crucible filter

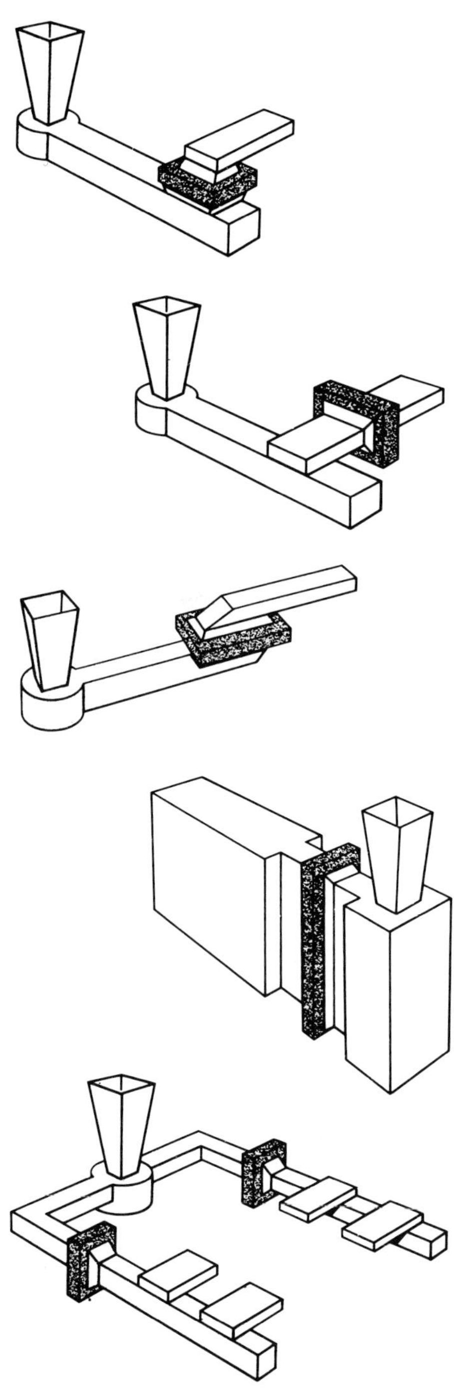

Figure 11.8. *Positioning of a filter close to the mold cavity.*

ing the pore walls and then being bonded in place. The three important mechanisms by which inclusions contact the pore wall in aluminum filtration are sketched in Figure 11.9. A particle is trapped by inertial forces if its weight prevents it from following the fluid flow line. It then simply deposits onto the filter wall. Particles may also directly intercept the pore wall (i.e., hit it), or they may be placed there by hydrodynamic effects resulting from the lower liquid velocity near the wall, which can cause particles to rotate into the pore surface.

When particles contact the internal filter surface, they are held in place by electrostatic forces of the Van der Waals type. These bonding forces are very weak, and particles or clusters of particles may become detached and washed from the filter by sudden increases in flow rate, mechanical vibration of the filter, or by backwashing.

Smaller inclusions (<20μm) are removed by the deep bed mechanism, which becomes more important as the pore size of the filter decreases, the flow path becomes more tortuous, and the filter thickness increases. Although all filters will exhibit some deep bed characteristics, particularly early in the filter life, this mechanism is most important in bonded particle and bed filters.

Cake Filtration

Cake filtration (Fig.11.10) occurs on the filter surface, and is essentially a sieving action. Initially, particles with sizes larger than the filter pore size are trapped on the inlet side of the filter. The spaces which remain between these trapped particles are finer than the original pores of the filter, and so smaller particles can now be stopped. Eventually a cake of trapped material builds up, and this filter cake itself acts as a filter. In cake filtration, the function played by the original filter is only to initially trap the largest foreign particles and then to act as a support mechanism for the filter cake. Of course, eventually the filter cake will become so thick that it will unduly impede liquid flow and the filter will become clogged.

Both cake and deep bed filtration occur in modern ceramic foam filters. The most effective filtration occurs during the deep bed period. Eventually, the filter becomes so clogged with inclusions that cake filtration predominates. When this stage is reached, the filter is essentially plugged, and will no longer pass molten aluminum.

11.5   Filter Efficiency

The efficiency of a filter refers to its ability to effectively remove inclusions of a given size, and is usually defined as:

$$\frac{\text{Inlet concentration of inclusions} - \text{Outlet concentration of inclusions}}{\text{Inlet concentration of inclusions}} \times 100\%$$

Some attempts to analytically predict efficiency in aluminum filtration have been made [1,7], and these show that efficiency depends on a complex inter-relationship between several factors as summarized in Table 11.3.

In Figures 11.11-11.13, the effects of these variables are shown on the calculated efficiencies, during the initial stages of deep bed filtration before cake build-up occurs. Close to 100% removal of large, dense inclusions can be achieved at appropriately small flow velocities. This value decreases to something of the order of 20% for small light particles at melt velocities in excess of 1 cm/sec. Filter manufacturers often cite 90-95% efficiency levels, but these are under truly optimum conditions only. Real efficiencies may be considerably less. Measured filter efficiencies have been

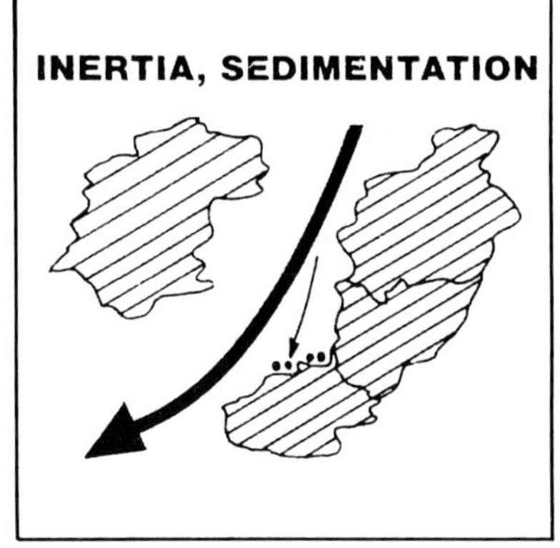

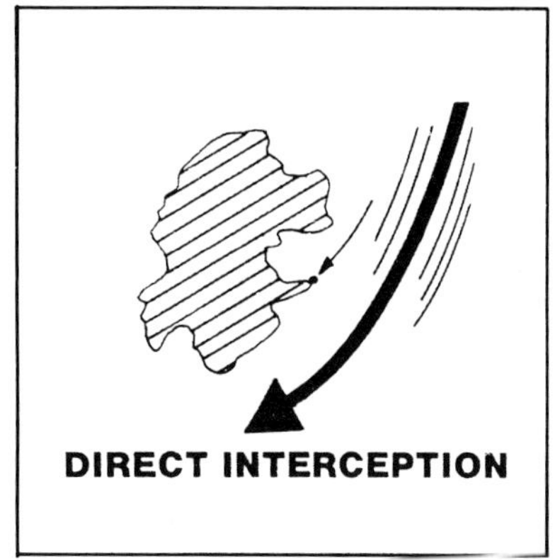

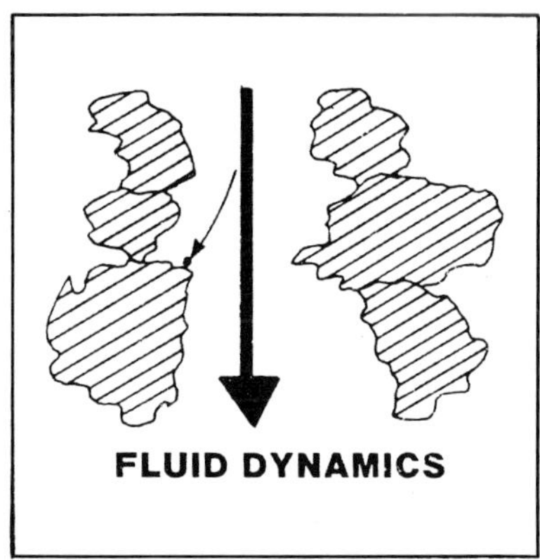

Figure 11.9. Mechanisms of particle trapping within a filter. (Illustration reprinted with permission of the Metaullics Systems Div., The Carborundum Co., Niagara Falls, NY.)

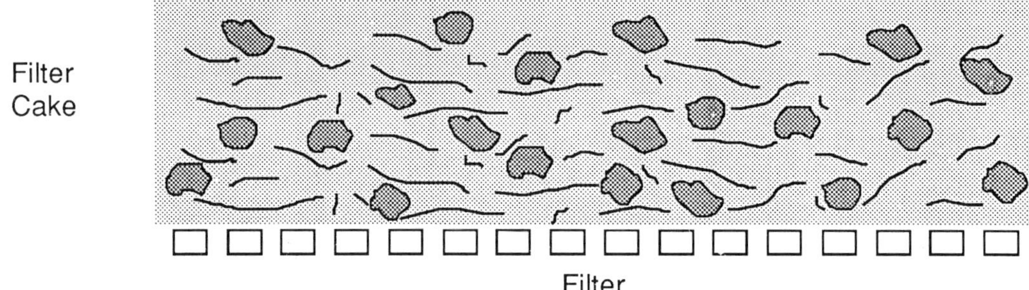

Figure 11.10. Cake filtration.

reported in a few cases and they show reasonable agreement with calculated data.

11.6   Filter Sizing

The incorporation of a filter in the mold running system adds a considerable resistance to flow, necessitating certain modifications to the runner system. If no changes are made, or if these are insufficient, the mold cavity will fill too slowly, and cold shuts or misruns may result. The first step in determining the size of filter needed is to calculate the mass of metal which is required to pass through the filter. The data in Table 11.4 gives the approximate quantities of liquid aluminum which can pass through a ceramic foam filter before it blocks, i.e., before the cake becomes so thick it severely restricts flow. Of course, these are average values; dirtier melts will pass less, cleaner melts, more. On this basis the area of filter required can be determined. The filter itself will be incorporated into a flared volume of the runner as illustrated in Figure 11.14.

Running systems incorporate a choke which determines and controls flow into the mold cavity. If no enlargement of the choke is made, the filter may restrict flow to such an extent that it acts as the choke. Usually, some modification to the runner will be necessary, and a convenient way to determine this is to appreciate that the filling time of the mold with a filter should be approximately the same as without a filter. A value of 90% is reasonable, and using the data of Figure 11.15, the ratio of filter area to choke area can be calculated.

Other factors which determine filling time are the metallostatic head (sprue height) and pouring temperature. Small increases in either of these parameters will aid mold filling in the presence of a filter. For example, flow rate increases as the square root of the head pressure. Small increases in pouring temperature of from 10–50F (18–90C) are often recommended if filters are used.

Table 11.3. Filter Efficiency

Factor	Efficiency	
	high for	low for
inclusions	large dense	small light
filter characteristics	large internal surface high ppi	small internal surface
liquid flow	low flow rate steady flow rate	high flow rate turbulent flow in filter medium non-steady (pumping) flow

Table 11.4. Quantity of Aluminum which Passes through a Ceramic Foam Filter before Blockage

Filter Size	Quantity	
(ppi)	(lb/in^2)	(kg/cm^2)
10	15	1.1
15	10	0.70
25	5	0.35

1.7   The Measurement of Melt Cleanliness

While melt filtration has been shown to be an effective method for cleaning liquid aluminum alloys, the actual quantitative assessment of melt cleanliness is a complicated matter. The need for a rapid and inexpensive technique to determine inclusion size, distribution, and type is obvious. Such a method would allow the effectiveness of filtration in furnace or melting crucibles to be controlled, and would allow changes in melting and pouring procedures to be evaluated in terms of their effects on melt cleanliness. At the moment, no such simple and inexpensive technique has been developed that seems suitable to a wide spectrum of foundry operations. As has often been the case, the primary aluminum producers have led in activity in this area, and have achieved some success in devising methods within are suitable to their large-scale continuous operations. Most methodologies to quantify melt cleanli-

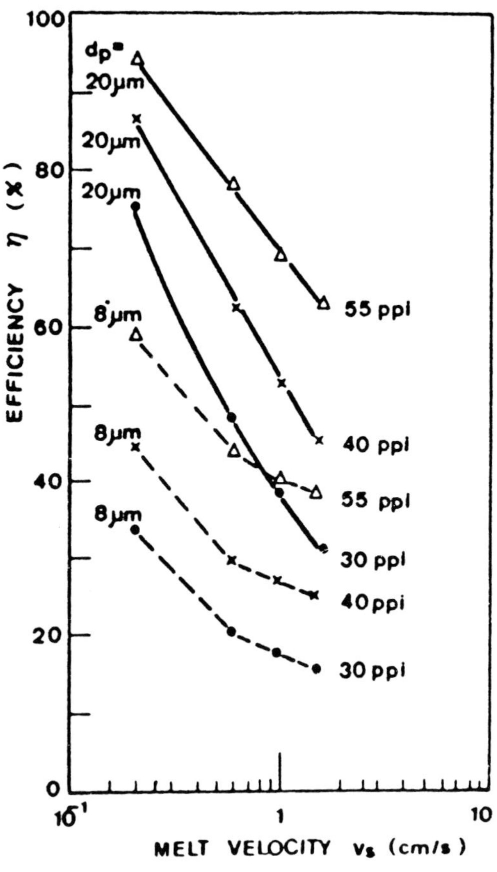

Figure 11.11. Calculated filter efficiencies as a function of melt velocity for various inclusion diameters (d_p) and filter sizes. A 5 cm head of aluminum is assumed[1].(Reprinted with permission from *Light Metals 1985*, edited by H.O. Bohner [1985], The Metallurgical Society, 420 Commonwealth Drive, Warrendale, PA 15086.)

ness are based on chemical, metallographic, or physical properties of the melt, in either the liquid or solid form.

Chemical Methods

While bulk chemical analysis might seem a simple way to determine the inclusion content, it is of very limited usefulness, since it tells nothing about the number or shape of the inclusions. For example, an aluminum alloy can be analyzed for oxygen content, but as seen in Table 11.1, it could contain both aluminum oxides and spinels in particle or film forms, and a wide range of sizes.

Metallographic Methods

An inclusion count on a polished metallographic sample would seem to be a relatively simple way to determine inclusion types and size distributions. The difficulty here is that reasonably clean melts contain relatively few dirt particles per unit volume, perhaps 20 particles per ml., $20\mu m$ and larger. In order to obtain statistically significant data, it is necessary to view a large number of

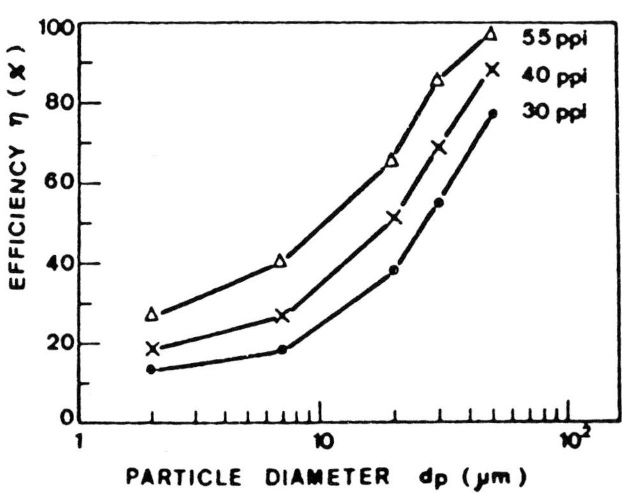

Figure 11.12. Calculated filter efficiencies as a function of inclusion diameter for three filter sizes. A 5 cm head of aluminum is assumed; inclusion density is 4.5 g/cm³, flow velocity 1 cm/sec [1].(Reprinted with permission from *Light Metals 1985*, edited by H.O. Bohner [1985], The Metallurgical Society, 420 Commonwealth Drive, Warrendale, PA 15086.)

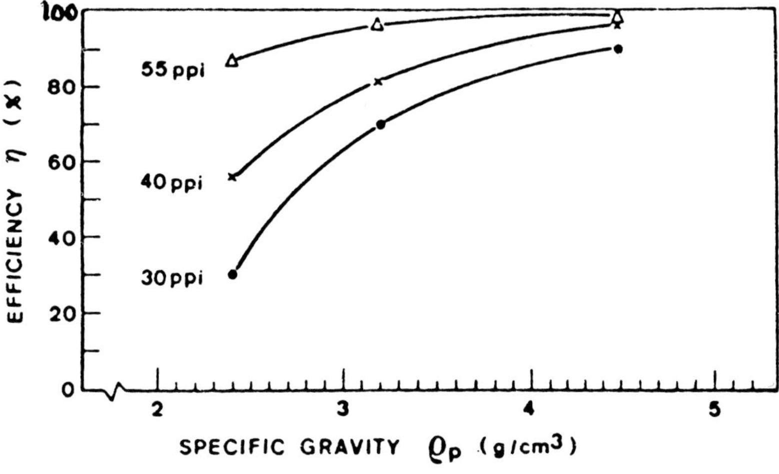

Figure 11.13. Calculated filter efficiencies as a function of inclusion density for three filters of 5 cm. thickness. Particle diameter is 50 μm and flow velocity is 0.6 cm/ sec[1]. (Reprinted with permission from *Light Metals 1985,* edited by H.O. Bohner [1985], The Metallurgical Society, 420 Commonwealth Drive, Warrendale, PA 15086.)

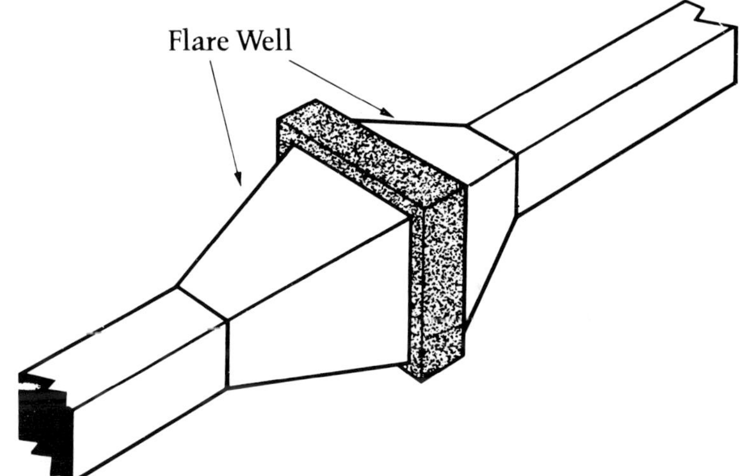

Figure 11.14. Flared filter well.

samples, and the method becomes cumbersome and impractical.

For this reason, preconcentration techniques have been developed. One of these, known as the PODFA method (Porous Disc Filtration Apparatus), has been used by Alcan for many years, and consists of forcing, under pressure, a five pound sample of liquid through a porous filter disc. Filtration is by cake formation and the cake is examined metallographically for inclusion type and size. As with any filtration process, the results obtained depend on the filter type and metal flow rate. This type of test is certainly not an on-line test, since more than one hour is typically needed for sample preparation, and considerable skill is required for proper and consistent interpretation of the metallographic results. Other methods of concentrating inclusions have been used, such as hot centrifuging[8], but this too requires time and skill in interpretation.

A somewhat simpler metallographic technique reported by Hedjazi[4], was to examine the fracture surface of a 3 mm thick tensile specimen cast from

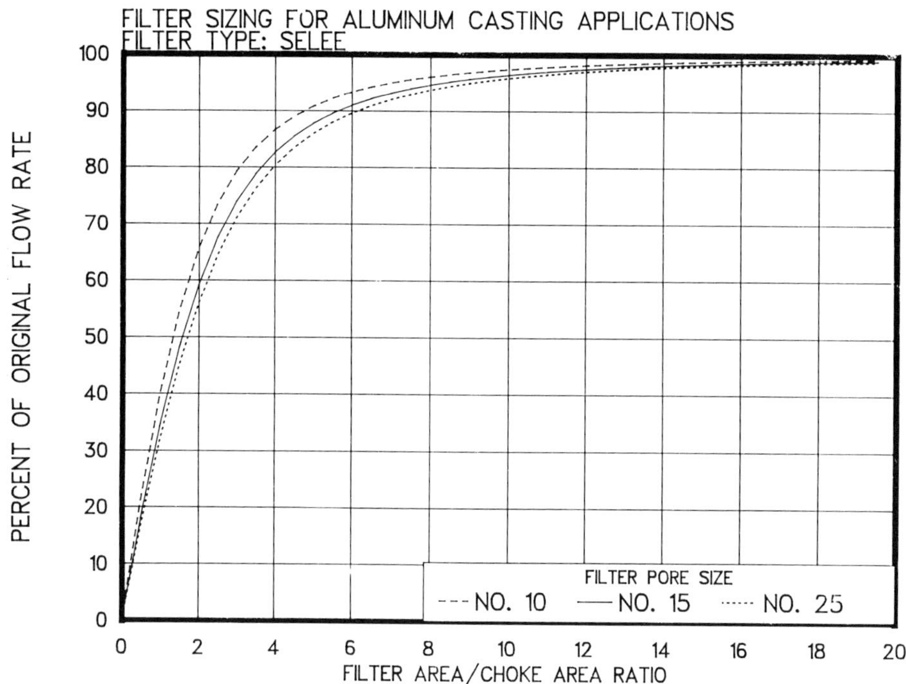

Figure 11.15. Variation of percentage of flow rate without a filter as a function of filter area/choke area ratio (courtesy Selee Corporation).

the melt in question and then rapidly broken in tension. Inclusions >50μm could be seen and counted at a magnification of X5; finer particles required examination at X100, and polarized light could be used to assist in identification.

Recently, there has been renewed interest in fracture tests to assess melt cleanliness. While they are of limited value in quantifying melt cleanliness, these tests can quickly demonstrate the presence or absence of large contaminants.

Physical Methods

The two most promising techniques based on a physical method which seem applicable to on-line testing of large quantities of liquid are based on ultrasonics and resistive pulse measurements. The ultrasonic technique, developed by Reynolds[9], uses a titanium probe to measure the attenuation and the discontinuity of an ultrasonic signal. An increase in either of these two factors is indicative of a decrease in melt quality. While this is an on-line technique, one of the major disadvantages of ultrasonic methods is their inability to detect the smaller inclusions. The detection limit of these techniques is probably 50 μm or larger.

Resistive pulse methods are metallurgical applications of the Coulter Counter, long used to detect fat particles in blood. The LIMCA System

(Liquid Metal Cleanliness Analyzer)[10], measures the change in resistivity (as indicated by a voltage change) of a thin tube of liquid metal as a non-conducting inclusion passes through it (Fig. 11.16). A resistive pulse or an increase in potential along the tube length results from the passage of one inclusion. The novelty of the technique is its application at high temperatures, and the use of high currents necessitated by the high conductivity of liquid metals. Currently the LIMCA is used in several aluminum smelters. Numbers of inclusions are readily measured, while information on size distribution depends on detailed signal analysis. Cost, coupled with the need for a skilled operator, will probably limit interest in LIMCA to only the largest foundries. It is nevertheless the most promising of sophisticated techniques for on-line inclusion measurement.

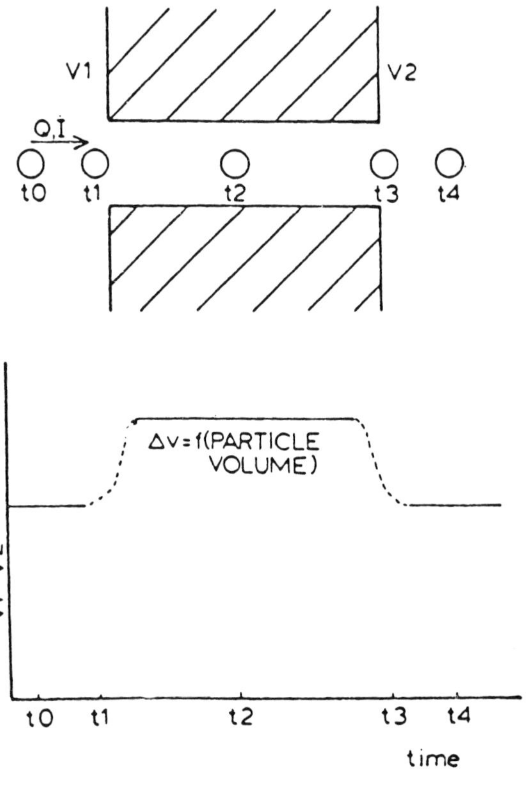

Figure 11.16. Principle of the LIMCA system for measuring melt cleanliness[10] An inclusion passing through a capillary increases the voltage difference (V_1-V_2) measured between the ends of the capillary.

Several other methods of quantifying liquid metal cleanliness have been studied. These include machinability tests and studies of the quality of a machined surface. Of all the methods investigated, perhaps one of the most promising for a quick overall test of cleanliness is the reduced pressure gas test described in Chapter 9[11]. The dependence of gas bubble nucleation on foreign particles in the melt has already been discussed, and it is known that dirtier melts will result in more porous castings. Hence, the porosity level of samples with similar dissolved hydrogen contents solidified under vacuum should be indicative of the inclusion content. Reliable use of this technique would almost certainly require an independent measurement of the hydrogen content, say by a recirculating gas measurement.

None of the techniques described here has been developed for foundry use, and many of them have never been applied to any great extent to foundry alloys. As seen, they range from the highly sophisticated to the quite simple, and it is reasonable to expect that the future will see detailed studies of these and other methods oriented towards foundry application as the need for a quick, shop floor test of melt cleanliness becomes more evident.

References

1. Gauckler, L.J., M.M. Waeber, C. Conti and M. Jacob-Duliere. "Industrial Applications of Open Pore Ceramic Foam for Molten Metal Filtration," *Light Metals 1985* (1985) pp. 1261-83.
2. Micks, F.W. and V.J. Zabek. "Identifying some Common Problems with Aluminum Castings," *AFS Transactions,* 82 (1973), pp. 38-42.
3. Jorstad, J.J. "Understanding Sludge," *Die Casting Engineer,* 30 (1986), pp. 30-36.
4. Hedjazi, D. J., G.H.J. Bennett and V. Kondic. "Removal of Non-Metallic Inclusions and Their Assessment in Al-Alloy Melts," *British Foundryman,* 68 (1975), pp. 305-309.
5. Chien, K.H., T.Z. Kattamis and F.R. Mollard. "Cast Microstructure and Fatigue Behaviour of a High Strength Aluminum Alloy," *Met. Trans,* 4 (1973), pp. 1069-76.
6. Groteke, D.E. "A Starting Point for A Foundry SPC Program: Uniform Metal Quality," SAE Congress Paper 850476, (1985).
7. Eckert, C.E., R.E. Miller, D. Apelian and R. Mutharasan. "Molten Aluminum Filtration: Fundamentals and Models," *Light Metals 1984* (1984) pp. 1281-1204.
8. Mollard, F.R., J.E. Dure and W.S. Peterson. "High Temperature Centrifuge for Studies of Melt Cleanliness," *Light Metals 1972* (1972) pp. 483-500.
9. Mansfield, T.L. and C.L. Bradshaw."Ultrasonic Inspection of Molten Aluminum," *AFS Transactions,* 93, (1985) pp. 317-22.
10. Guthrie, R.I.L. "On the Detection, Behaviour and Control of Inclusions in Liquid Metals," presented at: Foundry Processes: Their Chemistry and Physics An International Symposium, G.M. Research Laboratories (Sept. 21-23 1986).
11. Brondyke, K.J. and P.D. Hess. "Interpretation of Vacuum Gas Test Results for Aluminum Alloys," *Trans. Met. Soc. A.I.M.E.,* 230 (1964) pp.1542-46.

Chapter 12

Treatment with Cover and Cleaning Fluxes

12.0   Introduction

Flux treatment is a widely used term when applied to aluminum foundry technology, and in its most general sense is used to describe any external chemical operation performed on the melt. Simple degassing techniques such as nitrogen or argon treatment are often described as fluxing operations. Complex chemical fluxes can be used to degas, modify or grain refine melts, and these have been discussed in the appropriate chapters of the book. In this chapter the use of chemical fluxes to protect the melt, to deoxidize or clean the melt, and to clean the dross will be described. In addition, some of the problems associated with the use of chemical fluxes will be discussed.

At the outset, it should be made clear that it is not our intention to describe in detail all possible flux chemistries. A bewildering array exists from several international manufacturers, and new compositions are constantly finding their way into the marketplace. A visit to a flux manufacturer will reveal hundreds of different starting materials which are mixed in various proportions to form the commercially available compositions.

Of the over 150 flux compositions used in aluminum metallurgy, many are based on the NaCl-KCl system, with approximately equal amounts of each constituent. To this basic mixture are added other salts, either to increase the effectiveness of the NaCl-KCl flux, or for special purposes. Among these additives are: CaF_2, NaF, AlF_3, $AlCl_3$, $NaAlCl_4$, $KAlCl_4$, K_3AlF_6. Many fluxes fall within the system $NaCl-KCl-AlCl_3-NaF-KF-AlF_3$, but only recently has there been any detailed investigation of the reactions taking place within this system[1]. In most cases, the properties of fluxes or their behavior in contact with liquid aluminum are understood only scantily, if at all. It is little wonder, therefore, that flux treatment of aluminum alloys remains for the most part an empirical art.

12.1   Cover Fluxes

Cover fluxes are mixtures of various salts which form a molten layer on the surface of the liquid bath, protecting it from oxidation and reaction with moisture in the air. They isolate the liquid from the atmosphere, and in no way are they intended to add any new elements to the melt. A large number of compositions are available from several suppliers, but all cover fluxes contain alkali chlorides in various proportions and some also contain various fluorides.

The composition of a cover flux is determined by its melting point. Fluxes which are liquid at operating temperatures are usually desired, and a simple example of the influence of composition on the melting point of some very commonly used fluxes is seen in Figure 12.1. In this diagram, which represents the liquidus curve for the NaCl-KCl system, it is clear that the lowest melting point occurs at approximately equimolar quantities of NaCl and KCl. These fluxes would be molten at the surface temperatures of liquid foundry alloys. Increasing either the NaCl or the KCl content raises the melting point, until at very high concentrations of either component fluxes which do not melt at usual operating temperatures are obtained.

Of course, most fluxes are more complex than simple binary mixtures and higher order phase diagrams must be consulted. The effect of adding NaF to the simple NaCl-KCl flux described earlier is demonstrated in Figure 12.2, which gives the liquidus surfaces of the NaCl-KCl-NaF system. It is seen that the addition of NaF serves to raise the melting point.

In order to provide the reader with an idea of the range of possible cover flux compositions, a few taken from the work of Abreu[2], are listed in Table 12.1, and melting point ranges in terms of compositional ranges are presented in Table 12.2. Only the compositional ranges of the major constituents are listed in these tables, and so the mixtures do not add to 100%. The wide range in possible flux chemistries and melting points is evident.

While many fluxes can be used with a wide range of alloys, it is important that those containing

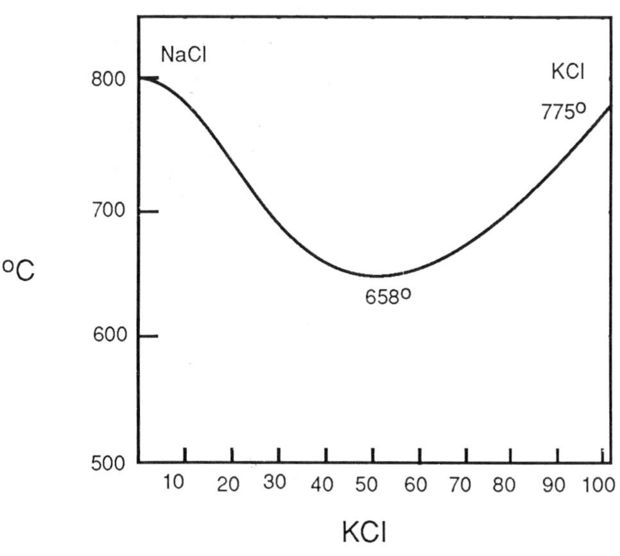

Figure 12.1. Variation of the melting point of NaCl-KCl mixtures with composition.

Table 12.1. Compositions of Some Typical Cover Fluxes

Component	Approximate Wt. Pct. of each Component in Typical Fluxes								
	#1	#2	#3	#4	#5	#6	#7	#8	#9
$NaCl$	10-50	85	20-40	35	24	4	30		53
KCl	30-60		26-30		34	13	40	20	16
NaF	10-30				4	13			
Na_3AlF_6 (cryolite)			20	50		13			20
CaF_2	15								10
KF			2				15		
$CaCl_2$						60			
$MgCl_2$					38			60	
Na_2CO_3				15			15		
MgF_2								20	

sodium salts not be used with alloys having more than about 2 wt.pct. magnesium. Magnesium will replace sodium in the salt according to a reaction of the type:

$$\text{Na-halide} + \text{Mg} \longrightarrow \text{Mg-halide} + \text{Na}$$

with the result that some magnesium is lost to the dross, and the alloy picks up sodium which may be deleterious to its mechanical properties.

For magnesium containing alloys, fluxes based on magnesium chloride or carnalite ($MgCl_2 \cdot KCl$) are used. While those fluxes may, in principle, be applied to all aluminum foundry alloys, they have the severe disadvantage that the magnesium containing salts are more highly hygroscopic, and the addition of hydrogen to the melt may be associated with their use unless very careful storage procedures are followed.

12.2   Cleaning Fluxes

Cleaning fluxes are designed to remove particles of alumina, aluminum oxide, from the melt. The active ingredient here is sodium fluorosilicate, Na_2SiF_6, which was originally thought to remove the aluminum oxides by a chemical reaction to produce silica, *viz.*:

$$6Na_2SiF_6 + 2Al_2O_3 \longrightarrow 4Na_3AlF_6 + 3SiO_2 + 3SiF_4.$$

The Treatment of Liquid Aluminum-Silicon Alloys

Table 12.2. Melting Point Ranges of Some Cover Flux Compositions

| | Melting Point Range (°C) | | | | | | | |
Constituent	590-610	550-625	530-720	600-680	580-620	750-760	600	655
	Compositional Range (wt. pct.)							
NaCl	40	60-70	55-65	30-45	45	30	15	
NaF	4-10	5-10		5-20	5-10			
KCl	34-51			65-35	30-56			
$AlCl_3$					6-10			
AlF_3	5-15							
Na_3AlF_6		6-10	20-30					
$BaCl_2$		14-25	5-25					
CaF_2						10		15
$CaCl_2$							80	85

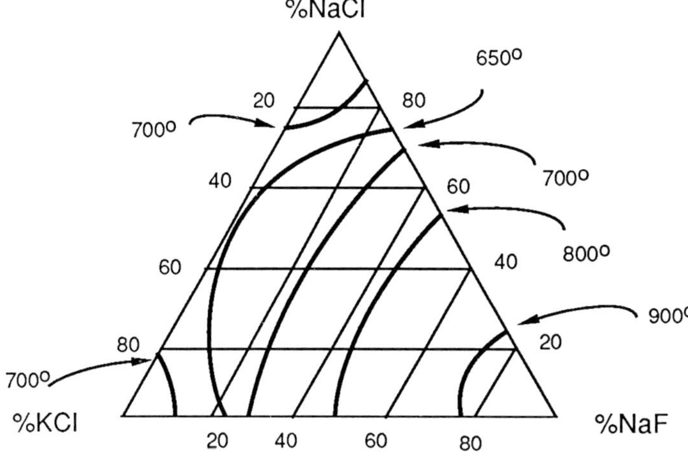

Figure 12.2. Liquidus surface of the KCl-NaCl-NaF salt system.

It now appears, however, that the cleansing action is due to a surface energy effect. The sodium fluorosilicate "coats" the interface between Al_2O_3 and the unreacted aluminum, which occurs at the center of the oxide particle. The oxide is thus stripped away, mechanically enveloped by the flux, and floats into the dross. The mechanism is illustrated in Figure 12.3.

While oxide removal is a problem common to all aluminum foundry alloys, there are some special cases in which a cleaning action is required, for example, the removal of sodium or calcium. This is accomplished by using a base flux of simple composition such as NaCl-KCl and adding to it a reactive component. If sodium removal is desired, $MnCl_2$ is added, and sodium is removed by the reaction:

$$MnCl_2 + 2Na \longrightarrow 2NaCl + Mn$$

To eliminate calcium, AlF_3 can be added to produce the reaction:

$$3Ca + 2AlF_3 \longrightarrow 3CaF_2 + 2Al$$

Magnesium can be similarly removed according to:

$$3Mg + 2AlF_3 \longrightarrow 3MgF_2 + 2Al.$$

It is clear from this discussion that the action of cleaning fluxes requires a reaction, either mechanical or chemical, between the active ingredients of the flux and the element or compound which it is desired to eliminate. The basic

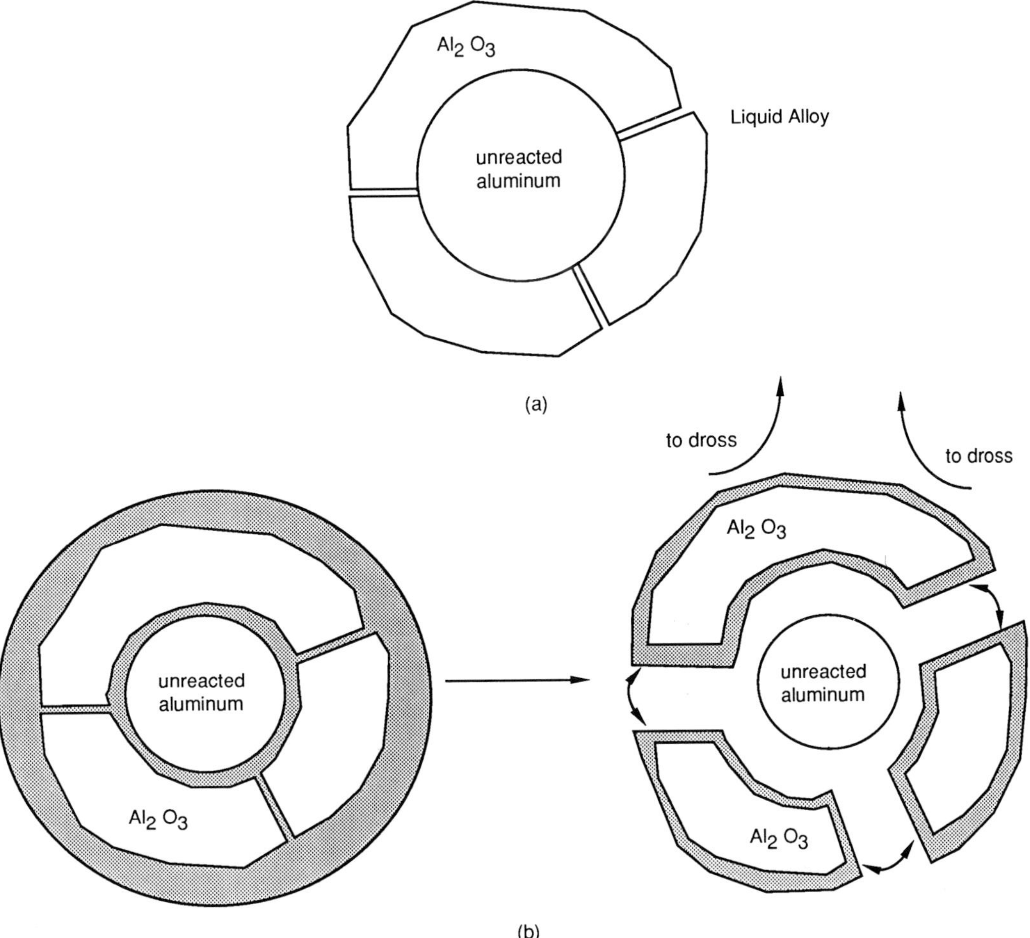

Figure 12.3. The cleansing action of Na_2SiF_6: (a) before addition of Na_2SiF_6; (b) after addition of Na_2SiF_6.

low melting point flux carries the reactive constituent, and in order to provide effective reaction kinetics, it is essential that the flux be mixed in with the bulk of the liquid alloy. This process can result in a significant metal loss through formation of a wet dross, i.e., one in which a large quantity of alloy becomes physically entrapped. This alloy may be present as fine solid or liquid particles which are extremely difficult to separate from the dross.

In order to eliminate this problem and to produce a powdery dry dross, cleaning fluxes most often contain small quantities of reactive oxidizing compounds such as sulphates or nitrates. These compounds trigger a series of exothermic reactions which increase the temperature of the dross to a point where the larger trapped aluminum particles coalesce into large liquid droplets. These droplets, because of their mass, fall out of the dross back into the bulk metal leaving a clean dry dross and increasing metal yield. A schematic illustration of this process is provided in Figure 12.4.

The exothermic reactions have been reported by Strauss[3] to occur in a series, beginning with oxidation (by the sulphates or nitrates contained in the flux) of the finest aluminum particles in the dross to alumina. This exothermic reaction raises the dross temperature to values in excess of 600C, whereupon further reaction of aluminum particles with fluorides in he flux occurs according to:

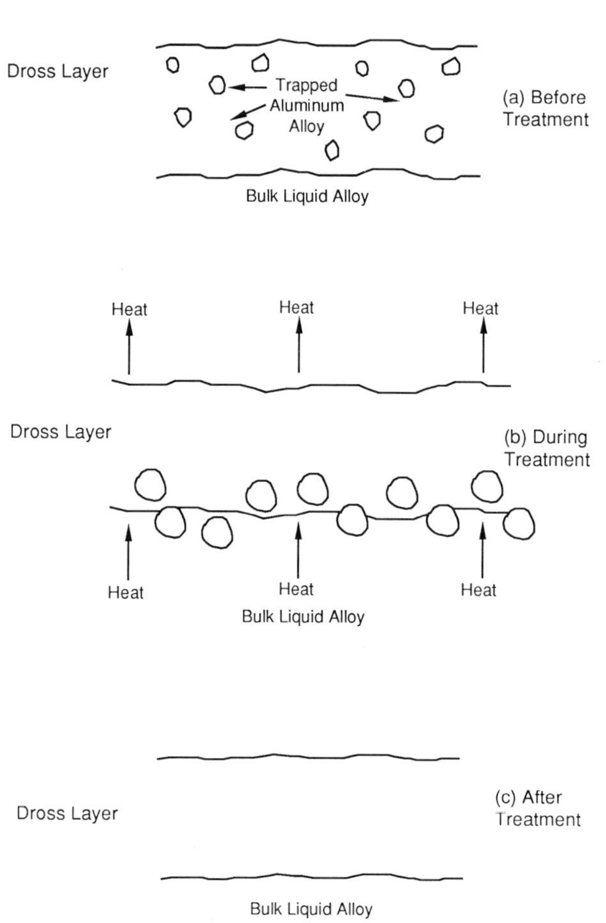

Figure 12.4. Dross cleaning by means of an exothermic reaction.

$$AlF_3 + 2Al \longrightarrow 3AlF \qquad (1)$$
$$3AlF + 3O \longrightarrow Al_2O_3 + AlF_3 + Heat. \qquad (2)$$

INJECTION PROCESS COMPONENTS

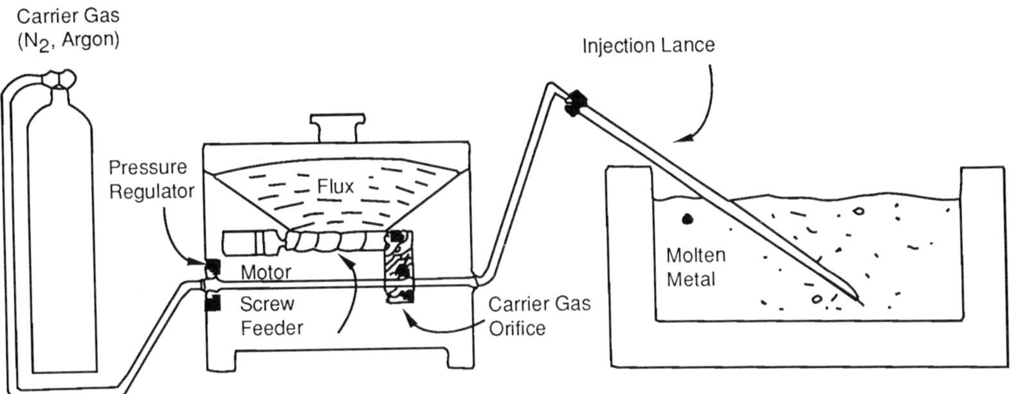

Figure 12.5. Schematic representation of flux injection process.

The product of reaction (1) is an unstable gas, AlF, which immediately reacts with oxygen from the atmosphere to form more AlF_3 gas. The reaction is thus self-perpetuating until all of the fine aluminum in the dross is consumed.

12.3   Flux Injection Process

Classical methods for using fluxes have involved the mechanical stirring of the flux powders into the melt. This fluxing method is extremely inefficient in that the physical contact between flux and liquid alloy, which is so essential to proper functioning of the flux, is difficult to achieve in a reproducible and effective way. In addition, inclusion of molten alloy in the dross is encouraged, and the excess agitation of the bath may actually enhance hydrogen gas pickup.

Recently, a more automated method of flux addition has been developed which encourages good flux-liquid contact—increasing efficiency and decreasing treatment time.[4] In the flux injection process (Fig.12.5), flux powders are blown into the melt using an inert carrier gas, which may be nitrogen or argon. Addition is via a lance, and so it is possible to blow the flux into the bottom of the bath. A flux powder dispersed in this manner will float to the surface, reacting chemically throughout the entire depth of the bath. It is therefore possible to contact significantly more of the alloy than by using classical means.

Several fluxes are available for use with this method, and in just one operation, it is possible to clean, degas, modify and grain refine. In addition, fluxes for the refinement of primary silicon in hypereutectic alloys have been developed.

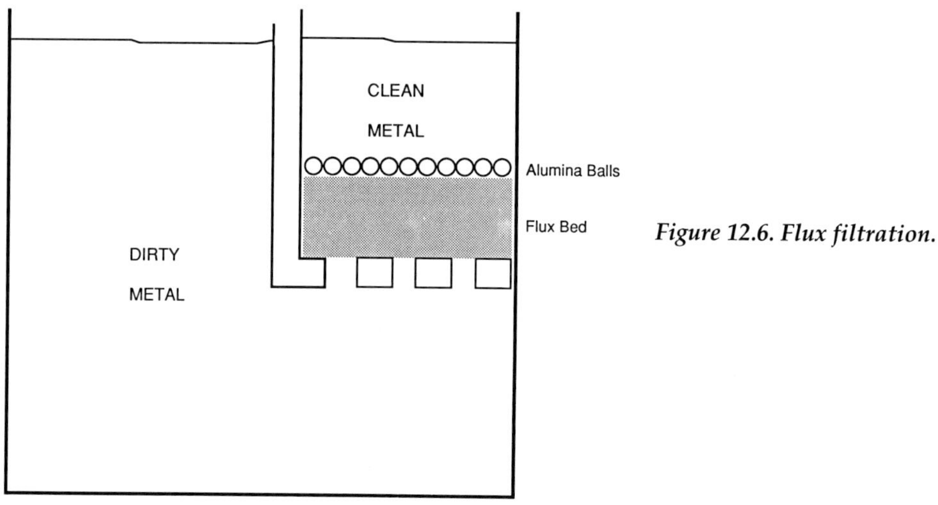

Figure 12.6. Flux filtration.

12.4   Flux Filtration

This process, illustrated in Figure 12.6, is similar to crucible filtration described in Chapter 11, with the exception that a bed of flux particles is used as the filter medium instead of a pore-type filter. Metal is ladled from the filter section of the crucible, and as this is done new metal is forced to pass through the flux bed. The bed, 10-15 cm in depth, may be a mixture of chlorides and fluorides which serves to deoxidize the melt as previously described. Two cleaning actions result: the physical filtration caused by the bed and the active surface deoxidation by the flux itself.

12.5   Some Problems Related to Fluxes

Flux Inclusions

As seen above, cleaning fluxes depend for their action on an intimate contact between flux and metal. Consequently, flux powders must be stirred or blown into the melt, and ultimately complete separation of the two may not occur. In principle, fluxes should float to the melt surface and go into the dross. This process requires an adequate flotation time, which will vary with furnace size and bath depth. Proper metal temperature is also important to allow flux flotation, since baths which are too low in temperature will be excessively viscous, and this will increase the difficulty of separating flux from metal. Temperatures in the range 750C-780C (1350-1400F), are usually recommended.

If complete flux-metal separation is not achieved, flux particles will remain in the bath, and if not removed by subsequent filtration (see Chapter 11), they will exist as non-metallic inclusions in the final cast product. Figure 12.7 shows a typical flux inclusion in a die cast alloy. Flux inclusions, like all other types, can lead to enhanced porosity and to a decrease in mechanical properties as discussed in Chapter 11.

Unlike many other inclusion types, flux inclusions can also lead to markedly enhanced corrosion. Fluxes are salts which are extremely reactive with water. Although they may not be obvious on a freshly machined surface, any contact with moisture will cause the salt crystals to grow on the surface of the casting, creating a freckled-white appearance. A simple non-destructive test is to soak a casting in tap water overnight. If flux inclusions are present, the surface corrosion will be very evident.

Flux-Master Alloy Interactions

Since salt fluxes contain reactive elements, notably halides, they can chemically interact with additions made to the melt for other purposes. For example, strontium modifiers will react to form the very stable strontium halides, i.e., ($SrCl_2$ or SrF_2). Strontium present in this chemically combined form is unable to act as a modifier, and in this case, flux treatment and modification are incompatible. The solution here is to completely remove all traces of the flux before strontium addition. The order of melt treatment is important, and flux treatment should always precede modification.

Hydrogen Pick-up

Many of the salts which are combined to form foundry fluxes are extremely

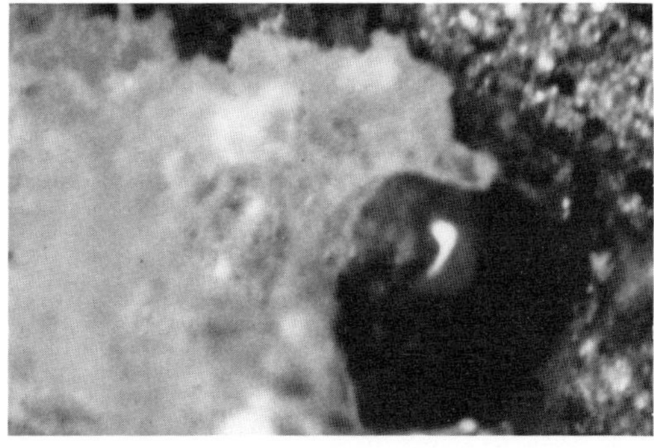

flux inclusion (x40)   eggshell-shaped flux inclusion (x43)

Figure 12.7. Typical flux inclusions[5].

hygroscopic, and will naturally pick up moisture from the air. If this occurs, use of the flux will add hydrogen to the melt, actually increasing melt hydrogen rather than decreasing it. Therefore, it is extremely important that fluxes be stored in sealed containers which are impermeable to air and in which a minimum of free space is allowed. Manufacturer's instructions on storage should be followed rigidly.

Flux Disposal

Disposal of fluxes is a problem of major environmental concern. Most are now dumped in landfills where the salts leach into the groundwater. Considerable effort is being expended by the industry into the recycling of salt fluxes so as to eliminate, or at least minimize, the need for their disposal.

References

1. Sorrell, C.A. and J.G. Groetsch. "Subsolidus Compatibility in the System NaCl-KCl-AlCl$_3$-NaF-KF-AlF$_3$," *J. Amer. Ceramic Society*, 69 (1986) pp. 333-38.
2. Abreau, I.A. "Products Used in Treating Metallic Aluminum Baths," *AFS International Cast Metals Journal*, 2, (1977), pp. 57-71.
3. Strauss, K. *Applied Science in the Casting of Metals*, Pergamon, (1970), pp. 253-56.
4. Fugua, J.M. "An Advanced Injection System for Aluminum-Silicon Alloys," *AFS. Transactions*, 95, (1987), pp. 635-42.
5. Bruner, R.W. "Basics on the Fluxing of Aluminum," *Die Casting Engineer*, (November/December 1986).

Chapter 13

Non-Destructive Microstructure Control

13.0   Introduction

Two of the melt treatment processes discussed in earlier chapters, grain refinement and eutectic modification, exert a profound effect on the cast microstructure. Each treatment is by no means foolproof, for grain refiners and modifiers can fade, and modifiers may react negatively with other elements in the melt to produce unmodified, or at best only partially modified structures. In addition, the use of improper treatment procedures may result in incomplete dissolution of the additive in the melt, leading to microstructures which are far from ideal.

Metallographic examination of the casting will reveal any deficiencies in the melt treatment process, but by then the improperly treated metal will have been cast, and scrap, or lower quality castings produced. Clearly it is preferable for the effectiveness of a grain refinement or modification treatment to be assessed before the metal is poured. One obvious method is to cast a small sample and to examine it metallographically. This is, however, time consuming, and the melt quality may well deteriorate during the time required to prepare and examine the sample. For example, in the case of sodium modification, considerable fading can occur in only a few minutes. In addition, many foundries do not possess either a metallographic facility nor personnel skilled in metallography.

In recent years, two techniques have been developed which are capable of providing a rapid evaluation of melt quality before casting: thermal analysis and the measurement of the electrical conductivity of the solid alloy. While the principles of thermal analysis have been known for a hundred years, it has only been with the development of microprocessor technology that equipment has become available which is both robust enough to operate in the foundry environment, and which can provide the required speed of analysis. This technology was first developed in the iron and steel industry, but is now used in several hundred aluminum foundries around the world.

The use of electrical conductivity to characterize cast microstructures is more recent, and so far is not used commercially. It is, however, more inexpensive than thermal analysis, and may be simpler to use in certain applications. The principles and use of each of these methods are outlined in this chapter.

13.1   Principles of Thermal Analysis

Simple Cooling Curves

In thermal analysis the temperature of a solidifying sample is recorded as it cools from the completely liquid state, through the solidification range, to become completely solid. The resultant plot of temperature versus time is the basic output of any thermal analysis. The shape of this so called "cooling curve" will vary depending on the phases which are produced during the solidification process. Whenever a solid phase forms during freezing, heat is evolved, and the rate of cooling is decreased. In some cases, for example when pure metals or eutectics freeze, the rate of cooling becomes zero until the freezing process is completed. If this occurs, a plateau will be evident on the cooling curve.

Some idealized examples of cooling curves are presented in Figure 13.1. A solid body, i.e., one in which no solidification occurs (Figure 13.1a) cools by Newtonian cooling. The temperature drops rapidly at first, and then more slowly. It gradually approaches the ambient temperature, but no abrupt changes in rate of cooling are evident.

Pure metals, such as aluminum, freeze at a unique temperature, and their cooling curves are characterized by a plateau as in Figure 13.1b. Freezing begins at (a) and is completed at (b). Both the liquid which is present before time (a) and the solid present after time (b) cool in a Newtonian fashion. The temperature of the plateau is the freezing temperature of the metal. In the case of pure aluminum this would be 660C (1220F).

Solid solution alloys (Figure 13.1c) freeze over a range of temperatures. At the beginning of freezing (the liquidus temperature, point [a]), latent heat is evolved and the cooling curve is typical of an Al-Cu alloy in which no eutectic is present.

Although eutectic alloys contain more than one solid phase, they exhibit cooling curves similar to those found for pure metals (Figure 13.1d). Eutectic solidification begins at (a) and ends at (b). The temperature of the plateau is the eutectic freezing temperature which for an Al-Si binary eutectic would be 578C (1072F).

Alloys which are either hypo- or hypereutectic (Figure 13.1e) exhibit primary phase solidification over a range of temperatures followed by eutectic freezing. Their cooling curves are combinations of those typical of a solid solution alloy and a eutectic. Primary freezing begins at the liquidus temperature (point a) and continues to the eutectic plateau (b-c). Solidification is complete at (c).

Real Cooling Curves

While the general shape of the idealized cooling curves presented in Figure 13.1 is typical of those found for real

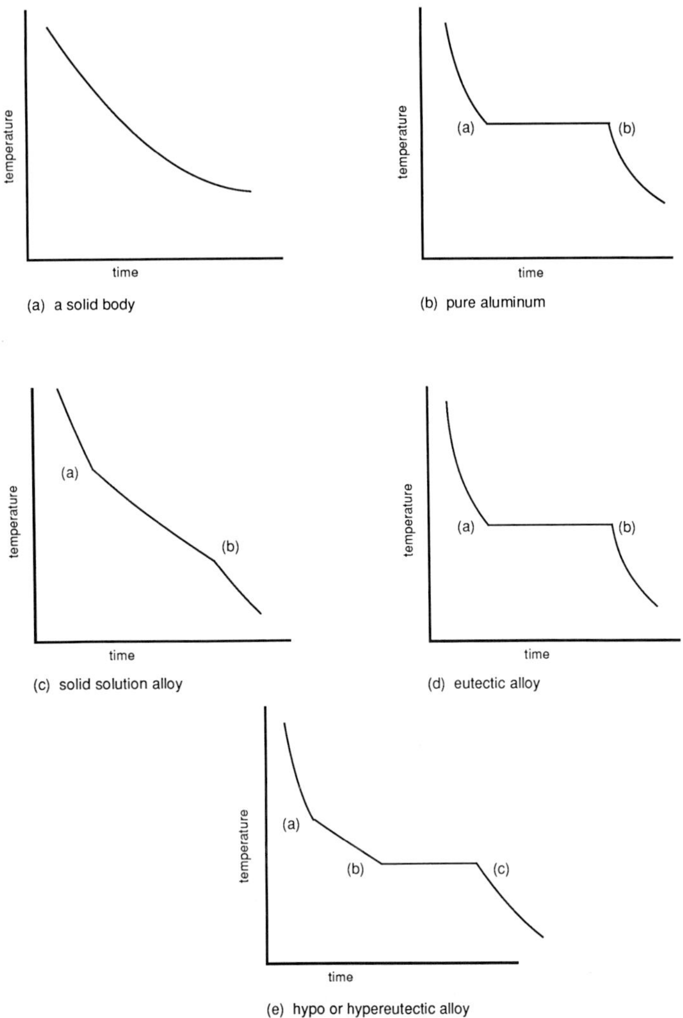

Figure 13.1 Some ideal cooling curves.

alloys, there are two important differences when real materials are considered. Solidification reactions usually require a measurable undercooling (or supercooling) in order to cause the freezing process to begin. This undercooling, which represents a thermodynamic driving force, appears on a cooling curve as a drop in temperature below the equilibrium temperature for the reaction. Once the solidification reaction begins, latent heat is evolved and the temperature rises to a value close to the equilibrium freezing temperature. This reheating process is known as recalescence. An undercooling with its accompanying recalescence is usually associated with the solidification of both primary phases and eutectics. In Figure 13.2 we have redrawn the cooling curve typical of an off-eutectic alloy (Figure 13.1e) to show the existence of undercooling. $\Delta\Theta$ is the undercooling associated with the formation of the primary phase and $\Delta\Theta_E$ is the undercooling associated with the beginning of eutectic solidification. We shall see later that the

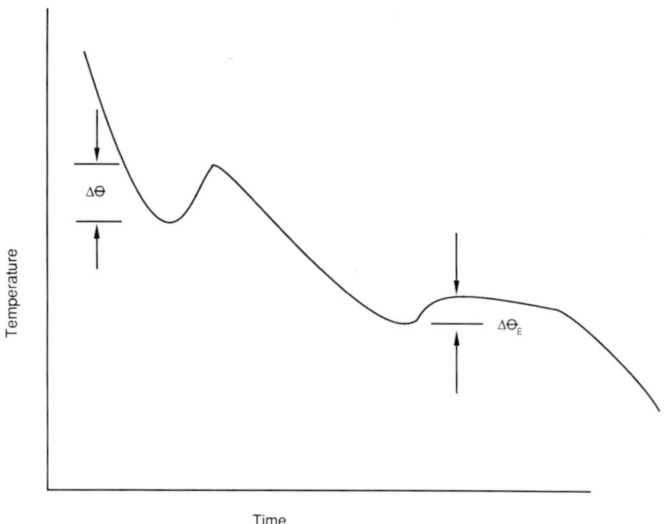

Figure 13.2. The real cooling curve of an off-eutectic alloy.

magnitude of these undercoolings is important in the assessment of the as-cast microstructure.

A second significant difference between the simple cooling curves of Figure 13.1 and those found for real alloys results from the chemical complexity of actual commercial alloys. Curves such as those shown in Figure 13.1c, d or e are true only for binary alloys. Most casting alloys are multicomponent and contain several solid phases, all of which can have an effect on the shape of the cooling curve. Real cooling curves can therefore be considerably more complex than those of Figure 13.1. An example is given in Figure 13.3 of a cooling curve obtained on a 356 alloy containing 6.7% Si, 0.44% Fe, 0.30% Mn, 0.35% Mg and 0.01% Ti. The alloy is basically hypoeutectic, and the cooling curve reflects the general shape for such an alloy. However, the minor elements present lead to subtle changes. For example, close inspection reveals a change in slope at point 3 at a temperature of 550C. This corresponds to the formation of a ternary eutectic composed of the phases Al-Mg$_2$Si-Si.

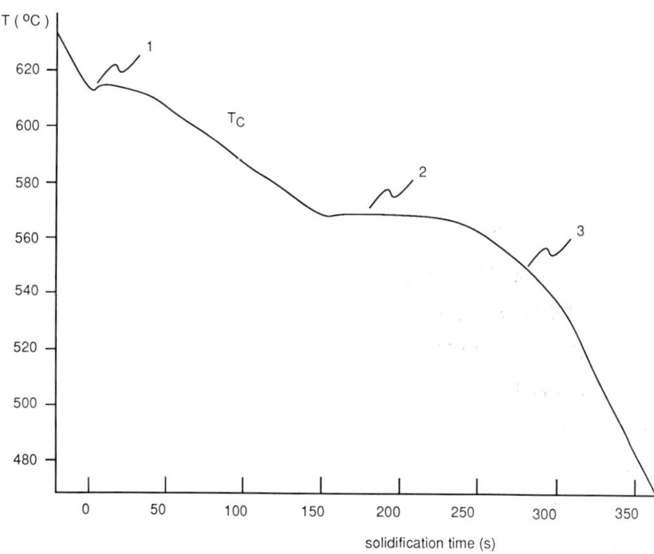

Figure 13.3. Cooling curve of a 356 alloy[1]:
1) primary phase nucleation;
2) binary Al-Si eutectic plateau;
3) Al-Mg$_2$Si-Si eutectic.

The Derivative Curve

A much clearer indication of the inflection points present on a cooling curve can be obtained by taking the first derivative of the cooling curve and plotting this versus time. This value, dT/dt, is the slope of the curve, and such a derivative curve is a powerful tool in the use of thermal analysis. The calculation is painstaking if done by hand, but software is readily available to calculate derivative curves.

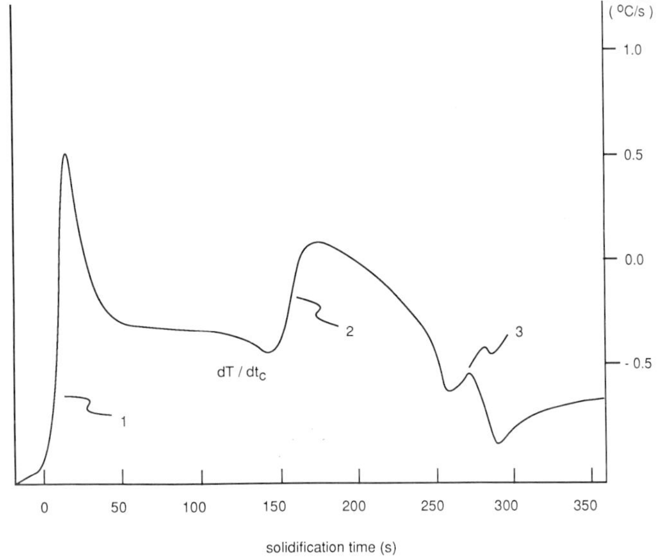

Figure 13.4. The derivative of the curve given in Figure 13.3[1].

The derivative of the curve given in Figure 13.3 is presented in Figure 13.4. An inflection point, or point of slope change, appears as a peak on the derivative curve. Thus the point of primary phase nucleation is peak 1; the Al-Si eutectic nucleation is peak 2; and the formation of the ternary Al-Mg_2Si-Si eutectic is peak 3. The main benefit of the derivative curve lies in its ability to magnify the important slope changes which are found on the cooling curve. Its value in emphasizing the ternary eutectic in the 356 alloy example is obvious.

Thermal Analysis Equipment

Several devices for the thermal analysis of aluminum alloys are available commercially. These consist of a sampling cup, a microcomputer for data acquisition and analysis, and appropriate software. The sampling cup is of a simple design, an example of which is shown in Figure 13.5. A thermocouple is fixed in the center of the cup with the bead at approximately the mid-point. Cups are either of shell-molded sand or thin-walled steel, with a shell molded

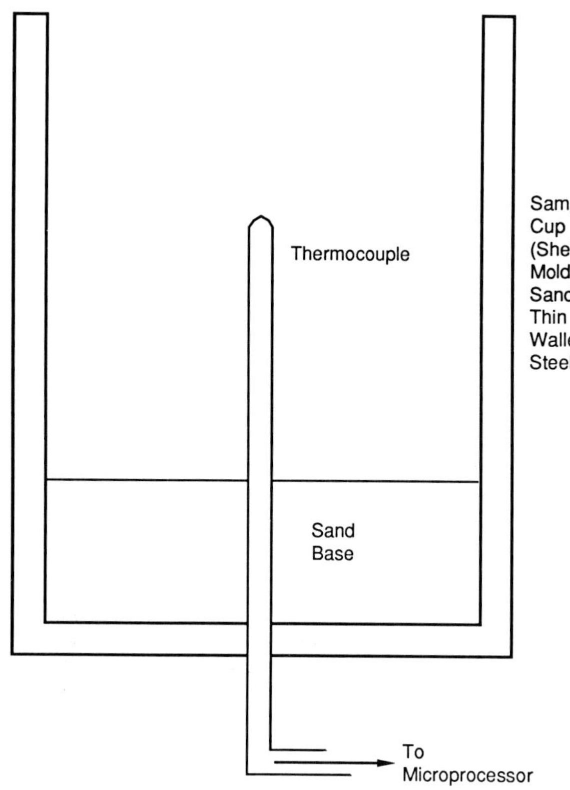

Figure 13.5. A typical sampling cup used for thermal analysis.

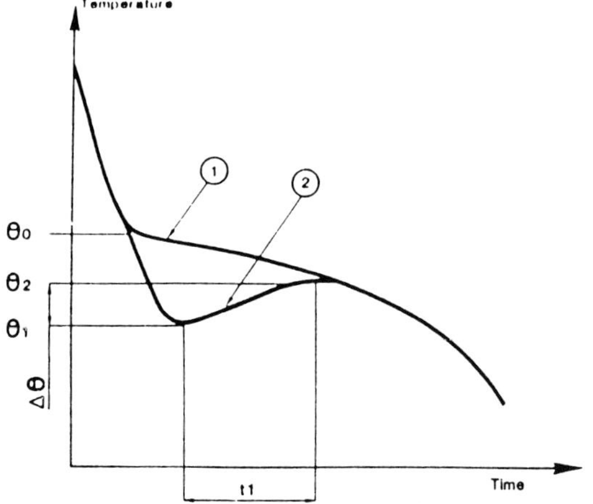

Figure 13.6. The cooling curve at the beginning of solidification[2]:
curve 1: a grain refined alloy (fine grain size);
curve 2: an unrefined alloy (coarse grain size);
Θ_0 = temperature at start of freezing of a well-refined alloy;
Θ_1 = temperature at start of freezing of an unrefined alloy;
$\Delta\Theta$ = $\theta_2 - \theta_1$: the apparent supercooling
t_1 = period of apparent supercooling.

sand base. In either case, solidification is relatively slow, requiring 6-8 minutes.

A major difference between the various testers lies in the approach to the analysis of the cooling curve. Some use a complex algorithm to assess the microstructure, while others employ only a single parameter. In all cases, however, information on the grain size and extent of eutectic modification can be obtained. Analysis of the cooling curve is performed automatically by the microcomputer, and the operator is presented with the results of this analysis in a simple form.

13.2 Thermal Analysis Control of Grain Size

The determination of grain size by thermal analysis utilizes that portion of the cooling curve associated with the beginning of primary solidification (Figure 13.6). In melts which contain a large number of nuclei (grain refined melts, for example), there will be little barrier to nucleation, and the cooling curve will appear as curve 1 in Figure 13.6. This type of curve usually denotes a fine grained material. If few nuclei are present, then a measurable undercooling will be necessary to start solidification of the primary grains, and a curve such as 2 will be observed. Larger values of $\Delta\Theta$, the apparent supercooling, are found when the grain size is large, and $\Delta\Theta$ values which approach zero are typical of fine grained materials.

In Figure 13.7, we show the average measured grain diameter as a function of $\Delta\Theta$ for an A356 and an A319 alloy. Grain size values were measured on round samples cast into a permanent mold with an open top. The bottom of the

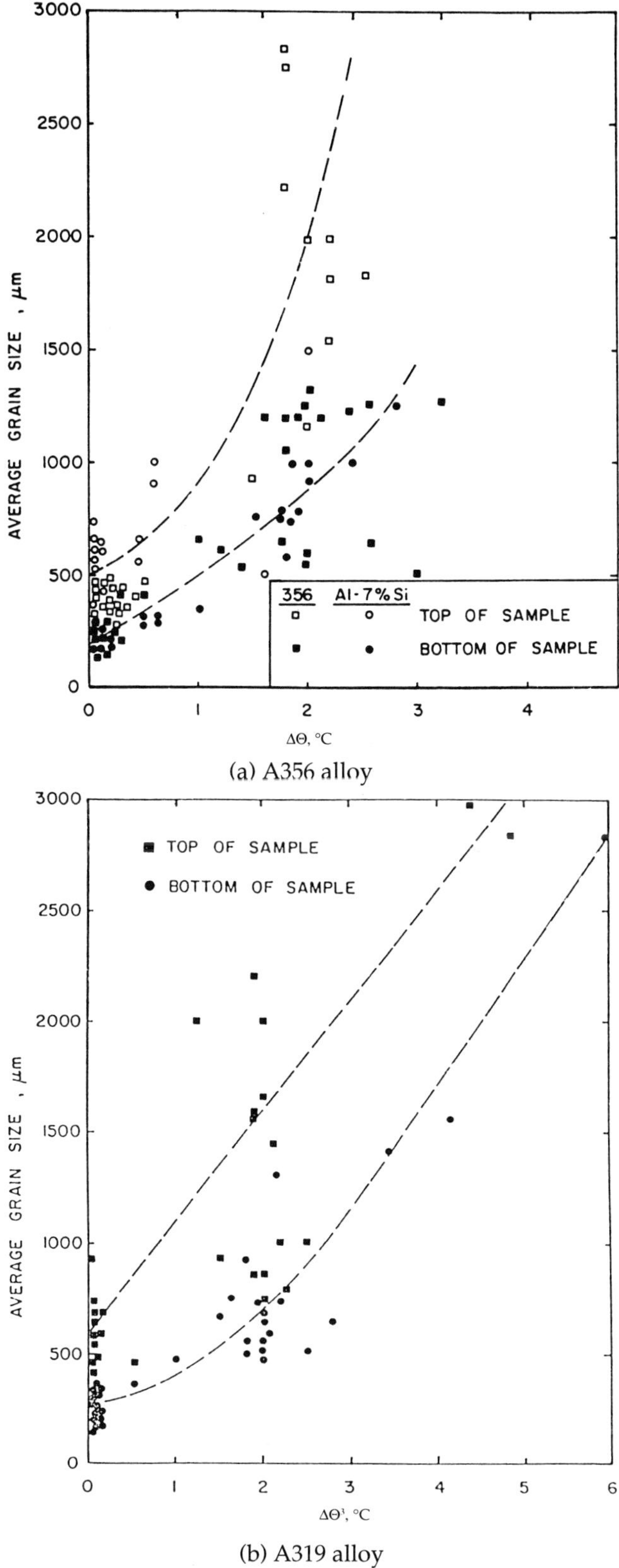

Figure 13.7. Average grain size as a function of $\Delta\Theta$ [3].

sample froze more rapidly than the top, and so the grain size is finer and is typical of sand castings. Values of $\Delta\Theta$ were obtained from thermal analysis samples cast at the same time as the grain size samples. Although there is a large amount of scatter, particularly at large grain sizes, it is clear that the measurement of $\Delta\Theta$, the apparent supercooling, can give an indication of grain size. In the simplest form of thermal analysis testing, the supercooling is measured and compared to a pre-determined calibration to obtain a grain size number which is similar, but not identical, to the ASTM grain size. This type of analysis is used in many foundries because measurement of the apparent supercooling is easy and straightforward.

A better indication of grain size, particularly at coarser sizes, can be obtained by taking into account the period of apparent supercooling which is shown as t_1 in Figure 13.6. This is useful because at large grain sizes, the undercooling may not be particularly large, but it tends to last for an extended period of time. This is due to the time factor required for nucleation mentioned in Chapter 8. The relationship between apparent supercooling, period of supercooling, and grain size is shown in Figure 13.8 for a hypoeutectic Al-Si alloy. More sophisticated thermal analysis testers use an algorithm based on data such as this to determine the grain size.

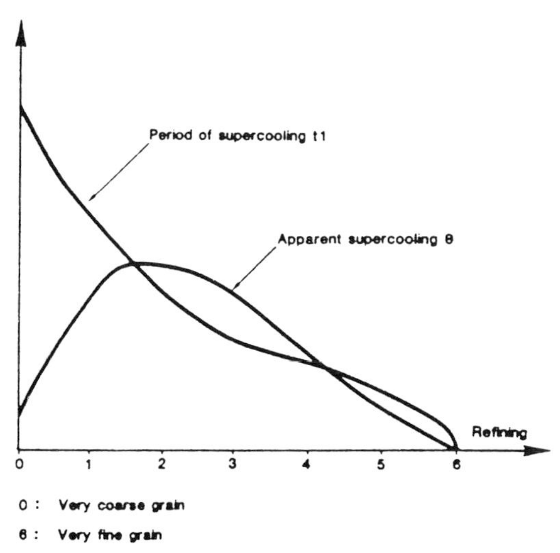

Figure 13.8. Variation of grain size with both apparent supercooling and period of supercooling[2].

13.3   Thermal Analysis Control of Eutectic Modification

Effects of Modification on the Cooling Curve

When an alloy is modified, three features of the cooling curve may be changed: the temperature of the eutectic plateau, the undercooling required to start eutectic freezing, and the time duration of this undercooling. These are indicated schematically in Figure 13.9. With modification, the eutectic temperature is depressed, the undercooling for nucleation of the eutectic is increased, and the period of this undercooling lengthens.

The feature most used in thermal analysis control of modification is the depression of the eutectic temperature. In an A356 alloy, full modification will lower this temperature by 6-8°C from the unmodified state (Figure 13.10). Since the eutectic temperature is easy to measure, it is most often employed by foundries to assess whether or not a melt is properly modified. This quantity is usually called the ΔT, and an example of its

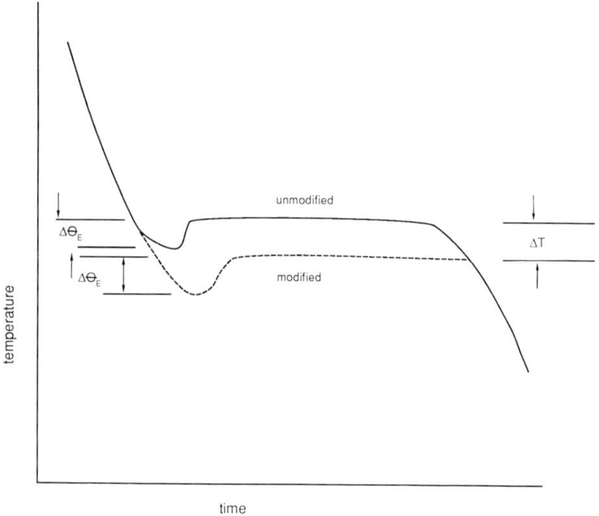

Figure 13.9. A comparison of the eutectic regions of the cooling curves of modified and unmodified alloys.

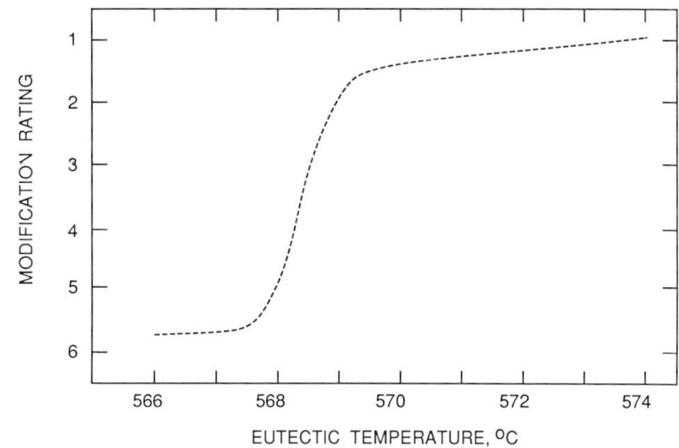

Figure 13.10. Relation between modification rating and the eutectic temperature for an A356 alloy[3].

calculation is given in Table 13.1 for structures which vary from non-modified to over-modified.

More sophisticated approaches to modification control are possible and are useful in some circumstances. If temperature alone is used as the criterion for proper modification, it is not very easy to detect overmodified structures, since the greatest temperature change occurs in the unmodified to modified transition. Consideration of the undercooling required to start eutectic freezing can, however, be used. Some experimental results taken from reference 4 on an A356 alloy are presented in Table 13.1. With modification, the undercooling, $\Delta\Theta_E$, increases and then falls off as the structure becomes overmodified. Values for overmodification are considerably lower than those found with unmodified structures, and so it is possible to distin-

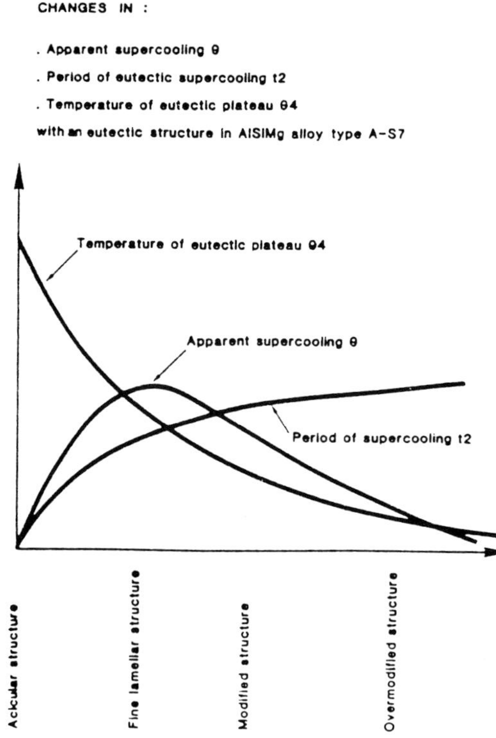

Figure 13.11. Relationship between eutectic structure and eutectic temperature, apparent eutectic supercooling and period of supercooling[2].

guish between the two.

An even greater sophistication is to couple the period of this undercooling with both the eutectic temperature and the undercooling. These relationships are shown in Figure 13.11 for a 356 type alloy, and are employed in the logic system of the most advanced thermal analysis testing devices.

The Effect of Alloy Composition

Assessment of the degree of modification by thermal analysis requires a comparison of the cooling curves for the modified and unmodified alloys. The differences in eutectic temperature or undercooling are not large, and the entire process is complicated by the fact that the cooling curves are influenced by normal variations in alloy chemistry. The problem can best be appreciated by considering the operational sequence of a thermal analysis test. If the eutectic temperature depression is used as the basis of analysis, a base eutectic temperature for the unmodified alloy must be programmed into the tester. The test is then run, and values from the modified alloy are compared to this temperature. The base temperature is determined

Table 13.1. Variation of Eutectic Undercooling with Structure in an A356 Alloy[4]

Structure	Eutectic Nucleation Temp. T_C (°C)	Eutectic Growth Temp T_E (°C)	Undercooling $\Delta\Theta_E = T_E\text{-}T_C$ (°C)	Undercooling $\Delta T = T_E$ (nonmodified) $-T_E$(°C)
non-modified	573.7	576.2	2.5	0
under-modified	569.1	573.4	4.3	2.8
modified	568.8	572.2	3.6	4.0
modified	568.5	571.9	3.4	4.3
modified	568.5	571.7	3.2	4.5
over-modified	570.2	571.7	1.5	4.5
over-modified	571.7	572.5	0.8	3.7

Table 13.2. Variation of the Eutectic Temperature of Several European Foundry Alloys Within the Normal Cu-Mg-Fe Specifications[5]

Alloy Designation	Minimum Cu-Mg-Fe Specification		Maximum Cu-Mg-Fe Specification	
	Unmodified (°C)	Modified (°C)	Unmodified (°C)	Modified (°C)
AlSi5Cu3Mg	558	548	555	545
AlSi6Cu2	563	553	559	549
AlSi7Cu3Mg	560	550	556	546
AlSi9Cu	568	562	562	560
AlSi9Cu3	568	559	564	557

by measurement or experience over a long period of time. If now a new alloy batch is used, the base temperature may change due to normal compositional variations within the alloy specification. Such changes can be 3-4°C, which are very significant when it is realized that the total change from the modified to unmodified condition may be only 6-8°C. Some measured variations of the eutectic temperature due to variations in Cu, Mg and Fe in various European alloys are summarized in Table 13.2.

Some attempts have been made to develop equations relating the eutectic temperature of an unmodified alloy to the composition. One such equation given by Mondolfo[6] is as follows:

$$\text{Eutectic temp. (°C)} = 577 - 12.5\%\text{Si} [4.43\%\text{Mg} + 1.43\%\text{Fe} + 1.93\%\text{Cu} + 1.7\%\text{Zn} + 3.0\%\text{Mn} + 4.0\%\text{Ni}].$$

This equation has been shown to be better than ±1°C in A356 alloys, and probably also applies well to other alloys where the total amount of elements other than Al and Si is less than 1 wt. pct. Thus, it should apply to alloys such as 355, 357, 359 and 360, but will not be valid for alloys in which the Cu, Fe, Mg or Zn levels are high, such as 319 or 384 [3].

Thermal analysis is a relatively new technology as applied to aluminum alloys, and a great deal of work remains to be done to establish the basic cooling curve behavior for all common alloys in the unmodified and modified states.

Applicability of Test Results to Real Castings

When thermal analysis is used to assess grain refinement or modification, it must be kept in mind that the test measures only the potential of the thermal analysis sample itself to be grain refined or modified. This sample is a small casting, and the results should be applicable to all sections of a casting which freeze at similar or faster rates compared to the test-piece. Modification is particularly sensitive to freezing rate, being enhanced at higher rates and decreased at lower rates. Most thermal analysis samples freeze relatively slowly, and represent worst case situations when applied to permanent mold and thin-walled sand castings. In such cases, if modification is indicated by thermal analysis, it should also be present in the casting. If, however, the casting contains sections which are thicker than the thermal analysis sample, some caution should be exercised, since these sections will freeze more slowly, and may not be fully modified.

Example of the Use of Thermal Analysis

An interesting example of the power of thermal analysis in the control of the structure of aluminum foundry alloys is presented in Figures 13.12 and 13.13. The two micrographs of Figure 13.12 were from two batches of the A356.0 alloy obtained from the same supplier. Each batch gave essentially the same chemical analysis, but clearly yielded different as-cast microstructures: one acicular and the other partially modified. While a chemical analysis was not sensitive enough to reveal the small compositional differences responsible for the two structures, thermal analysis accurately predicts the difference. The cooling curves

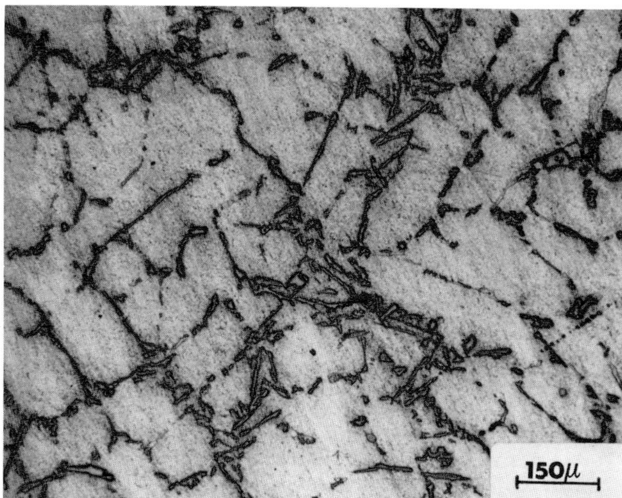

(a) batch 1

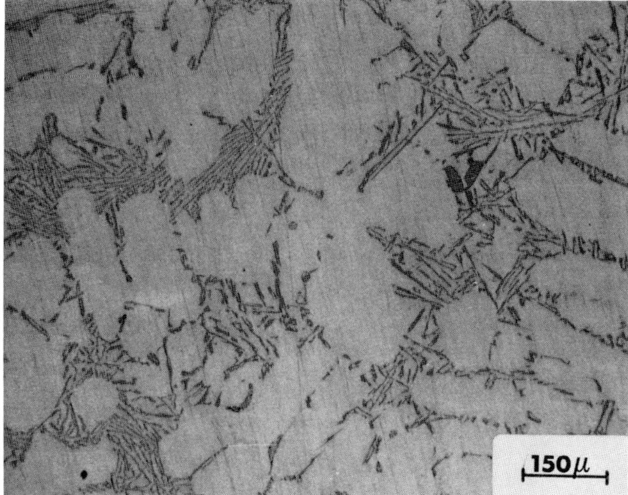

(b) batch 2

Figure 13.12. Microstructures of two batches of 356 alloy ingot.

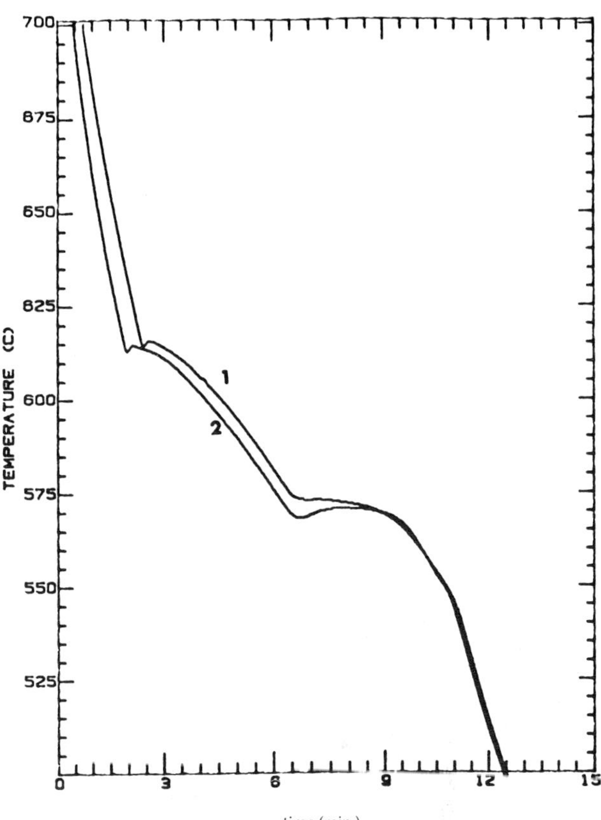

Figure 13.13. Cooling curves obtained from two batches of 356 alloy ingot.

of Figure 13.13 show, without a doubt, that batch 2 should yield a partly modified structure. This is indicated by the lowered eutectic temperature of batch 2 compared to batch 1. Producers of foundry ingots should develop the concept of guaranteed microstructure using thermal analysis, in addition to chemical analysis, as a quality control tool.

13.4 Electrical Conductivity Control of Modification

The electrical conductivity of Al-Si casting alloys is a function of the eutectic silicon morphology; thus this technique can be used to assess non-destructively the degree of modification. Silicon is essentially a non-conductor, and when present in a coarse, plate-like form in the non-modified eutectic, it acts as a serious impediment to electron flow and consequently decreases the overall electrical conductivity of the alloy. When modified, the silicon becomes finer and fibrous. This shape allows easier flow of electrons and results in a higher overall electrical conductivity. These considerations are represented schematically in Figure 13.14, while Figure 13.15 shows the variation of electrical conductivity of an A356 alloy with increasing strontium concentration. As the strontium level increases, the eutectic becomes better modified and the conductivity increases by about 10%.

Electrical conductivity is not difficult to measure. The simplest and most economical methods involve eddy current testing. A number of com-

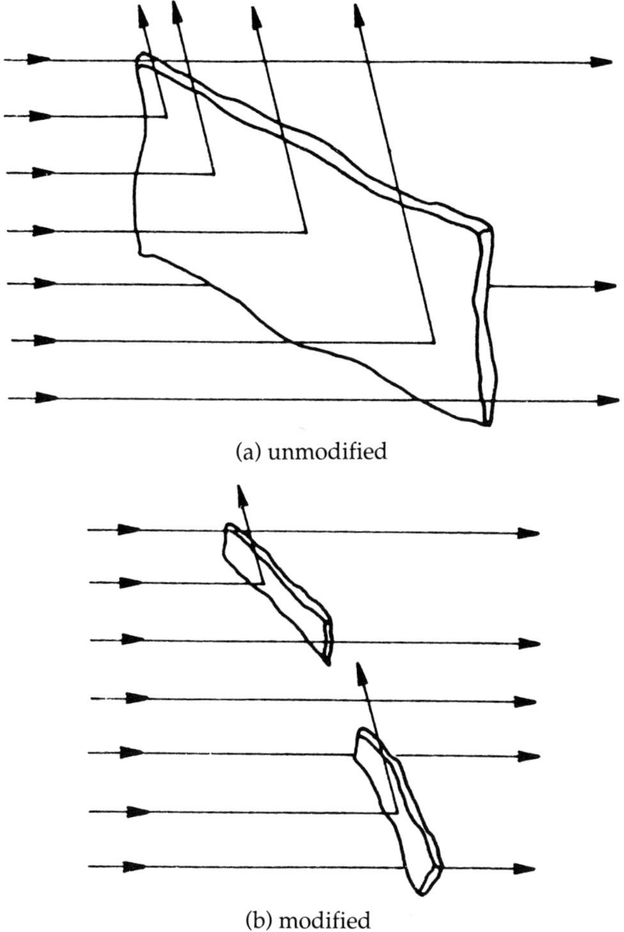

(a) unmodified

(b) modified

Figure 13.14. Schematic representation of the interaction between electron flow and silicon particles in the Al-Si eutectic[6].

mercial devices exist all of which operate on the same principle. A circulating electric current is induced in the alloy by means of an alternating magnetic field produced by an applied alternating current source. The interaction between the applied current and the induced eddy current is then measured and used to compute the electrical conductivity of the alloy. Small portable testing units are available, consisting of a digitized meter and a hand held probe. The depth of penetration of the eddy current field is a function of the frequency of the applied alternating current. Many devices are available with variable frequencies so that both thick and thin specimens may be tested. Depths of specimen penetration of up to 1 cm can be achieved at low frequencies. The meter provides a direct measure of the conductivity usually in terms of %IACS (percentage of the International Annealed Copper Standard).

Testing a melt for modification potential by this method is quite simple and straightforward. A button, similar to a spectrochemical sample, is cast into an insulated metal or graphite mold. After a predetermined time, usually 5-10 minutes, this button is quenched in water and the conductivity is measured. Such samples are taken both before and after the modification treatment, and the conductivities are compared. If the increase in conductivity is about 10%, the melt is well modified. Changes of only about 5% usually indicate partial modification, while no change is indicative of an unsuccessful modification treatment. These values are not meant to be absolute, and can vary from alloy to alloy and as a function of the sample mold. Some initial calibration of the technique for each alloy is advisable.

A particular advantage of the electrical conductivity method is that it is independent of the effect of compositional variations which plague ther-

Table 13.3. Electrical Conductivity Changes with Microstructure in a Sodium Treated A356 Alloy

Sodium Concentration (wt. pct.)	Microstructure	Electrical Conductivity (% IACS)	Percent Change From Unmodified Acicular Structure
0.0003	acicular	33.3	0.0
0.041	overmodified	35.2	6.0
0.012	well modified	35.9	8.1
0.009	well modified	36.5	9.9
0.007	partially modified	35.0	5.4
0.003	lamellar	34.1	2.7

mal analysis. Comparison is made on samples taken from the same melt before and after modification treatment. Any effect of batch-to-batch variation in alloy chemistry is therefore eliminated.

When electrical conductivity is used to assess modification, it is critical that the comparison always be made on samples cooled from the casting temperature in exactly the same way. Electrical conductivity is very sensitive to the quantity of alloying elements retained in solid solution, and this will vary depending on how the test button cools to room temperature. The effect is illustrated in Figure 13.16. Samples which are air cooled exhibit a higher electrical conductivity than those which are quenched, because the alloying elements have more time to precipitate out of solid solution (mostly as Mg_2Si in this 356 alloy example). Note that the shapes of the curves are identical for the two types of samples. It does not matter if the samples are quenched or air cooled, but it is important that all test buttons be processed in the same manner.

An example of the use of electrical conductivity to follow microstructure evolution in a sodium treated melt is provided in Figure 13.17 and Table 13.3. An excessive sodium addition was made to this melt so that overmodified structures were first produced. Sodium fade then occurred, and the structures be-

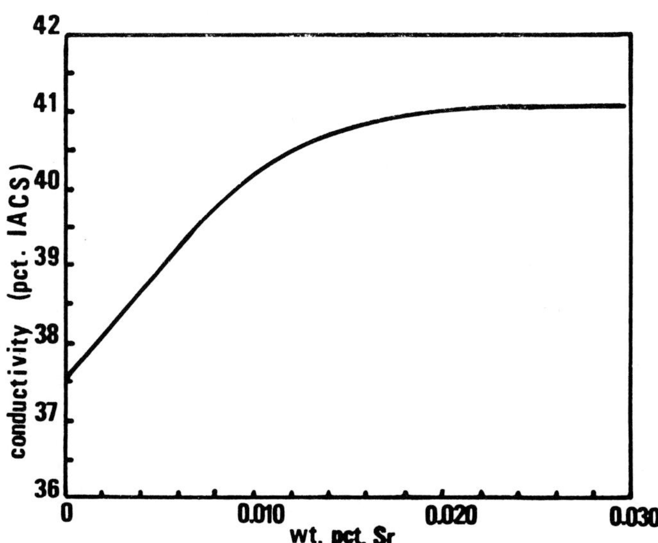

Figure 13.15. Variation of the electrical conductivity of a 356 alloy with strontium concentration.

came well modified and eventually quite undermodified. It is evident that electrical conductivity is a sensitive function of microstructure.

There are disadvantages to electrical conductivity measurements, however. The greatest of these is that electrical conductivity is not sufficiently sensitive to differentiate changes in grain size. If grain size measurement is required, then thermal analysis must be used.

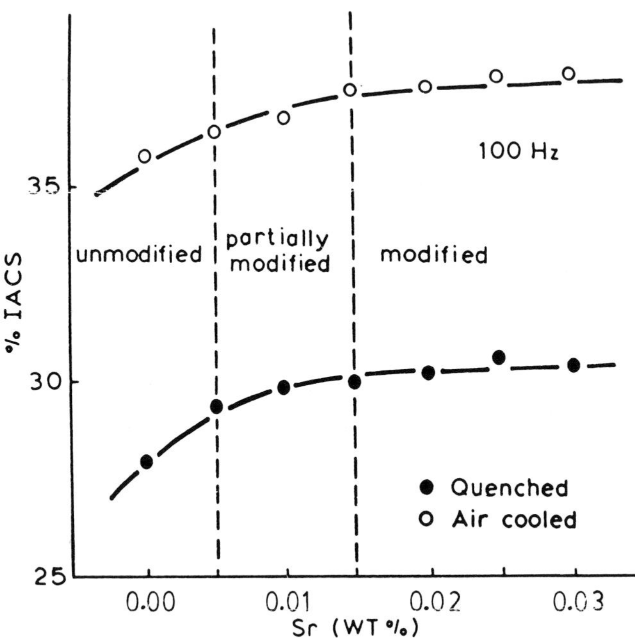

Figure 13.16. The effect of cooling rate in the solid state on electrical conductivity[7].

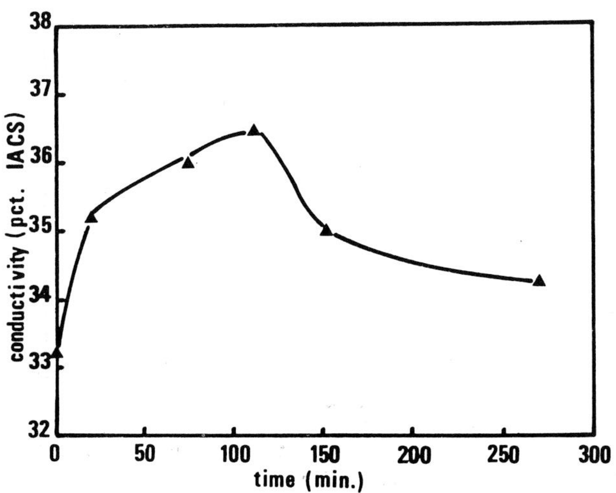

Figure 13.17. The change in electrical conductivity of samples cast from a 356 alloy melt treated initially with an excess of sodium.

13.5 Electrical Conductivity Measurement of Dendrite Arm Spacing

Although not currently in commercial use, there is evidence that electrical conductivity can provide a non-destructive measure of the dendrite arm spacing in Al-Si casting alloys[7]. As the dendrite arm size changes, so does the size of the eutectic pockets between them (Figure 13.18), and this can influence the electrical conductivity in a measurable way. There is a linear relation, such as shown in Figure 13.19, between the conductivity and the dendrite arm spacing. Establishment of this relation for a given casting raises the possibility of using conductivity to measure the dendrite arm spacing directly on a casting.

As noted previously, electrical conductivity is highly sensitive to alloying elements retained in solid solution. It is therefore essential that calibration curves such as those in Figure 13.19 be developed under cooling conditions which are identical to those found in the actual casting. One possible solution might be to use a production casting to develop the calibration curve.

Use of this method to determine the dendrite arm spacing requires that well developed pockets of eutectic structure exist. As a result, it is not possible to determine the arm spacing on non-modified alloys. This is illustrated in the curves shown in Figure 13.20, where the slope, and hence sensitivity, of the unmodified line is significantly less than its modified counterpart. Microstructurally, the reasons for this effect can be seen in Figure 13.21. In this 380 alloy example, it is clear that the dendrite arms are not well defined in the unmodified state. Changes in arm spacing in this type of microstructure result in much smaller electrical conductivity effects than do changes when the structure is modified and the arms are well defined by the regions of eutectic.

At present, this non-destructive measurement technique shows promise, but requires more development to demonstrate how it can be used on production castings. It has, so far, been shown to be workable on simple castings produced in modified 319, 355, 356, 357 and 380 alloys[7].

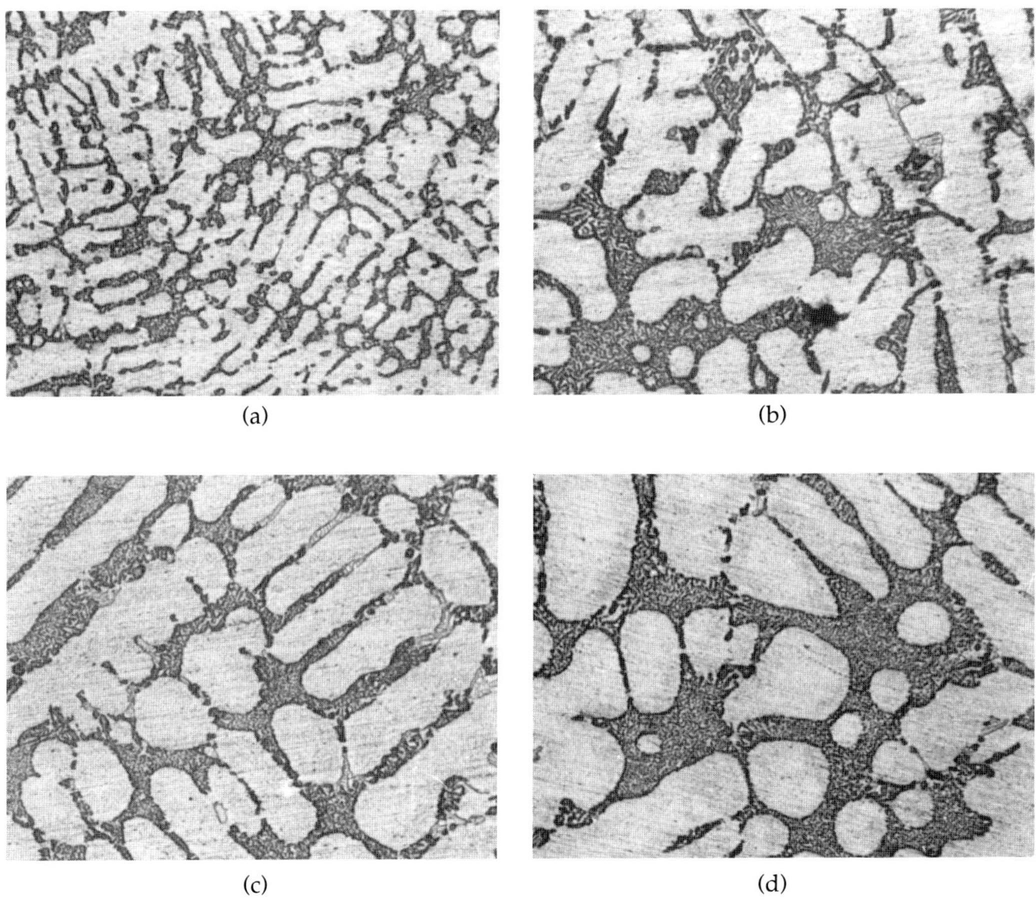

(a)   (b)
(c)   (d)

Figure 13.18. As the dendrite arm spacing changes, so does the size of the eutectic pockets and the electrical conductivity of the alloy. All photos are at 100X [7].

(a) DAS = 47 μm, %IACS = 39.0
(b) DAS = 86 μm, %IACS = 35.6
(c) DAS = 104 μm, %IACS = 33.4
(d) DAS = 117 μm, %IACS = 32.4

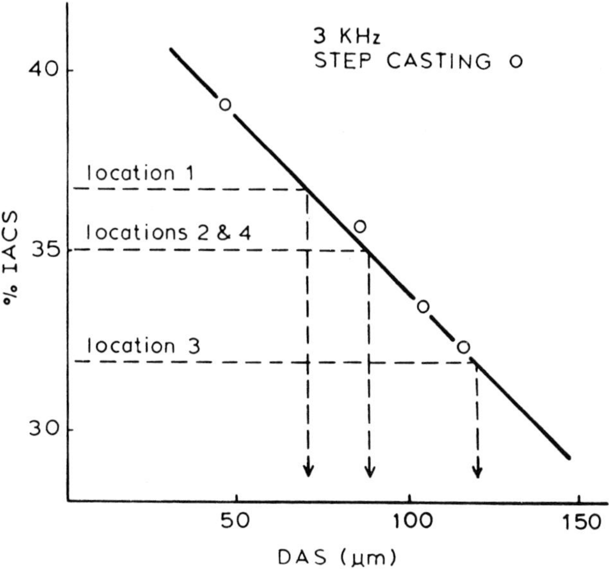

Figure 13.19. The relation between electrical conductivity measured as %IACS and dendrite arm spacing (DAS) in 356 alloy[7].

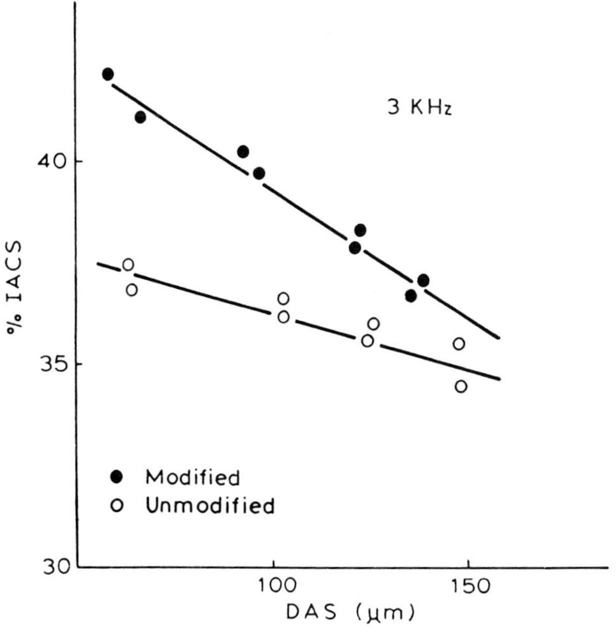

Figure 13.20. Conductivity variation with DAS in a modified and an unmodified alloy[7].

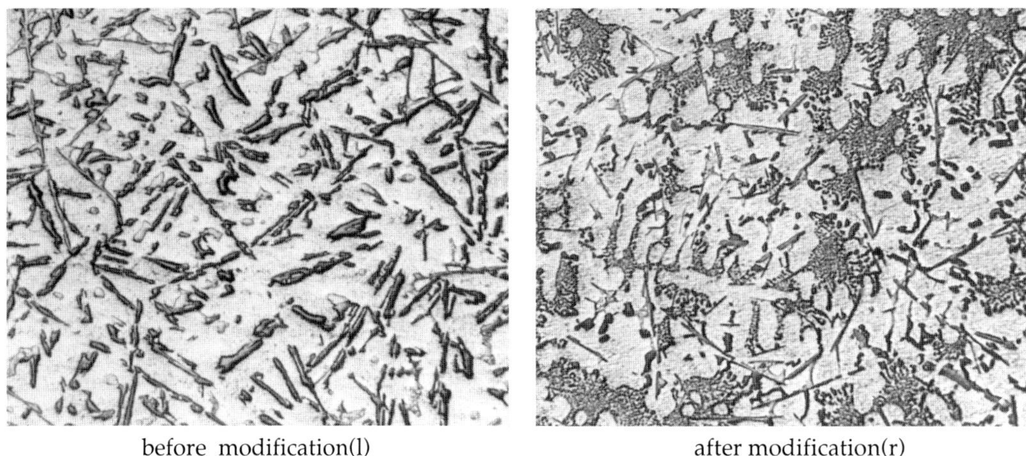

before modification(l)   after modification(r)

Figure 13.21. Microstructure of 380 alloy before and after modification(x100). It is the definition of the dendrite structure on modification which allows electrical conductivity to be used to measure the arm spacing.

References

1. Bäckerud, L.S. and G.K. Sigworth. "Recent Developments in Thermal Analysis of Aluminum Casting Alloys," *AFS Transactions*, 97, (1989), pp. 459-64.
2. Charbonnier, J. "Microprocessor Assisted Thermal Analysis Testing of Aluminum Alloy Structures," *AFS Transactions*, 92 (1984) pp. 902-22.
3. Apelian, D., G.K. Sigworth, K.R. Whaler. "Assessement of Grain Refinement and Modification of Al-Si Foundry Alloys by Thermal Analysis," *AFS Transactions*, 92 (1984) pp. 297-307.
4. Argyropoulos, S., B. Closset, J.E. Gruzleski and H. Oger, "The Quantitative Control of Modification of Al-Si Foundry Alloys Using a Thermal Analysis Technique,"*AFS Transactions*, 91 (1983) pp. 351-57.
5. Marino,F. R. Medana, M. Baldi, G. Silva. "Thermal Analysis in the Light Alloys Foundry," *Metallurgical Science and Technology* (June 1983) pp. 32-39.
6. Mulazimoglu, M.H., R.A.L. Drew and J.E. Gruzleski. "The Electrical Conductivity of Cast Al-Si Alloys in the Range 2 to 12.6 wt. pct. Silicon," *Metallurgical Transactions*, 20A (1989) pp. 383-89.
7. Argo, D., R.A.L. Drew and J.E. Gruzleski. "A Simple Electrical Conductivity Technique for Measurement of Modification and Dendrite Arm Spacing in Al-Si Alloys," *AFS Transactions*, 95 (1987) pp. 455-64.

Chapter 14

A Summary of Melt Treatment Processes, Techniques and Schedules

14.0   Introduction

In the previous chapters we have considered in detail the various melt treatment processes which can be applied to aluminum foundry alloys. The effects of these treatments on the structure and properties of the cast product have been stressed. Liquid metal treatment must be done in a rational manner and the processes applied in certain logical sequences. This necessity arises because some of the treatments are incompatible with others, and an incorrect treatment sequence can result in a complete negation of the otherwise beneficial effects of melt treatment. In this final chapter, we discuss processing schedules, beginning first with a brief summary of the various processes and then examining addition techniques and recommended sequences.

14.1 Melt Treatment Processes—A Summary

The treatment of liquid aluminum casting alloys may include one or more of the following processes:
- strontium modification of eutectic silicon;
- sodium modification of eutectic silicon;
- antimony refining of eutectic silicon;
- phosphorus refining of primary silicon;
- grain refinement;
- degassing;
- filtration;
- fluxing.

Strontium Modification

Strontium is now used as the main eutectic modifier. Strontium silicide (Sr_2Si) was employed extensively in the past, until pure strontium metal became readily available. Because of high impurity levels—mainly iron, phosphorous and oxides—the use of Sr_2Si-based master alloys has declined in recent years. Table 14.1 shows the various forms and compositions of strontium based alloys and master alloys produced by companies in Europe, North and South America, and Japan.

Strontium can be added at the foundry level by several different methods. Figure 14.1 describes the main avenues by which strontium can reach the liquid metal prior to casting. Pure strontium is used to manufacture the low-strontium master alloys shown in Table 14.1. The master alloys are added directly to the molten metal prepared in the foundry, or added to the liquid metal at primary or secondary smelters. Strontium premodified metal is delivered to the foundry in the form of solid ingots or liquid metal. Pure strontium or the high strontium Al-90%Sr alloy can be added directly to molten metal in primary or secondary smelters, without going through the master alloy stage. It is also possible to add the Al-90%Sr alloy directly to the metal prepared in the foundry without any intermediate steps.

Table 14.1 Form and Composition of Strontium Alloys and Master Alloys

Alloy No.	Composition	Sr Source	Form
A	Al-90%Sr	Pure	Rod, Can
B	Al-3.5%Sr	Pure	Rod
C	Al-5%Sr	Pure	Rod, Waffle
D	Al-10%Sr	Pure	Waffle, Buttons
E	Al-10%Sr-1%Ti	Pure	Rod
F	Al-10%Sr-14%Si	Sr_2Si	Waffle, Buttons

Sodium Modification

The limited solubility of sodium in aluminum has prevented the development of aluminum-sodium master alloys. The two most popular sodium products used to achieve modification are pure metallic sodium and sodium salt mixtures. Figure 14.2 shows the preferred routes to obtain sodium modification in the foundry.

Retention of sodium in the melt is the greatest difficulty associated with its use. The high vapor pressure results in significant losses at normal melt treatment temperatures, and during degassing with either inert or reactive gas mixtures. Pure metallic sodium is available in pre-packed vacuum cans. This form eliminates many of the handling difficulties, although the lowest cost method of addition is to purchase the element in bulk, store it under kerosene, and cut and add the required quantity.

Sodium salt mixtures can also be used. These are frequently of a quite low melting composition, and when added in a continuous permanent mold operation, they must be kept away from the active ladling area. Premodified ingots are prepared with high enough sodium levels to develop adequate modification after at least one remelt. If the metal is remelted more than once, or if the processing time is extended, then makeup additions of sodium will probably become necessary.

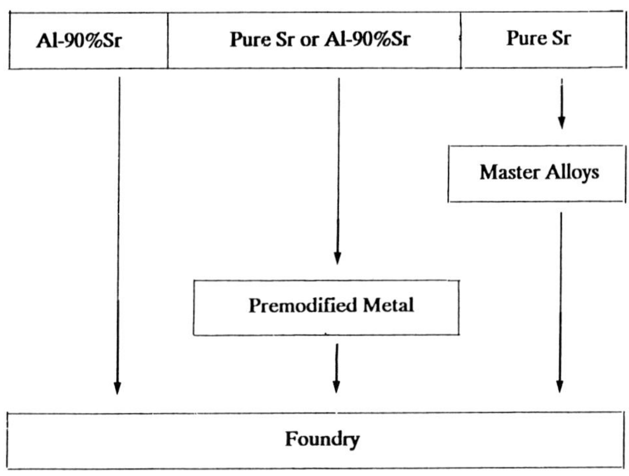

Figure 14.1. Introduction of strontium at the foundry level.

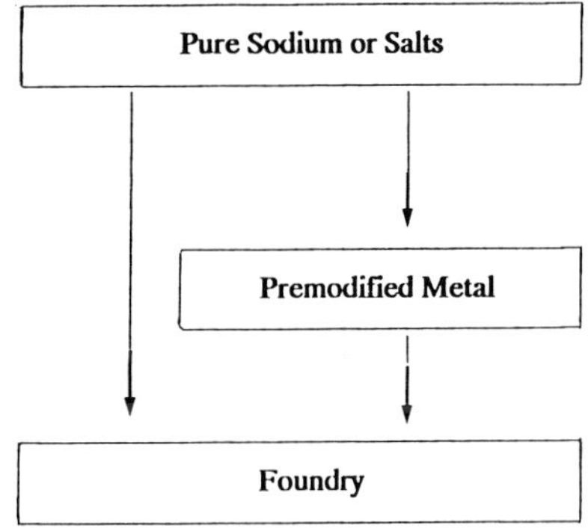

Figure 14.2. Addition of sodium at the foundry level.

Antimony Eutectic Refining

Antimony, in the form of pure metal or master alloys, is rarely added during melt treatment in the foundry. This is due mainly to the potential toxicity associated with antimony handling during addition. Reaction with hydrogen can lead to the formation of deadly stibine gas, which can also form if dross skimmed from the melt is allowed to come into contact with moisture. Figure 14.3 shows how antimony finds its way to the foundry. Pure antimony metal or a master alloy, generally Al-10%Sb, are added to the molten metal at the smelter. The favored method of obtaining antimony treated melts in the foundry is through the remelting of antimony containing ingots. The stability of antimony in the molten metal, and consequently its permanent effect, justifies the production of antimony treated ingots.

Phosphorous Refining

While strontium, sodium and antimony treatments are applied to hypoeutectic casting alloys, phosphorous is added to hypereutectic alloys to refine the

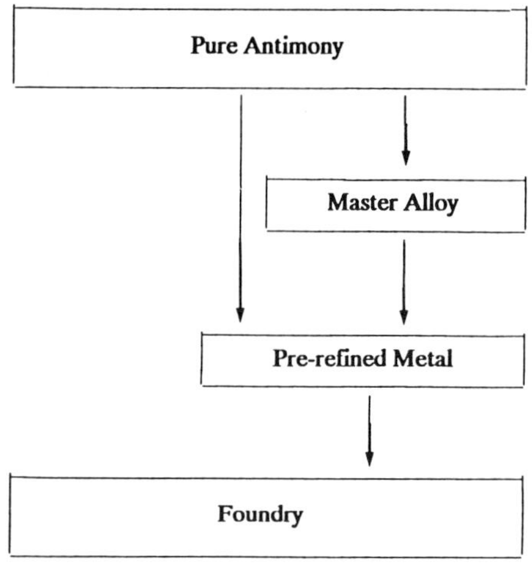

Figure 14.3. Addition of antimony at the foundry level.

primary silicon. Ingots delivered to foundries are normally pre-treated with phosphorous. As explained in detail in Chapter 7, the two major sources of phosphorous are salts containing red phosphorous and a Cu-8%P brazing alloy. Because of the limited effectiveness of phosphorous with time, it is necessary to retreat the molten hypereutectic metal in the foundry. If the copper content of the melt is not critical, the preferred form of phosphorous is the Cu-8%P alloy.

Grain Refinement

Grain refinement is carried out at both the smelter and the foundry levels. In order to facilitate grain refinement in the foundry and to obtain sounder ingots, titanium based products are added during the molten metal preparation in smelters. Since the effectiveness of grain refining is time dependant, it is generally necessary to repeat the treatment in the foundry. Titanium salt mixtures and titanium containing master alloys are the most common sources of grain refiners. Titanium sponge is often used as a smelter addition. Master alloys are aluminum based, and constitute the most widely used grain refiners. Binary Al-Ti and ternary Al-Ti-B master alloys contain up to 10% Ti. The ternary Al-Ti-B master alloy is available in different compositions, which vary according to the ratio Ti/B. As already discussed, the higher boron grain refiners appear advantageous when used in Al-Si foundry alloys. Master alloys are produced in various forms such as waffles, wires, buttons and briquets. Briquets have a high density and sink to the bottom of the melt, yielding a high recovery. Waffles, on the other hand, are more variable in performance and less effective than cut lengths of wire.

Degassing

To improve the quality of casting alloys, degassing in foundries as well as smelters is carried out just before casting. Because of the propensity of casting alloys to regas during remelting, degassing always has to be repeated in the foundry, especially when high quality castings are desired.

A wide range of products exists today to degas aluminum melts. Table 14.2 lists the most common degassing agents. The use of inert gases has increased in recent years, since the introduction of systems which produce finely dispersed bubbles, although mixtures of inert and reactive gases are still widely used. Pure, reactive gases such as chlorine have seen their use decline in recent years due to environmental concerns. Nevertheless, reactive gases or gas mixtures still find application in the removal of impurities such as sodium in primary smelters, and magnesium in secondary smelters which recycle beverage cans. Degassing with salt tablets (C_2Cl_6) is confined to small batch processes.

Table 14.2. Degassing Agents

Form	Constitution
Inert Gases	N_2, Ar
Reactive Gas	Cl_2
Gas Mixtures	N_2-Cl_2
	N_2-Freon
	N_2-SF_6
Salt	C_6Cl_6

Filtration and Fluxing

Molten metal filtration is usually the last step before casting, and comes immediately after degassing. Filtered metal is an important tool in the sale of primary and secondary ingots. Some specific benefits arising from the use of filtered ingot include: reduced melt loss, less gas pickup on remelting, and the elimination of hard spots.

The fluxing of liquid aluminum alloys has been practiced since the origin of aluminum casting. Fluxes are chemical mixtures which come in the form of powder or tablets, and are used generally in the initial stage of melting and alloying. Later in the molten metal preparation cycle, fluxes are also added for modification, refining, grain size control and degassing. As described in detail in Chapter 12, there is a wide range of such products manufactured by numerous companies around the world. Often, fluxes differ only slightly in composition and color, and their effectiveness is not well documented.

14.2   Melt Addition Techniques

The form and composition of the additive dictate the proper addition technique.

Strontium Modification

The low-strontium containing master alloys listed in Table 14.1 can be added to the surface of the melt because of their low reactivity in air. It has been shown that their dissolution rates increase with increasing melt temperature. In a normal melt temperature range, 700-750C, (1290-1380F) binary or ternary master alloys D and E will dissolve gradually because of their higher liquidus temperature, about 775C (1425F). The dissolution rate of master alloy F, containing 14% Si, is considerably slower because of its high melting point of 875C (1600F). While master alloys D and E will sink to the bottom of the melt, master alloy F will float on the surface due to its lower bulk density caused by a high porosity level. In this case, a gentle stirring should be applied to obtain good dissolution rates, and high strontium recoveries, especially at melt temperatures near 700C (1290F). Master alloys B and C melt readily if the molten metal is kept in the 700-750C (1290-1380F) range, and consequently complete dissolution is achieved in less than 10 minutes.

In rod form, master alloys B, C and E can be added into the launder by an automatic wire feeding machine. By adjusting the feeding speed to the melt temperature, master alloy composition, and final level of strontium desired, homogeneous, strontium-premodified ingots are produced in primary and secondary aluminum smelters.

The dissolution characteristics of the Al-90%Sr alloy are markedly different from the low-strontium master alloys. Dissolution rates are higher at low melt temperatures, 650-700C, (1200-1290F), than at higher melt temperatures, 700-750C (1290-1380F). The silicon content also affects the dissolution rates. An increase in the silicon content from 7% to 12% results in a lowering of the optimum temperature range from 650-700C (1200-1290F) to 625-675C (1160-1250F). To utilize the full potential of the dissolution characteristics of the Al-90%Sr alloy, three addition techniques have been designed: surface addition, cage addition and ladle addition. Figure 14.4 shows schematically the three addition techniques, and the optimum temperatures are given in Table 14.3. At low melt temperatures, the surface addition technique is recommended, since the Al-90%Sr alloy will float on the surface and dissolve exothermically.

After the dissolution period, which lasts 5 minutes, a gently stirring of the melt surface completes the addition. At intermediate melt temperatures, a closed-end, perforated cage is immersed into the melt, and stirring is carried out for a 5 to 10 minute period to achieve complete dissolution. At intermediate or high melt temperatures, addition technique no. 3 offers an elegant solution to obtain fast dissolution rates. In an intermediate step, a sufficient quantity of metal is drawn off in order to obtain a strontium level between 2% and 5% (Step A) in the ladle. The metal in this small ladle cools rapidly, and exothermic dissolution is achieved and completed in less than 5 minutes

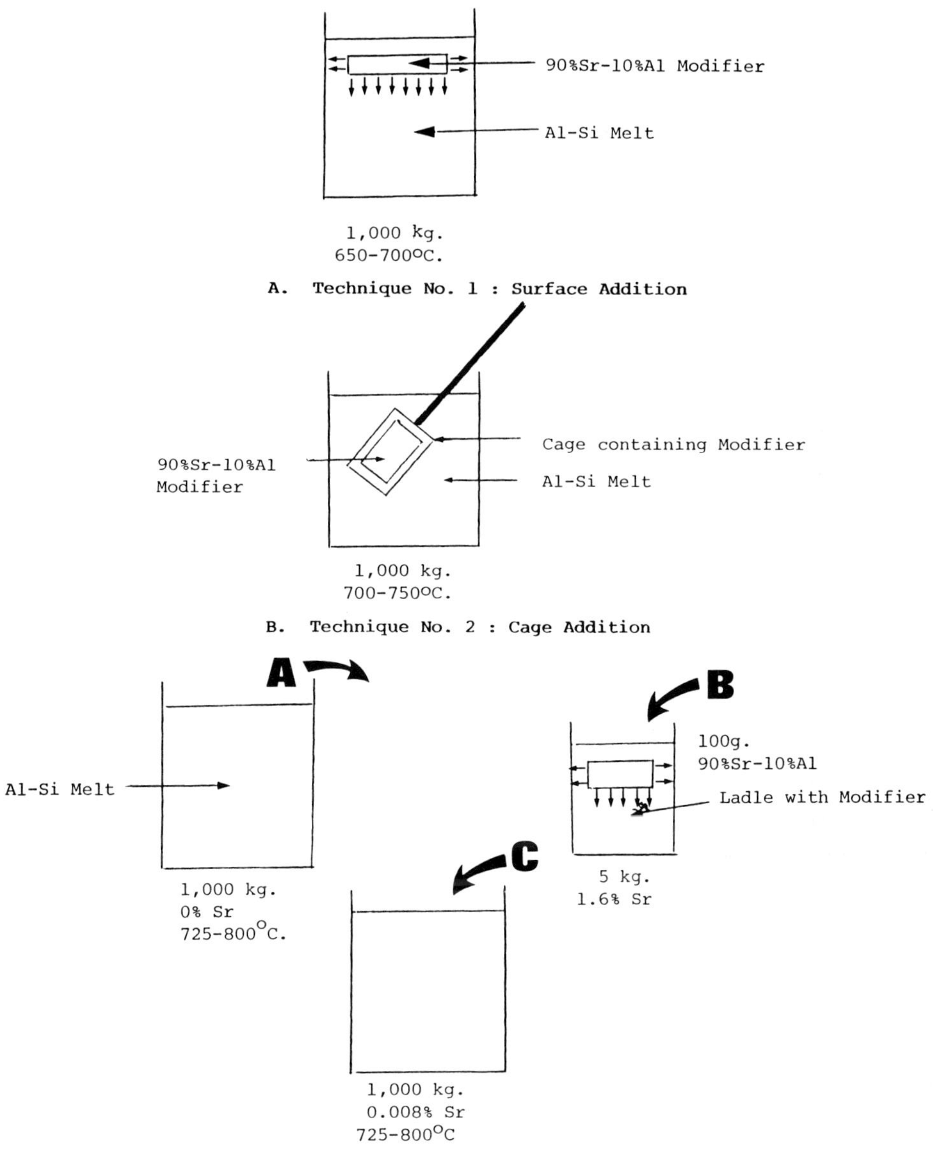

Figure 14.4. Addition techniques for Al-90% Sr alloy.

(Step B). The contents of the small ladle, which by now are in a semi-solid form, are then transferred to the melt, where instantaneous dissolution occurs (Step C).

By using the appropriate product form and addition technique, dissolution of strontium modifier is achieved in less than 10 minutes. High strontium recovery levels are obtained and are usually in the 90%–100% range. In addition to high recoveries and fast dissolution, strontium modification allows reproducible strontium levels as shown in Figure 14.5. The different successive treatments of an A356.0 alloy in a 500 kg transfer ladle

Table 14.3. Optimum Temperature Ranges for Al-90% Sr Alloy Addition

Technique	Melt Temperature Range (°C)	
	Alloy A356.0 (7% Si)	Alloy 413.0 (12% Si)
1: Surface Addition	650 - 700	625 - 675
2: Cage Addition	700 - 750	675 - 725
3: Ladle Addition	725 - 800	700 - 775

yield little variation in strontium content from one addition to another. The strontium levels vary from a minimum of 0.006% to a maximum of 0.009%. The final required level of strontium in the casting is a function of:

- Al-Si alloy composition;
- casting method;
- ingot quality (primary or secondary).

In Table 14.4 are summarized the strontium ranges to achieve the optimum mechanical properties. The main rules to be followed are:

- the lower the cooling rate the higher strontium content;
- lower alloy quality requires higher strontium content;
- the higher the silicon content, the higher the strontium content required.

Table 14.4. Recommended Strontium Content in Casting Alloys

Alloy	Alloy Quality	Sr Content (%) Casting Method		
		Sand or Investment Casting	Permanent Mold	Die Casting
A356.0	Primary	0.01–0.02	0.005–0.015	—
356.0	Secondary	0.015–0.025	0.01–0.02	—
319.0	Secondary	0.015–0.030	0.01–0.025	—
A413.0	Primary		0.015 –0.025	—
413.0	Secondary		0.020–0.030	0.015–0.025
380.0	Secondary			0.02 –0.03

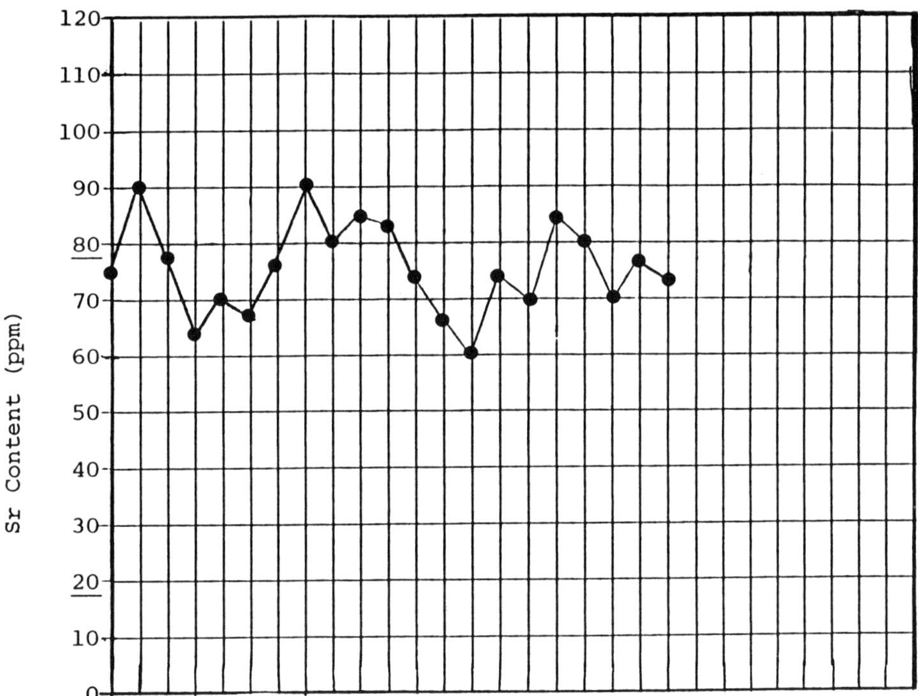

Figure 14.5. Strontium levels in different transfer ladles treated with Al-90% Sr.

Many foundries use strontium modification only to redistribute macroshrinkage to microshrinkage. In this particular application, strontium levels in the 0.005%–0.01% range are adequate to achieve the objective.

Premodified ingots contain strontium levels in a much wider range, from 0.02% to 0.05%. Such a wide range is necessary to accommodate the different melt treatment schedules in each particular foundry. Generally, premodified ingots contain higher strontium levels to compensate for losses during molten metal processing with reactive gases and fluxes.

Sodium Modification

Metallic sodium is added under the melt surface by using a plunger or a bell. The plunging tool must be thoroughly cleaned of accumulated salts and deposits. These are very hygroscopic, and can contribute to the production of hydrogen gas. Sodium fluxes in the form of solid tablets are added under the melt surface using a technique similar to that developed for metallic sodium. Fluxes in the form of powder are added to the surface and then stirred into the melt. In recent years, techniques have been developed to co-inject fluxes with an inert carrier gas through a lance.

Sodium addition is characterized by low metal recoveries, 20% to 50%. Significant dross formation can result if the plunging bell is not clean. In order to minimize melt losses, the dross is reprocessed to partially recuperate some

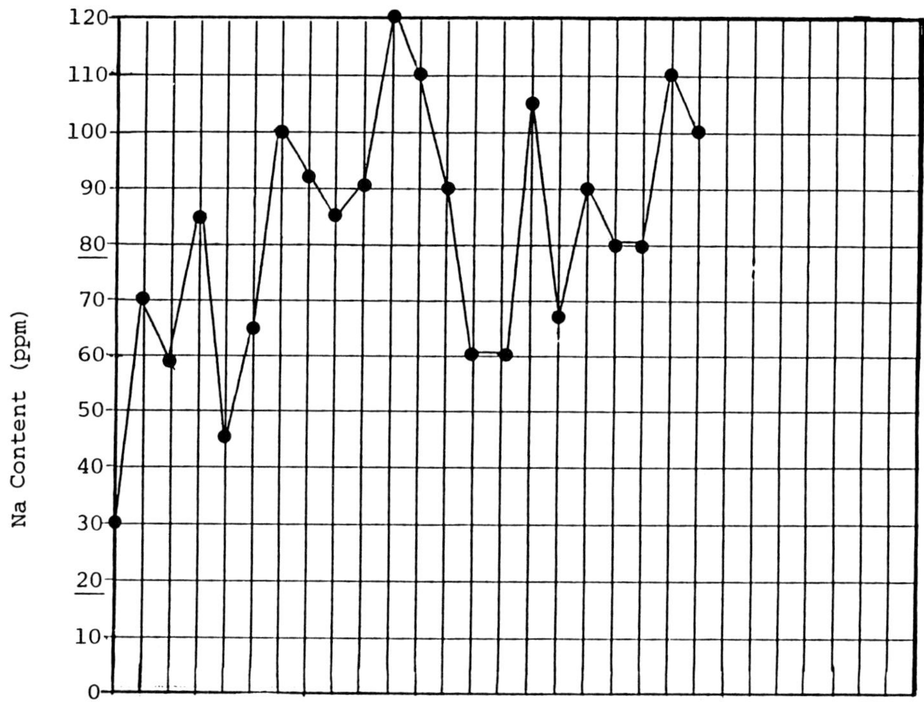

Figure 14.6. Sodium levels in different transfer ladles treated with metallic sodium.

aluminum metal. Sodium treatment of the melt is also associated with poor reproducibility. Figure 14.6 shows the sodium levels obtained in different successive sodium metal treatments of an A356.0 alloy in a 500 kg transfer ladle. From one ladle to another, the sodium fluctuates in a large band from 0.003% to 0.012%.

Sodium modification of Al-Si melts has declined in recent years. Its usage has become more and more confined to small foundries processing small batches. In large foundries, sodium is not well adapted for use, because constant attention has to be given to the melt composition in order to maintain the level within a narrow range. This is necessary in order to avoid under or over-modification. Sodium levels are kept in a range from 0.005% to 0.01% to achieve good mechanical properties and castability.

Antimony Refining

Antimony refining can be accomplished using pretreated ingots. There is practically no antimony loss during remelting or melt processing. In the case of recycled metal, it is difficult to eliminate antimony, and some secondary metal can contain up to 0.008% antimony.

Pretreated ingots normally contain high levels of antimony. For example, A356.0 alloys have levels in the order of 0.15%, while alloys with higher silicon levels (9% - 11%) require up to 0.30% antimony to achieve a fine lamellar structure.

Primary Silicon Refining

Phosphorous refining is the favored method to control and optimize the size of primary silicon. Remelting of pre-treated ingots or long holding periods prior to casting will necessitate the addition of phosphorous containing alloys (Cu-8%P) or fluxes in order to obtain small primary silicon particles. The Cu-8%P alloy is added directly into the melt, while fluxes are added to the surface and then stirred into the molten metal. Good silicon refinement is obtained with phosphorus levels between 0.003% and 0.015%. However, because of a loss in effectiveness during holding of the liquid metal, it is often necessary to add an incremental amount—on the order of 0.01%—in order to achieve optimum silicon refining.

Grain Refinement

There exist various forms and compositions of grain refiners. Master alloys in the form of waffles are added to the melt surface in foundries and smelters, followed by a slight stirring to permit good dissolution and a homogeneous dispersion of the nuclei. In smelters, automatic wire rod feeding machines permit the addition of master alloys into the launder. Titanium-based salt tablets are added through a plunger or a bell into the melt. After a reactive dissolution, the dross is skimmed before the next treatment stage.

In most cases, ingots delivered by smelters have been treated with grain refiners and contain from 0.05% to 0.1% titanium. Again, remelting and long holding periods cancel most of the grain refinement treatment. Therefore, the founder is usually forced to re-refine the grain prior to casting by increasing the titanium content to values in the order of 0.01% to 0.02%. Fortunately, thermal analysis is now available to allow easy determination of the need to add grain refiners.

Degassing

Inert gases, or mixtures of inert gases and reactive gases, are injected via a lance usually made of graphite or silicon carbide. Various lance designs, including open-end, perforated end and porous plug head, are available to achieve degassing. Recently, lances with rotary impellers or spinning heads have increased the degassing efficiency by a factor of 5 to 10.

Salt tablets are added by plunging the degassing agent into the molten metal. This technique is relatively inefficient and generates fumes and dross which have then to be removed from the melt surface. This method is now limited to small batch processes.

Vacuum degassing is more rarely used. This technique requires special equipment, and the time to process a batch of molten metal is relatively long. To increase the degassing kinetics, vacuum treatment is often associated with inert gas purging.

Special crucibles with a porous bottom have also been designed to degas melts. An inert gas is passed through the porous bottom to form a uniform dispersion of fine bubbles throughout the melt.

In order to obtain sound castings with a minimum of porosity, it is necessary to decrease the hydrogen content to levels in the order of 0.10 ml H_2/100 g of aluminum before casting. Such a level is adequate to obtain porous-free permanent mold castings. In order to produce sand or investment castings (which solidify at much lower rates) with low porosity levels, hydrogen levels from 0.05 to 0.10 ml H_2/100 g. of aluminum are recommended. Castings with heavy sections also require hydrogen levels below 0.10 ml H_2/100 g Al.

Fluxing

Fluxes have been designed to achieve multiple functions, such as removing impurities and degassing, or sodium modifying and degassing. Powders are added to the surface and stirred into the melt or injected with the help of an inert carrier gas. Tablets are usually plunged to the bottom of the melt with a bell shaped tool.

There is a tendency to add excessive amounts of fluxes based on the perception that they afford low treatment efficiency. Under normal conditions the addition of 0.1% fluxes should be enough to achieve the desired objectives.

14.3   Processing Schedules

Reactive interactions between melt treatment products have to be taken into account in establishing liquid metal processing flow charts. Locations where the various treatment processes are carried out are different for smelters and foundries. Other important factors in optimizing melt treatment processes include production rates, molding equipment and methods, alloy type, series size and final melt quality sought.

Smelter Metal Processing

Some typical smelter metal processing schedules and operations are illustrated in Figures 14.7 and 14.8. A typical simple schedule without grain refinement or modification is represented in Figure 14.7a. Degassing could be carried out with inert or reactive gases and tablets. Grain refiners in the form of master alloys and tablets would be added after fluxing (Figure 14.7b). If a rod grain refiner is used, then it is possible to refine the metal in the ladle with a wire feeding machine. In this case, grain refining is the last step before ingot

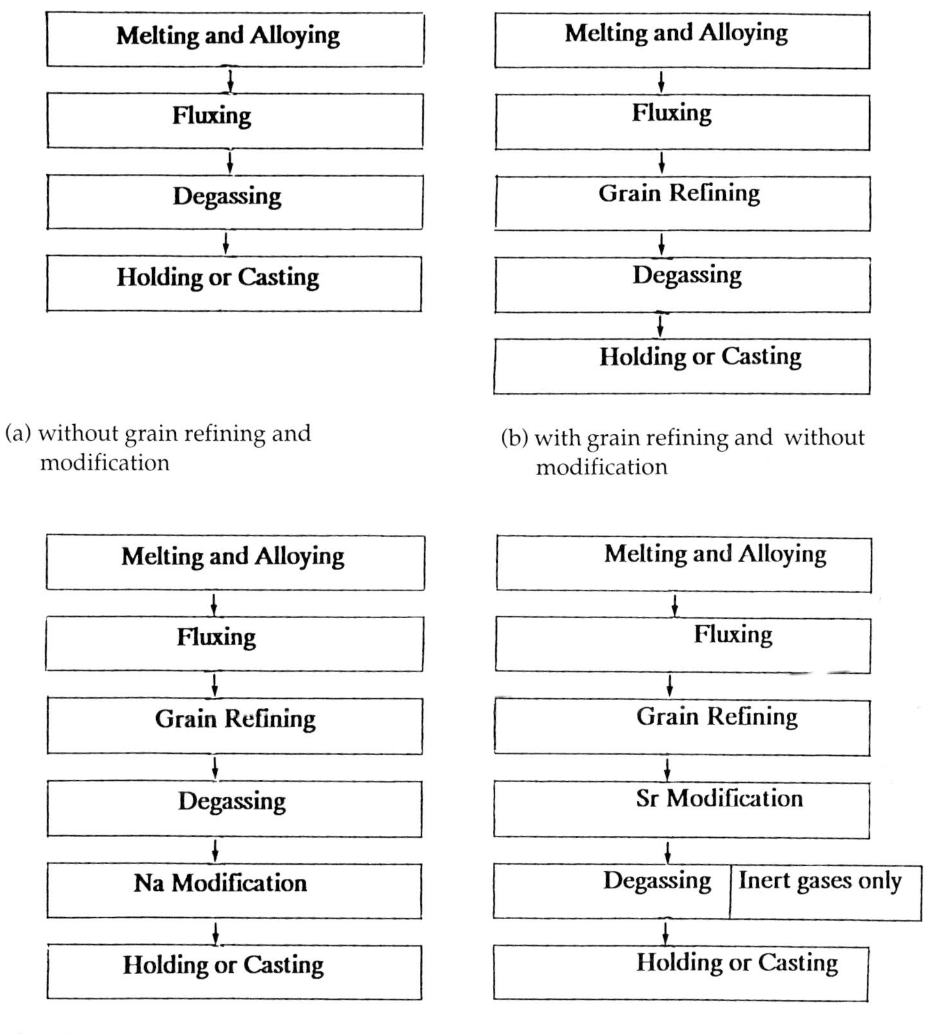

(a) without grain refining and modification

(b) with grain refining and without modification

(c) with grain refining and Na modification

(d) with grain refining and Sr modification

Figure 14.7. Typical smelter processing schedules.

casting. Sodium modification is usually carried out after degassing (Figure 14.7c), in order to avoid extra losses, or the addition of high levels of modifier to compensate for these losses. Again, grain refining can be performed in the launder if the appropriate material and equipment is available. Strontium modification can be carried out before degassing if inert gases are used (Figure 14.7d). If reactive gases are used the strontium modification should be carried out after degassing to avoid modifier losses. It should be noted that strontium reacts readily with fluoride or chloride compounds. Therefore it is recommended that the metal surface be cleaned by removing the dross before

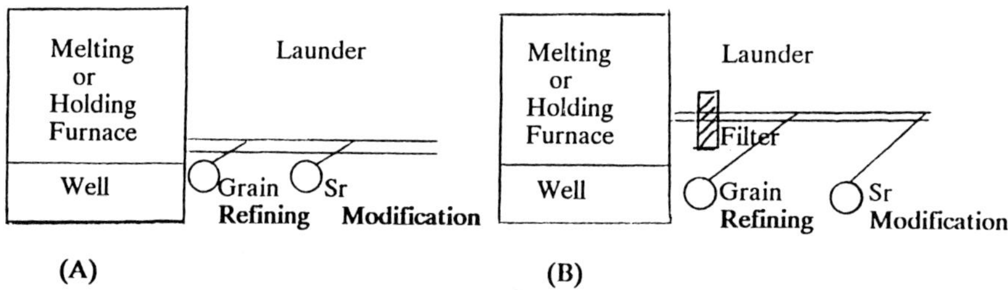

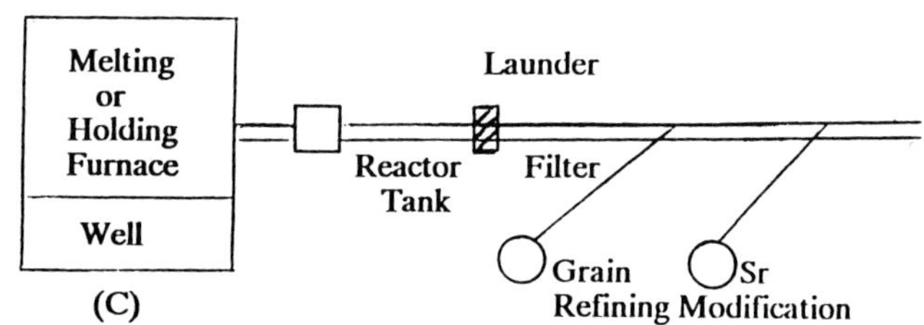

Figure 14.8. Typical smelter processes.

strontium additions are made.

Both strontium modification and grain refining can be conducted in the launder. With sodium modification this is not a recommended practice, again due to high sodium losses.

Smelters often add a filter in the launder to clean the metal before the ingot casting. Such an arrangement is illustrated in Figure 14.8. Grain refinement and strontium modification is often done after filtration (Figure 14.8b), but may be done before.

In Figure 14.8c, the use of a special reactor tank for melt degassing outside of the furnace is illustrated. The degassing efficiency of such a system is greatly superior to a lance system, since it is possible to use a rotary impeller in the tank. Strontium modification can still be done in the furnace, but then inert gases must be used in the reactor. If grain refinement and strontium treatments are carried out in the launder, it is possible to use reactive gases in the reactor tank. Sodium modification is not well suited to such a system because of the heavy modifier losses which occur in the reactor tank during degassing with either inert or reactive gases.

Foundry Metal Processing

A simple layout of a high capacity production foundry is given in Figure 14.9. A variety of mold types is shown, but for each casting method most liquid metal processing would be done in the transfer ladle, although some fluxing and degassing could occur in the melting furnace as well.

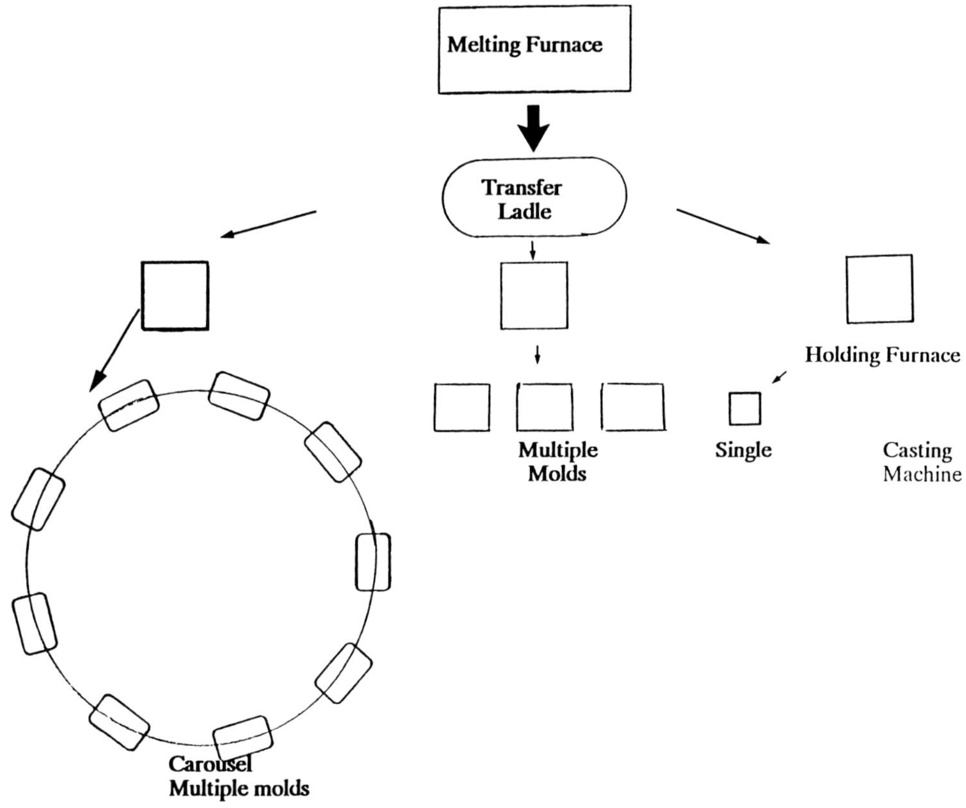

Figure 14.9. Typical foundry processing; automatic casting.

Figure 14.10 shows the different melt treatment schedules including grain refinement and modification in the transfer ladle. Metallic sodium is added under the melt surface which is covered with a flux. After dissolution the dross is skimmed off (Figure 14.10 a. If a modifying flux is applied, then it is wise to choose a flux which combines both fluxing and modification.

After degassing with inert or reactive gases, sodium is added into the melt either in the form of metal or flux (Figure 14.10b). Grain refining is usually carried out before sodium modification (Figure 14.10c). Strontium modification, as with sodium modification, is generally preceded by a flux treatment (Figure 14.10d). Because of strontium's affinity for chloride or fluoride compounds, the dross must be removed from the melt surface before modification.

Degassing can be done before or after strontium addition depending on the type of degassers used (Figure 14.10e). Strontium modification is performed first if degassing is done with inert gases such as nitrogen or argon. In the case of reactive degassing using either gas mixtures or fluxes, strontium addition should follow degassing. It is definitively not recommended to use reactive degassing agents after the addition of strontium. This will result in large strontium losses and almost no hydrogen removal.

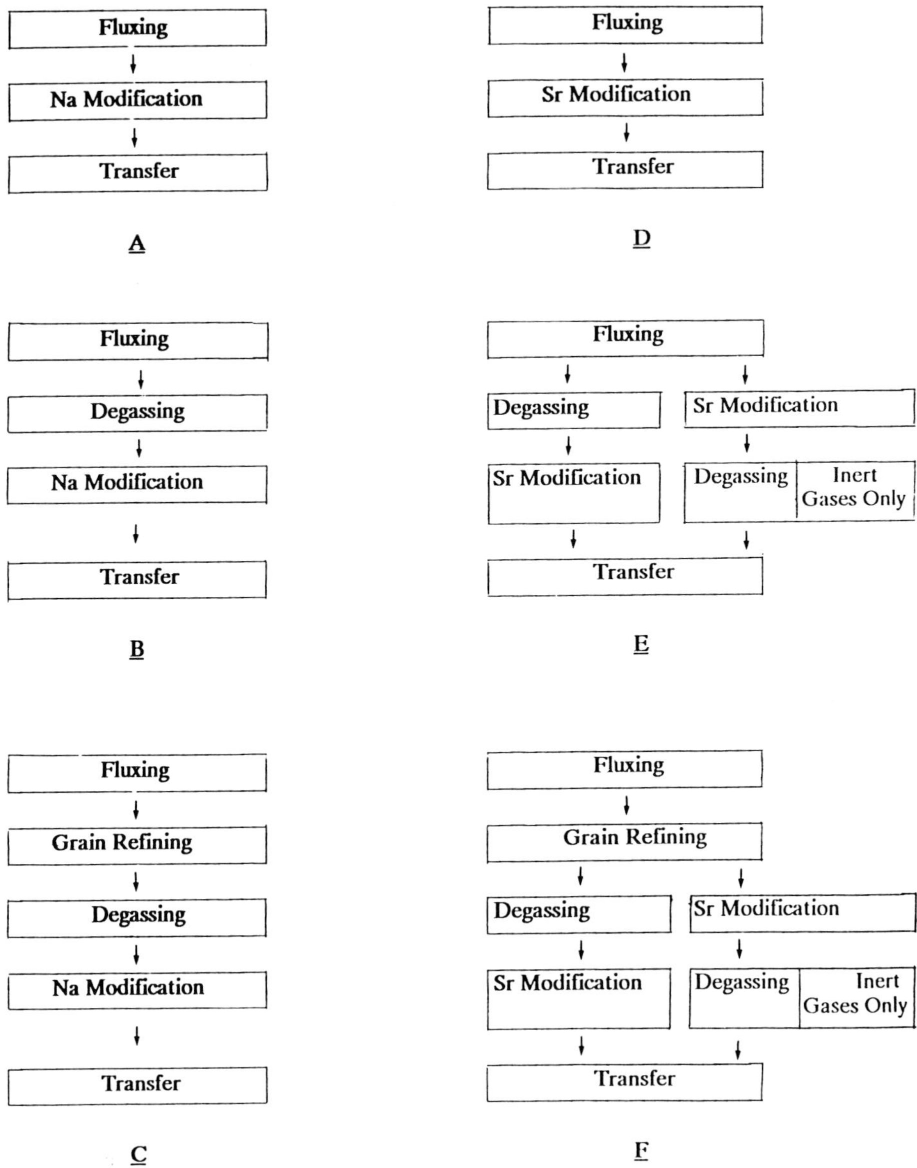

Figure 14.10. Typical foundry processing schedules in the transfer ladle.

Grain refinement of modified melts (either sodium or strontium) is preferable after fluxing (Figure 14.10f). Nevertheless, metallic master alloys can be added after strontium treatment, while reactive grain refiners should always be added before strontium treatment in order to avoid loss of modification.

The treatment schedules detailed in Figure 14.10 are applicable not only to high production foundries, but also to smaller operations using crucible furnaces and manual pouring methods. In this case treatments may be done in the crucible. The quantities of metal involved may be smaller, but the melt chemistry remains the same, as does the sequence of performing the treatments.

Index

-A-

A201 alloy, influence of purge gas composition
    on degassing of, 180
A206.0 alloy, 18
A356.0 alloy, 4, 16, 18, 35, 224-25
    antimony-sodium interactions during the
        treatment of, 100-101
    as-cast microstructure of, 26
    changes in dissolved hydrogen level
        following additions of metallic sodium
        to, 63
    effect of cooling rate through the
        solidification range and strontium level
        on the eutectic structure of, 43
    effect of pressure on porosity in, 158
    electrical conductivity control of, 225-28
    evolution of microstructure in, as measured
        by modification rating for Na- and Sr-
        modified, 55
    mass transfer coefficients for hydrogen
        regassing of, 62
    mechanical properties of, with different
        modifiers, 82
    microstructures of, 21, 78
    modification rating of strontium modified, 97
    overmodification of, 45
    Quality Index of, 81
    recovery of low-strontium, high-aluminum
        master alloy in, 38
    reduced pressure test of, 161-64
    relation between modification rating and
        the eutectic temperature for, 221
    samples subjected to reduced pressure test,
        160-61
    slumping and contraction in, 68
    sodium modification of, 81
    sodium-strontium treatment of an, 124
    strontium modification of, 17-19, 42, 49, 88,
        103-104
    Tatur Test measurements on, at constant
        hydrogen concentration, 66
    tensile properties and radiographic ratings
        for, modified with sodium, 84, 132
    variation of eutectic undercooling with
        structure for, 222
    variation of pore size in cast, for strontium
        modified and unmodified samples, 71-73
    variation in the Quality Index with different
        cooling rates, 80-81
A357.0 alloy, 4, 16, 20
    microstructures of, 21-22

A390.0 alloy
    chill-cast, 109
    effect of strontium additions to a
        phosphorus-refined, 122
    macrosegregation in sand cast, 113
    strontium additions to, 120-22
α-AlFeSi, 14
Aerospace
    aluminum-copper alloys used in, 18
    applications for aluminum castings, 8
Age hardening, 78. *See also:* Heat treatment
Al-Cu phase diagram, 17
Al-4.5%Cu-0.25%Mg alloy, 17
Al-Mg2Si, pseudo binary phase diagram of, 17
Alcan, 199
ALPUR, 172
Aluminum alloying elements, 1
    selection of, 4
    solid solubility of, 2
Aluminum Association's designation of
        aluminum casting alloys, 3-4, 13
Aluminum casting(s), 32. *See also:* Aluminum
        foundry alloys; Cast microstructure
    of automobile wheel, 10-11
    current market for, 5-9
    effect on modification treatment due to the
        temperature of, 117
    future market for, 9-11
    high pressure die casting of, 5, 7
    modification and porosity in, 57-73
    permanent mold production of, 5
    of pistons, 11
    porosity of, 150, 152-57
    production, as a percentage of total castings
        (by country), 3
    production in Europe, 7
    property variations within, 83-85
    solid solubility of alloying elements in, 2
    unchilled, 81
    U.S. production of, since 1945, 6-7
    used in aircraft, 8-9
Aluminum-Copper alloy(s), 16-17, 18
    solubility of various, 144-46
    use of grain refiners in, 19
Aluminum foundry alloy(s). *See also:*
        Aluminum casting(s); Aluminum-
        Copper alloy(s); Aluminum-Magnesium
        alloys; Aluminum-Silicon alloy(s); Cast
        microstructure(s) *and* Eutectic modification
    advantages of, 4-5
    alkali fluoride fluxing of, 25
    alloying elements, 1
    binary, 14

castability of, 4
characteristics, 4
chemical modification by sodium, strontium
    or antimony of, 31-44
commercial importance of, 1-5
composition, 222
degassing of, 169-84
designation, 3-4, 13
filtration of, 185-202
gassing of, 143-68
heat treatment of, 4, 78-80
machinability of, 4
mechanical properties of, 4
processing schedules for, 244-48
properties of, 4-5
properties of modified, 75-94
recommended strontium content in, 240-41
refinement of, 107-26
solubility of, 144-46
sources of hydrogen pickup for, 145-49
summary of melt treatment processes,
    techniques and schedules for, 233-48
treatment of, with cover and cleaning
    fluxes, 203-12
U.S. tonnage shipped of, 1986, 1
used in the transportation industry, 3, 10
Aluminum-Lithium alloys, 145
Aluminum-Magnesium alloys, 17
    phase diagram of, 18
    use of grain refiners in, 19
Aluminum metal matrix composites, 11
Aluminum-Silicon alloy(s), 13-23
    alloyed with Mg, 19-20
    characteristics of, 15
    composition of common, 14
    effects of phosphorus on, 50-57
    equilibrium binary Al-Si phase diagram of, 16
    eutectic modification of, 25-55
    fluidity of, 92-93
    foreign particles in, 130
    hot tearing tendency of, 91-92
    as percentage of the total aluminum cast
        parts produced, 13
    refinement and modification in
        hypereutectic, 107-26
    relationship between UTS, elongation, YS
        and Quality Index for a sample, 76
    replacement of aluminum-copper alloys by,
        16-17
    resistance to hot cracking in, 19
    solubility of various, 144-46,
    tensile properties of, 75-85
    200 series of, 127
    treatment with alkali fluoride fluxes of, 25
    typical applications of, 17-18
    use of grain refiners in, 19
Alusuisse, 6

Antimony, 34, 81
    change in concentration of, measured near
        the top of A356 sodium treated melts,
        101
    eutectic refining using, 235-36
    interactons with strontium, 96-98
    phosphorous interaction with, 53
    reaction with sodium and strontium, 42, 95-
        105
    refining, 242
    sodium contamination of melts refined
        with, 95
    treatment and fatigue strength, 89
ASTM
    classification of radiographs by, 83
    grain size classification, 220
Australia, A356.0 alloy cast wheel production
    in, 17-18
Automobile(s)
    fuel savings resulting from the use of
        aluminum castings in, 6, 8
    future of aluminum in, 10-11
    wheels, 10-11

-B-

Beryllium additions to Al-Mg alloys, 17
Binary aluminum silicon alloys, 14
    microstructures of, 19
Blistering, effect of porosity on, 156-57
Bonded particle filters, 191

-C-

C355.0 alloy, 18
Cake filtration, 194-95
Canada
    percentage of aluminum castings used in
        autos in, 8
    use of 380 alloys in, 18
Cast iron, thin walled, 11
Cast microstructure(s)
    basic elements of an Al-Si, 128
    common factors which determine, for cast
        aluminum alloys, 18-23
    effects of chemical modification on, 77
    evolution of, with time, 55
    non-destructive control of, 213-32
Cast parts as a percentage of total aluminum
    products, 3
Casting alloy designation, 3
Ceramic foam filter, 191, 197
Charge materials as sources of hydrogen
    pickup, 148
Chemical modification by sodium, strontium
    or antimony, 31-32

Chlorides, free energies of formation of, 182
Chlorine
    affinity for sodium of, 47, 49
    degassing, 171-72
    in situ formation of, 179
    used to remove magnesium, 182
Chilling, grain refinement by, and the effect on dendrite arm spacing, 131-32
Chills, 70-73
Cleaning fluxes, 205-209
Complex intermetallics, 14-15
Cooling curves, 214-17
    effects of modification on, 220-22
Cooling rate(s), 152-54
    chemical grain refinement, modification and, 141
    effect on the size of primary silicon, through the solidification range, 114-15
    variation of the DAS of a 356 alloy with different , 131
    variation in modification at different, 80-81
Coulter counter, 200
Cover fluxes, 204-205
Crucibles, as source of hydrogen pickup, 147
Crystal(s), twinning in a silicon, 28

-D-

Deep bed filtration, 192-94
Degassing. *See:* Melt degassing
DAS (Dendrite Arm Spacing), 127-28, 131-32
    electrical conductivity measurement of the, 229-32
Density variation versus distance from graphite chill for sand cast bars, 70
Die castings, projected future demand for aluminum, 9
DMC, 172
Ductility, effects of iron and beryllium additons on, 16

-E-

Engine blocks, 11
England, use of aluminum engines in, 11
Europe, aluminum casting production in, 6-7
Eutectic, 220-28
    alloys, 4-5, 14, 214
    displacement caused by modification or high freezing rates, 108
    in hypereutectic alloys, 109-10
    modification, 18-19, 25-55, 220-25
    silicon morphology, 110
    structure in pressure die casting, 19
    structures observed in the presence of both strontium and antimony, 96
    undercooling, 220-22

-F-

Fading. *See:* Modifier fading
Fatigue properties, 89
    effect of porosity on, 155
    filtration's effect on improved, 188-89
Ferrous casting shipments, U.S., from 1978, 2
Filter(s), 237. *See also:* Melt cleanliness
    bonded particle, 191
    ceramic foam, 191, 197
    efficiency, 195-200
    operation of, 192-95
    placement, 192
    sizing, 196-97
    types and placement of, 190-92
Filtration, 185-202
Fluidity, 92-93
    effect of grain refinement on, 140
    improvement following filtration, 190
Fluorides, free energy formation of, 183
Flux(es), 173, 177, 237, 244
    cover and cleaning, 203-12
    as source of hydrogen pickup, 147
    treatment, 46-47
FM Process, 11
Ford Motor Co., 6
413.0 alloy, 14
443.0 alloy, 14
Fracture toughness, 85, 88-89
France, 7
    use of aluminum castings in, 10-11
Freezing rate variation with type of modifier, 44
Freon-12-nitrogen mixtures, 172, 179-81

-G-

Gassing
    of aluminum foundry melts, 143-68
    and sodium treatment, 75-76
Gating and risering of Al-Mg alloys, 17
Grain(s)
    definition of, 127-28
    size, chilling affect on, 131
    thermal analysis control of, 219-20
Grain refinement, 19, 27-42, 125-26, 153, 236, 243
    chemical, 132-37
    effect of chilling on dendrite arm spacing and, 131-32
    effect on properties of, 118-20, 137-41
    interrelationship between chemical, cooling rate and modification, 141

and modification in hypereutectic alloys,
    107-26
principles of, 128-130
Graphite lance, 172-73

-H-

Hall-Heroult process, 1
Heat treatment
    of castings, 4, 13
    properties of some alloys that have
        undergone, 79
    some properties of alloys which have not
        undergone, 80
    T5/T6, of 319.0 alloy, 14
    of 357.0 alloys, 16
Hexachloroethane, 173, 179
High pressure die casting(s), 5, 16
    use of 380.0 alloy in, 18, 23
    use of 390.0 alloy in, 18
    use of 413.0 alloy in, 14
    growth in Japan of, 6
    production of aluminum, 7
    rapid cooling in, 19
Holding time and remelting, 116
Hot tearing
    of aluminum-copper alloys, 16-17
    effect of grain refinement on, 138-39
    effect of porosity on, 155-56
    tendency, 91-92
Hydrogen
    analysis, 166-67
    avoiding pickup of, 148-49
    concentration in the melt, 153-54
    measurement techniques, 157-67
    changes in levels of melt, following
        additon of Sr, 58-60
    changes in levels of melt, following
        additon of Na, 63
    gas control, 69-70
    methods of adding, 175
    pickup, 145, 148, 211-12
    solubility in liquid aluminum, 57-58, 143-45
    sources of, 146-49
Hypereutectic casting alloys, 18, 21-23, 107,
    109-10
    feathery eutectic structure found in, 112
    fluidity and feeding of, 118
    range of microstructures seen in the
        modification of, 39-41
    refinement and modification in, 107-26
    tensile strength variation of unrefined and
        phosphorus refined, 118
    390.0 alloy, 18
    wear properties of, 119-20

-I-

IACS (International Annealed Copper
    Standard), 226
Impact properties and fracture toughness, 85,
    88-89
Impeller degassing. *See:* Rotary impeller
    degassing
Impurity(s)
    induced twinning, 31
    present in the melt, 42
Inclusions, 186-90, 197-201
Inert gas purging, 170-73, 179-81
Intermetallic compounds
    effect of grain refinement on the
        distribution of, 139
    in modified alloys, 52
ISO designation, 85
Isopores, 139
Italy, 7, 10

-J-

Japan, use of aluminum castings in, 6-7, 10

-L-

Lamellar structures, 50
Lance degassing, 172, 177-78
LIMCA (Liquid Metal Cleanliness Analyzer)
    System, 200
Lost foam process, 10
Low pressure casting, 16, 18

-M-

Machinability
    of aluminum alloys, 4
    effects of silicon fineness on, 16
    expense related to tool life, 90-91
    filtration's improvements in, 189
    improvements following phosphorus
        treatment, 118-19
Macroshrinkage, 65, 157
    microporosity used to combat, 175
Magnesium
    effect of additons of, on the Quality Index,
        77
    removal using chlorine, 182
    used in engine blocks, 11
McGill University, studies on regassing in the
    presence of strontium at, 59
Mechanical properties of aluminum alloys, 4-5
    improvements in, from grain refinement,
        140-41

Melt cleanliness, 197-201
Melt degassing, 169-84, 236-37, 243-44
   agents, 237
   avoiding loss of strontium when
     performing, 69-70
   chemical changes during, 181-84
   effect on phosphorous modification of, 114-
     116
   efficiency, 175-79
   methods, 170-75
   purge gas effect on, 180
   and regassing, 59-62
   rotary impeller, 173-74, 177-78, 181
Melt regassing, effect of strontium
     modification on, 59-62
Metal matrix composites, 11
Mg_2Sb_2Sr, 97-98
Mg_2Si precipitates, 16
Microshrinkage
   effects of modification on, 39-44
   formation in aluminum-copper alloys, 16
Microstructure. *See:* Cast microstructure
MINT, 172
Modification, 64. *See also:* Overmodification,
     Refinement
   changes in freezing rate due to, 44
   changes in tensile properties due to, 77-80,
     83-85
   effects on microstructure from, 39-44
   effects of, on machinability, 90-91
   effect on thermal shock properties from, 90
   electrical conductivity control of, 225-228
   of heavily chilled castings, 81
   hot tearing tendency following, 91-92
   interrelationship between, chemical grain
     refinement and cooling rate, 141
   and melt hydrogen, 58-63
   and porosity, 57-73, 83
   Rating (M.R.), 39
   and refinement in hypereutectic alloys, 107-
     26
   sodium, 57-73, 83, 108-109
   strontium, 17-18, 42, 80-81, 88-89, 103-104
Modifier(s), 11, 17, 29. *See also:* Antimony,
     Phosphrus, Sodium *and* Strontium
   additions, 33, 154-55
   amounts used of, 42-43, 50-51, 80-81
   common commercial, 34
   fading, 46-50, 81-83, 116
   interactions of, 95-105
   overcoming negative interactions of, 102-
     104
   treatment temperature and casting
     temperature, 117
   type and tensile properties, 81-83

M.R (Modification Rating), determination, 39

-N-

NaOH formation on the surface of metallic
     sodium, 61
Nickel addition to 332.0 alloy, 18
Non-destructive microstructure control, 213-32
Nonferrous casting shipments, U.S., 1977-86, 2

-O-

Overmodification, 44-46

-P-

Pacz, Aladar, 25, 32
Pechiney, 50, 52
Permanent mold casting, 5, 7, 9
   cooling rates and use of eutectic modifiers
     in, 19
Phosphorus, 116, 122
   additives, types of, 112-113
   effect on modifiers of, 50-55
   effects of, on aluminum casting alloys, 50-57
   improved machinability of Al-Si alloys with
     treatment of, 16, 118-19
   level, with sodium modification, 54
   optimum concentration of, 114-15
   reactions with strontium, 120-21
   refinement, 110-17, 235-36
   treated ingot, holding time and remelting
     of, 116
Piping reduction resulting from modification,
     65-67
Piston(s), 11, 18
   castings, addition of phosphorus to, 50
   surface roughness of machined, 119
PODFA (Porous Disc Filtration Apparatus),
     199
Processing schedules, 244-48
Pore(s)
   due to gas and shrinkage, 149-52
   size, 67
Porosity
   avoiding, in modified castings, 68-73
   design criteria affecting, 71, 73
   distribution, 63-68, 139
   effect of filtration on, 189-90
   effect on properties of, 155-57
   effect on tensile strength of, 16
   modification to overcome, 57-73, 83
   pinhole suface, 149, 151
   pure gas, 149
   quantitative assessments of the amount of,
     66-67

in unchilled castings, 81
Porous plug degassing, 177-78
Pressure tightness, effect of porosity on, 155

-Q-

Quality assurance, 4-5, 8
Quality Index, 80-81
    determination of the, 76-77
Quantitative reduced pressure test (Severn
    Science), 162-64
Quench modification, 29-31

-R-

Radiographs, ASTM classification of, 83
Rating system for modified microstructures,
    39-41
Reactive gas treatment for element removal,
    181-84
Recirculating gas techniques, 58, 165-66
Reduced Pressure Test (Straube-Pfeiffer Test),
    158-62
Refinement. *See also:* Modification
    in hypereutectic alloys, 107-26
    selenium, 125-26
Refractories as a source of hydrogen pickup,
    148
Rotary impeller degassing, 173-74, 177, 181
    effect of melt temperature on, 178

-S-

Sand casting, 5, 9
Schedules for melt treatment processing, 244-
    48
Scrap cycle, mixing of antimony, strontium
    and sodium in a national, 95
Selenium refinement, 125-26
Severn Science Technique (Quantitative
    Reduced Pressure Test), 69, 162-64
Shrinkage
    coefficient, 4
    distribution, 64
    porosity patterns of cast alloys, 63-67, 149-57
    variation, with modification, 64
Sievert's Law, 164-65
Silicon
    coarse, caused by overmodification, 45
    contamination of Al-Mg alloys, 17
    effect of cooling rate through the
        solidification range in primary, 114
    effects on machinability with increasing
        fineness of, 16
    increasing amounts of modifier with higher
        concentrations of, 44

    lamellar, 3
    morphology of, 110-11
    particle sizes, 115
    refining of primary, 243
    schematic illustration of the solid-liquid
        interface of a solidifying crystal of, 29
    schematic representation of the growth of
        an acicular melt crystal of, 27
    shapes, primary, 123
    solidification of, in an Al-Si eutectic, 27-28
    transmission electron micrographs of, 30
    typical sizes of primary, 112
Slumping and contraction, measured by the
    Tatur Test, 66-68
Smelter metal processing, 244-46
SNIF, 172
Sodium, 18, 31-33, 81, 234-35
    additions causing reduced pipe volume, 65-
        66
    addition methods, 241-42
    additions through flux treatment, 46-47
    additions and their effect on mechanical
        properties, 75
    affect on porosity redistribution of,
        compared to strontium, 66-67
    antimony interactions with, 99-102
    changes in dissolved hydrogen in an A356
        melt after addition of metallic, 63
    changes in fluidity resulting from additions
        of, 92-93
    dissolution of, 34-39
    fading, 46-47
    fluoride treatment of Al-Si alloys, 25
    interaction with strontium, 102
    losses during degassing, 70
    and melt hydrogen, 60-63
    modification leading to the formation of
        primary aluminum dendrites, 108-109
    modification and porosity, 57-73, 83
    phosphorus interactions, 51, 54
    production of twins by pure, 29
    Quality Index changes resulting from
        modification by, 81
    reactions with phosphorus treated alloys,
        122-25
    treatment after degassing, 183
Solid solubility of elements in aluminum, 2
Solution treatment , 14-17
Stirring, effect on regassing due to bath, 60
Straube-Pfeiffer Test (Reduced Pressure Test),
    158-62
Strontium, 18-20, 31-33, 80, 234
    addition methods, 238-41
    additions causing reduced pipe volume, 65-
        66
    additions to a phosphorus-refined A390
        alloy, 122

alloys containing, 60
alloys low in, 37-38
and antimony interactions, 96-98
changes in fluidity resulting from addition
    of, 92-93
changes in porosity following additon of,
    compared to sodium, 66-67
changes in thermal shock properties
    following treatment with, 90
degassing techniques which do not remove
    large quantities of, 69-70
fading, 47-50
freezing rate changes following additions
    of, 44
hot tearing tendency following treatment
    with, 91
incubation period, 52-53
influence of alloy temperature on the
    dissolution of, 36, 39
interactions with antimony, 96-98
interactions with sodium, 102
and melt hydrogen, 58-60
modification of A356.0 alloy with, 17-18, 42,
    88, 103
modification of an antimony-containing
    A356.0 melt, 103-104
modification with different cooling rates,
    80-81
modification and fatigue strength, 89
overmodification, 45
reaction with chlorine, 49
reactions with phosphorus, 120-21
regassing following additions of, 59-61
used in 3HA alloy, 11
variation of the electrical conductivity of a
    356 alloy with the concentration of, 227
variation in porosity of cast A356, modified
    and unmodified following additions
    of, 71-73
Sub-Fusion hydrogen analysis, 166-67
Sulphur
    hexafluoride, 172
    refinement, 125
Summary of melt treatment processes,
    techniques and schedules, 233-48
Supercooling, 220
Surface appearance, 139
    effect of porosity on, 155
Sweden, use of magnesium automobile
    components in, 11

-T-

T4 treatment, 15-16
Tatur test, 64-66
Telgas, 59-60, 162-63

Temper designations, 13
Tensile property(ies), 75-78
    of binary and strontium-treated Al-17%Si
        synthetic alloy, 122
    effect of porosity on, 155
    effect of refinement on, 118
    filtration's affect on improved, 186-88
    and hardness of modified alloys, 86-87
    and modifier type, 81-83
    of 319.0 alloy, 10, 14
    variation of, with DAS for an A356 alloy,
        132
Tensile strength, 16-18
201.0 alloy, 18
319.0 alloy, thermal shock resistance of,
        following strontium treatment, 90
332.0 alloy, 16
355.0 alloy, 16
356.0 alloy, 16, 228
    cooling curve of a , 216, 231
357.0 alloy, distinguished from 356.0, 16
380 alloy, 11, 14, 18, 232
    addition of strontium to, 20
    different international designations for, 15
    microstructures of, cast by high pressure
        die, 23
    microstructures of, cast at a low cooling
        rate, 22
    tool life variation with iron content and
        modification in a, 90-91
390 alloy, 11, 16
    microstructures of, chill cast and treated
        with phosphorous, 23
    used in high pressure die casting of engine
        blocks, 18
393.0 alloy, 16
3HA alloy, 11
Thermal analysis
    applicability to real castings of, 223-24
    equipment, 217-18
    example of the use of, 224-25
    principles of, 214-18
Thermal shock properties, 90
    and titanium-boron grain refinement, 133-
        37
Tool life variation, 90
Transportation industry
    use of aluminum in the, by country, 10
Twin(s), 28
    impurity induced, 31
    measured spacings of, in several modified
        systems, 32

-U-

Unchilled castings, 81
United States, 6-10
    ferrous and nonferrous casting shipments
        from the , 2

-V-

Vacuum
    degassing, 173-75, 183
    fusion hydrogen analysis, 166-67
Volumetric shrinkage of aluminum alloys, 4

-W-

Wear properties, 119-20
West Germany (GDR), 6-7, 10-11
Wheels, 10-11